FLORICULTURE AND LANDSCAPING

FLORICULTURE AND LANDSCAPING

Editor

Gavino Merlo

scitus academics

Floriculture and Landscaping
Edited by **Gavino Merlo**

ISBN: 9781681179360
Library of Congress Control Number: 2017949074

SCITUS Academics LLC,
616, Corporate Way, Suite 2, 4766,
Valley Cottage, NY 10989

www.scitusacademics.com

Printed in United States of America.

Preface

Flowers have been considered as the symbol of grace and elegance and a feast for our eyes. Flowers have always remained an integral part of social fabric of human life. They are used on all religious festival occasions. Flowers are given as a birthday presents, wedding gifts or while meeting sick people and even at funerals. Floriculture is the aesthetic branch of horticulture which deals not only with cultivation of ornamentals, annuals, biennials, and perrnnial plants including potted ones but also their marketing. This also includes marketing for local, distant markets and export of cut flowers, live plants and their economic products like scents, oils and medicines, etc. Floriculture is expanding very fast as an industry world over. Floriculture has become a very popular commercial activity during last decade of the outgoing millennium. he economic importance of ornamentals has been increasing in many countries, and international demand has rapidly expanded. Cut flowers represent the largest segment of the industry followed by flowering pot plants, tree and nursery crops, flower bulbs, and other propagation material. Worldwide trends in floriculture are rapidly changing. Excess greenhouse capacity has developed in the Netherlands following reduced demand for Dutch cut flowers. The trend away from cut flower production has resulted in a shift to flowering pot plants.

This volume 'Floriculture and Landscaping' aims to present the latest trends and overview of the floriculture industry and reviews the product, production process, market segmentation and peculiar characteristics of floriculture. Overview of Regulations, code of conduct and Standards in Floriculture is presented along with health and environmental hazards.

Table of Contents

CHAPTER 1

Effect of Different Substrates on Survival and Growth of Transplanted Orchids (*Dendrobium Nobile* cv.) into Net House

[1]Abdul Aziz Mirani*, [2]Adel Ahmed Abul-Soad and [1]Ghulam Sarwar Markhand

[1]Date Palm Research Institute (DPRI), Shah Abdul Latif University, Khairpur, Pakistan.

[2]Horticulture Res. Institute, Agricultural Res. Center, Cairo, Egypt.

ABSTRACT

Orchids are considered the most beautiful and valuable cut flower and potted plants all over the world. Although, micropropagation has become very important nowadays to meet growing market demand, there are still some barriers that hamper the ultimate goal to achieve viable ex vitro plants. Acclimatization has remained very serious issue. The current study was carried out to investigate the impact of different substrates available in nature on the survivability and development of ex vitro plantlets under net house conditions. The step-wise protocol of acclimatization utilized plantlets with well-developed roots produced in MS rooting medium involved NAA at 3.0 mg l^{-1}. The substrates used for acclimatization were different mixtures of coco-peat, desert sand, bricks chips, saw dust, coal pieces and date palm chips. The treatment contained coco-peat: desert sand: perlite at ratio 1.0:1.0:0.2 (v/v/v) proved the best mixture with 100%

survival rate after 120 days of transplantation. Furthermore, addition of wood and bark chips to this best soil substrate as an upper layer improved the subsequent growth and development of plants. Number of new shoots and other growth parameters were recorded and the whole protocol is discussed in the current study.

Keywords: Acclimatization, coco-peat, hill sand, micropropagation, perlite, substrates.

INTRODUCTION

Orchids have an old history in terms of exploitation for therapeutic & ornamental purposes. They are considered as the most beautiful in ornamental plants due to their diverse shapes, colors, durable flowers and adaptability to room conditions. Their modern uses as cut flowers and potted plants have increased their demands in cut flower industries throughout the world. They account for 27% of the global cut flower production in terms of value (Kabiret al., 2012). The increasingdemand of orchid cut flowers and potted plants all over the world on different occasions has enticed the growers to adopt micropropagation technique to reach the market demands.

The micropropagation technique has been generally used worldwide for rapid and disease free propagation of several plant species but its broader use is restricted often by high percentage of morality or damage to plantlets during acclimatization (Deb and Imchen, 2010). This is a crucial step in micropropagation and is considered as difficult due to the variability of environmental factors from in vitro to ex vitro conditions. Generally, the conditions during in vitro shooting and rooting stage render plantlets to develop poor roots and shoots in terms of root hairs & cap and stomata & cuticle development respectively. The plantlets at this stage, if not dealt with expertise, become vulnerable to high percentage of damage and loss (Ortega-loezaet al., 2011). Júnior and Venturieri (2011) have reported that the germination phase in micropropagation of orchids has been fully mastered but the ex vitro phase is rarely acknowledged. They further stated that the ex vitro phase is the most difficult part of the orchids cultivation and the knowledge about the phase does exist but is kept secret (Colombo et al., 2005).

A variety of approaches such as hydroponic systems, different in vitro culture media and concentration of agar & sucrose in growth medium have been employed to reduce the loss of micropropagated orchid plantlets at acclimatization stage (Deb and Imchen, 2010). Beside these all approaches, the proper selection of soil substrates, correct shade management, irrigation and gradual lowering of humidity can solve the problems and reduce the losses during acclimatization (Deb and Imchen, 2010). The in vitro plantlets should be large and vital enough when transferring ex vitro onto an appropriate substrate and under plausible relative humidity (Lesar et al., 2012).

In the current study, an improved protocol using cheap and simple substances found in nature for acclimatization of in vitro grown orchid plantlets has been discussed.

MATERIAL AND METHOD

This study was carried out at the Biotechnology Lab., Date Palm Res. Inst., Botany Dept., Shah Abdul Latif Univ., Khairpur, Sindh, Pakistan during the period from2012 to 2013.

Substrate used

The protocol was divided into two phases: acclimatization and post-acclimatization (after 5 months of transplanting). During the first phase, 9 different substrate mixtures were tested to serve as soil beds of the in vitro plantlets under ex vitro conditions of the net house (Table 1, Figure 1). Thereafter, the soil mixture used for the second phase was categorized into a control and a treatment. The control (T3) mixture of second phase consisted of cocopeat, hill sand and perlite at 1.0:1.0:0.2 (v/v/v), whereas the treatment (T10) composed of two layers: the lower layer occupied 30% of the plastic bag volume having coco-peat, hill sand and perlite at 1.0:1.0:0.2 while the upper layer filled with wood and bark chips of different plants. It is worth to mention that hill sand is used to acclimatize Banana (Abul-Soadet al., 2012). This sand of hills found on the boundaries of „Khairpur" district, Pakistan. It is always mixed with small stones and has to be sieved to separate the small particles looking like sand. Its color is dull yellowish brown. The pH is 5.57 and Electric Conductivity (EC) is 1.03 (dS m-1).

All the substrates, except commercial coco-peat that was already sterilized, were treated by fungicide solution (Mancozeb + Metalaxyl (M/S Zhejiang Minngyun Industries Co., Ltd. Wenzhu China) before making mixtures. The soil mixtures were poured into small plastic pots of 15 cm long × 12 cm wide during the first phase and transferred into plastic bags 25 cm long × 15 wide during the second phase of acclimatization.

Plant material

The plantlets with uniform growth, 2-4 leaves and 2-3 roots from the in vitro rooting medium involved NAA at 3.0 mg l-1 were transferred to the net house to conduct the first trail. After then, the survived plants in acclimatization were randomly selected after five months of transplantation and shifted into larger plastic bags for post-acclimatization trail.

The acclimatization protocol

The experiments of acclimatization and postacclimatization phase were carried out under net house conditions and the following protocol was followed: The in vitro developed plantlets were rinsed with sterilized water to remove any media adhered to their roots.

To avoid any contamination, the plantlets were subsequently dipped into the fungicide solution at 2 g/l for 5 min. before implanted into the pots.

Immediately after potting, the plantlets were shifted under polyvinyl plastic sheet used as a low tunnel inside the net house to preserve the moisture and avoid drying.

The plantlets remained completely covered and undisturbed during the first week of the transplantation to maintain maximum percentage of relative humidity around the plantlets.

During the second week, the cover was slightly opened for 2 to 3 minutes every day to reduce relative humidity inside the tunnel.

During the third week, the cover used to be kept open for 5 minutes daily. Subsequently, the plantlets were completely uncovered at the end of fourth week of the transplantation.

The plantlets were misted twice a week during winter and twice to thrice a day during summer. The plants were watered regularly as per need according to the moistness of the substrates.

Recorded measures and statistical analysis of data

The number of surviving and dead plantlets was recorded after every 4 weeks from transplantation. The first and the second data during acclimatization stage were recorded after 30 and 70 days of transplanting

(DAT) respectively. While during post-acclimatization, these were recorded after 07 and 40 DAT respectively.

Table 1. The survival percentage of the ex vitro Orchids (Dendrobium nobile) transplanted on different soil mixtures after 120 days

Treatments	Coco Peat	Hill sand	Coal pieces	Date Palm chips	Brick chips	Saw dust	Perlite	Survival%
T^1	2	1	-	-	-	-	0.2	80
T^2	1	2	-	-	-	-	0.2	60
T^3	1	1	-	-	-	-	0.2	100
T^4	1	-	2	-	-	-	-	0
T^5	1	1	1	1	-	-	-	20
T^6	1	-	1	1	-	-	-	0
T^7	-	-	1	1	1	-	-	0
T^8	1	-	-	-	-	2	-	60
T^9	-	1	-	-	-	2	-	40

T1-9 stands for 1-9 different soil mixtures tested. Soil mixtures were prepared by v/v ratio.
"-" stands for "Nil".

Figure 1. Substrates used in soil mixtures during acclimatization phase; a. hill sand; b. brick chips; c. saw dust; d.date palm chips; e. coal pieces.

Each treatment was consisted of 6 replicates and each replicate (represented by a plastic pot/bag) contained 1 plantlet. The documented data was analyzed through ANOVA and the significance among treatment means was compared by using Standard Error & Duncan"s multiple range (DMR) test at $P \leq 0.05$ (Steel et al., 1997).

RESULTS AND DISCUSSION

During acclimatization phase

Micropropagation technique without a successful acclimatization is of no use even if it is most effective for in vitro shoot and root development. Deb and Imchen (2010) have reported that the broader use of micropropagation is mostly limited by high percentage of mortality or damage to plantlets during acclimatization. In the current study concerns the use of proper soil substrate and methodology to increase the percentage of survival and sustain healthy growth for the growing plants.

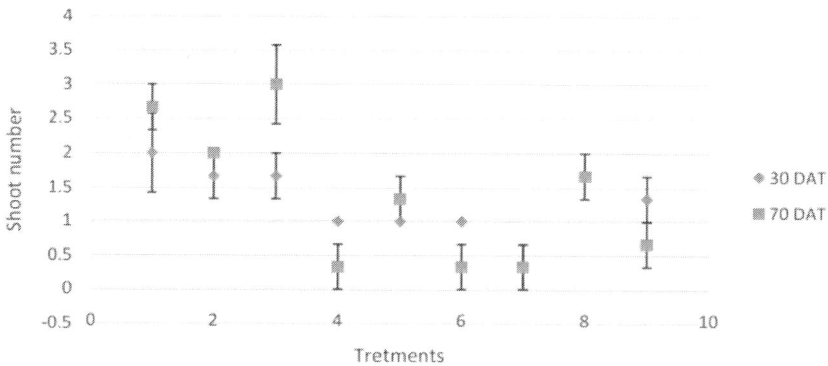

Figure 2. Effect of different soil mixtures on number of shoots during acclimatization.

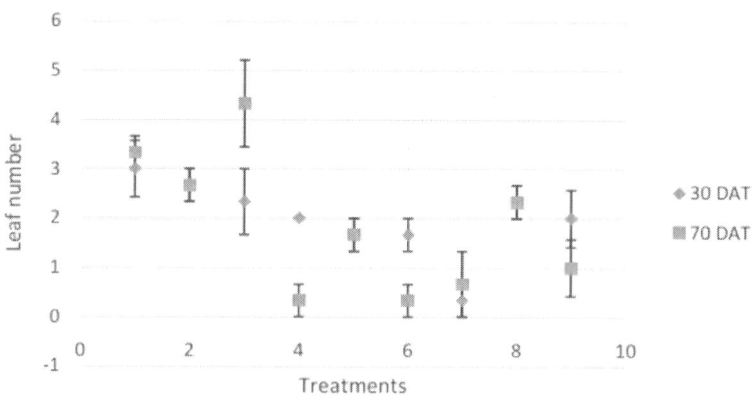

Figure 3. Effect of different soil mixtures on number of leaf during acclimatization.

The maintenance of maximum humidity and optimum temperature around the plantlets during initial days of transplantation is of utmost importance. During in vitro cultures, plantlets are grown under high humid conditions inside jars. Whereas in greenhouse or in filed, it is much lesser than the laboratory conditions (in vitro). Therefore, a rapid change in environmental conditions, especially in terms of temperature, humidity, CO2 concentration and irradiance, produces wilting and increases plantlets mortality rate (Pospíšilová et al., 1999).

In the present study, a maximum relative humidity of about 90-100% was maintained during the first week of transplantation by keeping the transparent tunnel closed. However, for a long time, the closing of tunnel increased temperature and maximized the relative humidity, which not only damaged the plantlets but also made them vulnerable to microbial contamination. Abul-Soad (2011) reported the same problem in date palm acclimatization, and suggested decrease relative humidity gradually after 3-7 days in order to reduce fungal infection and increase the adaptability of the plants to the ambient conditions. In the current study, the tunnel was regularly opened with increasing time duration after one week of transplantation to gradually lower down the relative humidity and temperature. It was observed that the plantlets under such tunnels remained healthy and free of contamination, wilting and any type of chlorosis.

It is reported that in order to establish a successful ex vitro transplantation, the plantlets of Phalaenopsis and Cattleya must achieve a growth stage with an appropriate sprout number, foliage area, root number and root length while they are in test tubes. Beside this, the selection of a proper soil substrate with low septicity, high aeration, permeability and acidity is the key condition for initiation of ex vitro growth (Diaz, 2010). The soil substrate beside the medium of mechanical support to plantlets also acts as storage pool for nutrients, air and water to roots.

The data presented in figure 2-5 showed that there was a significant difference among all the treatments (soil mixtures) in respect to shoots number/plant, leaves number/plant, leaf length and leaf width. The plantlets grown in T 3 (coco-peat, hill sand and perlite; 1.0:1.0:0.2) produced 100% survivability, noted after 120 DAT (Table 1). Therefore, the survived plants on this soil mixture (T3) were healthy and no any plant in all the replicates showed any type of leaf chlorosis or leaf fall (Figure 7a). However, in T1 , T2 and T8 , the survival rate was 80%, 60% and 60%respectively, noted after 120 DAT. While in rest of the treatments survival rate was very low, which decreased up to 0% in T4 , T6 and T 7 after 120 DAT. These results indicated the crucial impact of the soil mixture on successful transplantation process.

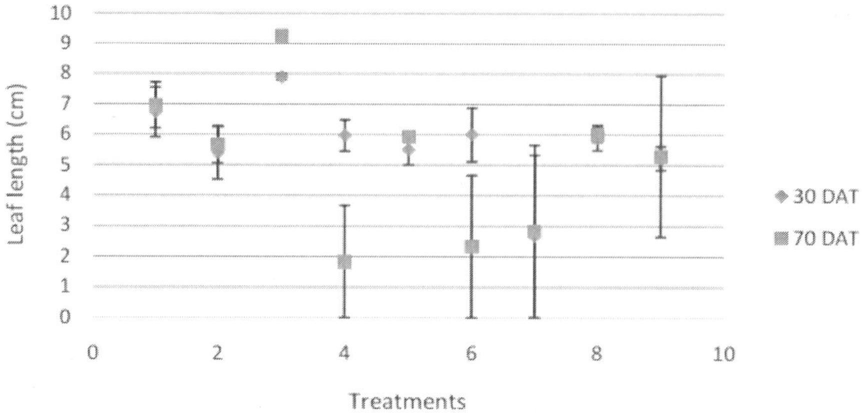

Figure 4. Effect of different soil mixtures on leaf length during acclimatization.

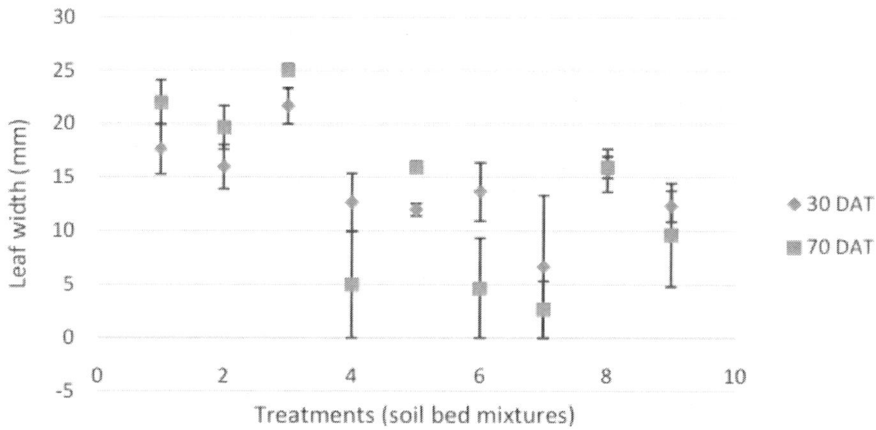

Figure 5. Effect of different soil mixtures on leaf width during acclimatization.

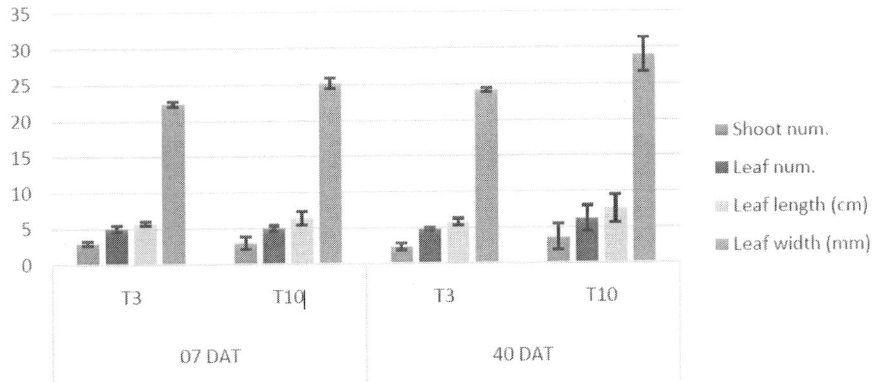

T^3 stands for control.
T^{10} sands for treatment.

Figure 6. Impact of improving soil bed composition on the further growth of the ex vitro survived Orchid during Post-acclimatization stage.

The number of shoots, leaves, leaf length and leaf width of the transplanted plants on T 3 and T1 were significantly better than other treatments (Figure 2-5). Nevertheless, the growth rate in T2 was not significant compared to T3 and T1 but it was relatively better than all other remaining treatments. The treatment number 8 (T8) although revealed very poor growth rate, its survival rate was comparatively better than the rest of the treatments except T3 , T1 and T2 . It was also observed that fungal contamination appeared in T8 and T9 only, which was perhaps due to high level of nutrient contents and unhygienic conditions in wood dust. The results in T8 indicated that the soil substrates having high water holding capacity increased survivability of in vitro grown plantlets in ex vitro net house conditions during acclimatization. However, there was a slow growth rate due to the fungal contamination.

The results showed that for acclimatization of in vitro grown orchids, the soil substrates having high water holding capacity and moderate aeration were the best. This explains why the treatment number 3 (T3) showed better performance among the rest of the treatments as it contained coco-peat, hill sand and perlite. The coco-peat is mostly used as a substitute of peat-moss and can absorb water up to 20% of its weight, which is suitable to improve water holding capacity of the substrate and nutrient contents at the initial stage. The hill sand is having relatively big size soil particles (Abul-Soadet al., 2012) which made good aeration and mechanical support to plantlets (Figure 1a). Whereas, the perlite is ground inert volcanic lava, which is expended at 800ºC, is also useful to provide aeration and permeability in any substrate mixture.

During post-acclimatization phase

Although in the previous stage of acclimatization stated above, there was a successful establishment of plantlets in terms of survival rate but the rate of shoot and root development was very poor and insignificant. The intention to establish this stage was to investigate the rate of vegetative growth of in vitro acclimatized plantlets on two different soil substrates (T3 and T 10) and in order to ensure the subsequent development of the acclimatized plants.

The results presented in figure (6) revealed that there was a significant difference between treatment (T10) and control (T3). The increase in growth rate of leaf length and width was significantly better in T10 as compare to T3 (Figure 7c). Moreover, the formation of new shoots, leaves and roots was also only observed in T 10 (Figure 8, left). It was observed that the in vitro developed leaves and roots had stopped growing further and new leaves and roots replaced them, which were very healthy and vigorous. Pospíšilová et al., (1999) reported similar results that in most of the plant species leaves formed in vitro do not grow further but are replaced by new leaves in ex vitro conditions.

On the other hand, it was observed that the high moisture contents and less aeration to roots for long time in T3 delayed their development and new root formation. Moreover, the in vitro formed roots were being damaged in T3 as showed drying out started from tips to base (Figure 8, right).

In general, the growth was very slow in terms of new shoot or leaf and root formation on T 3 during postacclimatization stage. Results in current study showed that during post-acclimatization stage, there was very rapid rate of new root and shoot formation produced on T 10 compared to T3 . It shows that the transfer of plantlets to more aerated substrate (T10) after their adaptation to ex vitro conditions (acclimatization) proved very significant for their subsequent rapid shoot formation and root development. These results were confirmed by Pospíšilováet al., (1999) and Van Huylenbroeck and De Riek (1995) who stated that during acclimatization of Spathiphyllum floribundum two different stages were observed; an adaptation period with slow shoot growth & root formation, followed by a period of fast growth of roots and shoots.

Figure 7.(a-c). Different stages of growth during acclimatization in the net house showing new shoots and leaf formation of orchid plantlets; a. acclimatization (T3 , 120 DAT); b. post-acclimatization (T10, 07 DAT); and c. post-acclimatization (T10, 40 DAT).

CONCLUSION

In the current study a protocol was developed for an economic and efficient acclimatization of in vitro grown plantlets to ensure subsequent growth of Orchid. The achieved results revealed that most of the ex vitro plants showed two stages of growth after successful transplantation: a stage of adaptation to ex vitro conditions associated with slow growth followed by a stage of rapid growth during their acclimatization. During the first phase a need of high humidity around the transplanted plantlets for few days which should be decreased gradually to normal ambient conditions. The best soil substrate having high water holding capacity, moderate porosity or aeration and low contamination rate composed of coco-peat: desert sand: perlite at ratio 1.0:1.0:0.2 (v/v/v). In the second phase (postacclimatization phase), rapid growth occurred with a necessity to change the soil bed composition into low water holding capacity and higher porosity or aeration. The current protocol can be applied for orchids as well as for other similar plant species.

REFERENCES

1. Abul-Soad, A.A.(2011): Micropropagation of date palm using inflorescence explants. In: Jain SM, Al-Khayri JM, Johnson DV (eds.) pp 91–118. Date palm biotechnology, Springer, Dordrecht.
2. Abul-Soad, A.A., Markhand,G.S. and Akhtar,N.(2012): Effect of different substrates on survival of transplanted Banana „Grand Naine" cultivar into greenhouse. In: Proc. XXVIIIth IHC – IS on Citrus, Bananas and Other Trop. Fruits under Subtrop. Conditions Eds.: J.-N. Wünsche and L.G. Albrigo, Acta Hort. 928: 131-138.
3. Colombo, L.A., Faria, R.T., Assis,A.M. andFonseca, I.C.B.(2005):Aclimatização de um híbrido de Cattleyaemsubstratos de origem vegetal sob doissistemas de irrigação. ActaScientiarum.Agronomy. 27 (1): 145-150.
4. Deb, C.R.andImchen, T.(2010): An efficient in vitro hardening technique of tissue culture raised plants. Biotechnol.1-5. Diaz, L.P., Namur, J.J., Bollati,S.A. andArce, O.E.A.(2010): Acclimatization of Phalaenopsis and Cattleyaobtained by micropropagation. Rev. Colomb. Biotecnol. 12 (2): 27-40.
5. Júnior, D.S. andVenturieri, G.A.(2011):Ex vitro acclimatization of Cattleyaforbesii and Laeliapurpurata seedlings in a selection of substrates. ActaScientiarum.Agronomy Maringá. 33 (1): 97-103.
6. Kabir, M.I., Mortuza,M.G. andIslam, M.O.(2012): Morphological features growth and development of Dendrobiumsp. Orchid as influenced by nutrient spray. J. Environ. Sci. Natural Resources. 5 (1): 309-318.
7. Lesar, H., Hlebec, B., Čeranič, N., Kastelec, D. andLuthar, Z.(2012): Acclimatization of terrestrial orchid BletillastriataRchb.f. (Orchidaceae) propagated under in vitro conditions. ActaagriculturaeSlovenica. 99 (1): 69-75.
8. Ortega-LOeza, M.M.,Salgado-Garciglia, R.,Gómez-Alonso, C. andÁvila-Díaz, I.(2011): Acclimatization of the endangered Mexican Epiphytic Orchid, Laeliaspeciosa (H.B.K.) Schltr. Eur. J. Environ. Sci. 1 (2): 48-54.
9. Pospíšilová, J., Tichá, I., Kadleček, P., Haisel,D. andPlzáková, Š.(1999):Acclimatization of micropropagated plants to ex vitro conditions. BiologiaPlantarium 42 (4): 481-497.
10. Steel, R.G.D., Torrie, J.H. andDickey, D.A.(1997): Principles and Procedures of Statistics: A biometrical approach, 3rd ed. McGraw Hill Book Co, New York.
11. Van Huylenbroeck, J.M.,DeRiek, J.(1995). Sugar and starch metabolism during ex vitro rooting and acclimatization of micropropagatedSpathiphyllum "Petite" plantlets. - Plant Sci. 111: 19-25.

CHAPTER 2

Within-Crop Air Temperature and Humidity Outcomes on Spatio-Temporal Distribution of the Key Rose Pest Frankliniella occidentalis

Hicham Fatnassi[1] *, Jeannine Pizzol[1] , Rachid Senoussi[2] , Antonio Biondi[1,3,] Nicolas Desneux[1] , Christine Poncet[1] , Thierry Boulard[1]

[1] *INRA, UMR1355, ISA, 400 Route des chappes, 06903, Sophia Antipolis, France,*

[2] *INRA, UR 0546, Biostatistique et Processus Spatiaux, F-84914, Montfavet, France,*

[3] *University of Catania, Di3A, Via Santa Sofia 100, 95123, Catania, Italy*

ABSTRACT

Frankliniella occidentalis (Pergande) is a key pest of various crops worldwide. In this study, we analyse the dependence of the infestation of this pest on spatially distributed micro climatic factors in a rose greenhouse. Despite the importance of this subject, the few existing studies have been realized in laboratory rather than in greenhouse conditions. However, recent progress on greenhouse microclimate characterisation has highlighted the strong indoor climate heterogeneity that may influence the within-crop pest distribution. In this study, both microclimate (air temperature and humidity) and thrips distribution were simultaneously mapped in a rose greenhouse. The measurements were sensed in a horizontal plane situated at mid-height of the rose crop inside the greenhouse. Simultaneously, thrips population dynamics were assessed after an artificial and homogeneous infestation of the rose crop. The spatio-temporal distribution of climate and thrips within the greenhouse were compared, and links between thrips infestation and climatic conditions were investigated. A statistical model was used to define the favourable climate conditions for thrips adults and larvae. Our results showed that (i) the air temperature and air humidity were very heterogeneously distributed within the crop, (ii) pest populations aggregated in the most favourable climatic areas and (iii) the highest population density of thrips adults and larvae were recorded at 27°C and 22°C for temperature and 63% and 86% for humidity, respectively. These findings confirm, in real rose cropping conditions, previous laboratory studies on the *F. occidentalis* climatic optimum and provide a solid scientific support for climatic-based control methods against this pest.

INTRODUCTION

Thrips (Thysanoptera: Thripidae) are important opportunistic pests of various cultivated crops around the world [1,2,3]. More specifically, *Frankliniella occidentalis* (Pergande) is a highly polyphagous phytophagous insect and it is considered among the most important pests of greenhouse vegetable and ornamental crops worldwide [4,5,6]. Native to North America [7], it is an invasive species in Europe. It was accidentally introduced into Europe in the early 80s, and reached France in 1986 [4–8]. It causes considerable damage to commercial flower crops, through direct feeding on marketable plant organs, i.e., flowers or flower buds, and by plant-viruses vectoring [3,9,10]. Damage related to *F. occidentalis* infestation varies according to the host plants. On rose crops such as necroses and deformations have been observed on young leaves, but it is mainly on flowers that effects are the most serious [11–12].

The most common practices used for controlling thrips pests rely on the chemical approaches [1,3,14]. Unfortunately, the intensive horticulture cropping activity selected many *F. occidentalis* insecticide-resistant populations [14,15,16]. Moreover, the widespread insecticide use is often associated with multiple human health [17] and environmental issues [18, 19, 20]. This exhorts for developing environmentally sound and sustainable management pest strategies restraining the use of insecticides [6–21]. For this purpose, a thorough knowledge on pest biology and ecology in its environment is strongly needed [22,23,24]. In particular, air temperature and humidity conditions closely surrounding the thrips habitats are known as important factors influencing their populational biology.

Several laboratory studies have shown that various *F. occidentalis* life history traits, such as development duration, longevity, sex-ratio and fertility, can largely depend on the climatic conditions tried out in laboratory [25,26,27]. Thereby, Robb and Parella [28] showed the longevity and fertility dependence of *F. occidentalis* on air temperature, finding that the highest longevity occurs at 20°C whereas the highest fertility is achieved between 20 and 27.2°C. However, all these life-table studies were carried out in strictly controlled laboratory conditions, and not in the actual cropping settings, in which the climate is often very heterogeneous [29]. This heterogeneity has been recently demonstrated in several studies evaluating the greenhouse climate variations using Computational Fluid Dynamics (CFD) modelling [30,31,32,33]. This modeling approach details the airflows and heat exchanges by numerically solving the equations governing the air dynamic in the greenhouse.

In this context, the deepening of our knowledge concerning the relation between fluctuating micro-climate cropping conditions and thrips population dynamic responses is crucial in any attempt of pest management. So, our study aimed at characterizing/quantifying the effect of air temperature and humidity on the spatio-temporal distribution of *F. occidentalis* populations in a protected rose cropping system.

MATERIALS AND METHODS

Experimental greenhouse and climatic measurements

The experiment was performed during the summer of 2008 at the facilities of the INRA research center in Sophia Antipolis, South-Eastern France (43° 36' 44.9'' N latitude, 07° 04' 40.4'' E longitude and 125 m altitude). We used a compartment of about 40 m^2 (6.2 m wide and 6.2 m long) in a rose experimental glasshouse. The compartment was naturally ventilated by roof opening oriented east-west and equipped with an insect-proof screen.

15 individual air temperature and humidity measurements were set up at equal distance throughout the greenhouse area in a horizontal plane at the level of the rose buds (1.3m height) (Fig 1) [30,31]. The thermo-hygrometer sensors (model EE06, Intertechnique, Plaisir en Yvelines, France) were connected to two data loggers (model CR 23, Campbell, Shepshed, United Kingdom) on which data were transferred every second. The average

values were recorded every 15 min before being downloaded and processed in a computer for displaying air temperature and humidity contours plots using 3DField software.

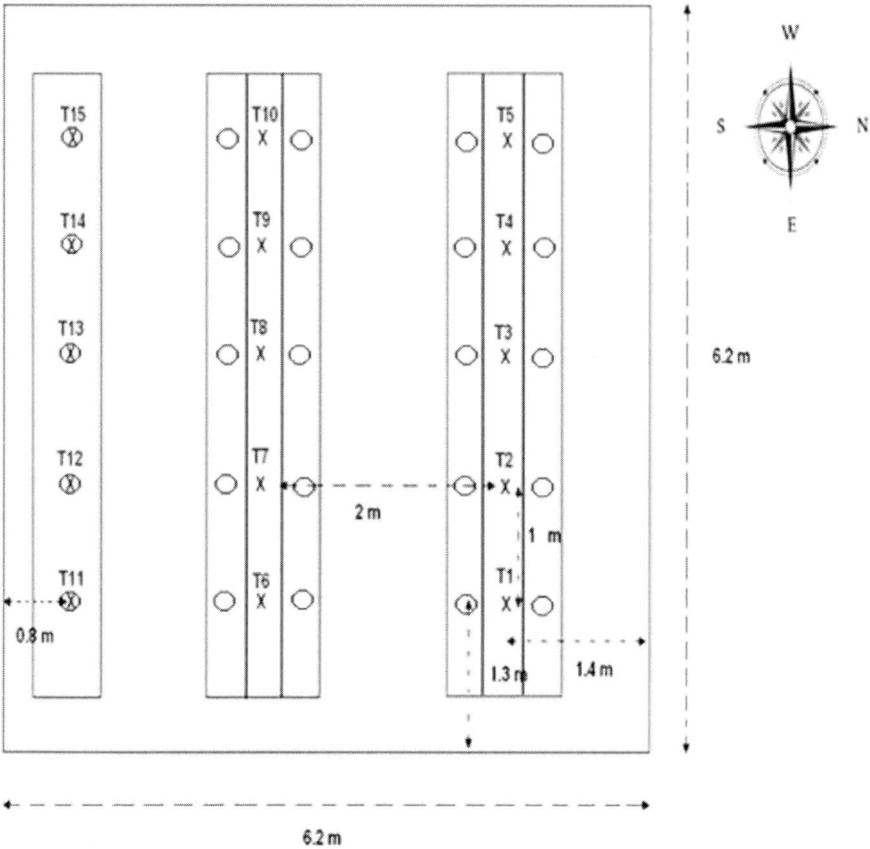

Fig 1. Layout of temperature and humidity sensors (x) and sampled plants (o) locations inside the experimental greenhouse.

Biological materials and samplings

The plants used in the experiment were young pest- and pesticide-free rose plants, *Rosa hybrida* L. of the thrips-susceptible variety Aubade (*Nirp International*), obtained using in vitro synthesis cultivation. Plants were grown on new rock wool slabs in single rows. Five rows of 25 rose plants (25±2 cm high) were transplanted on May 15, 2008 (week 21), see Fig 1 for the layout and details of the greenhouse. Routine plant management was carried out, such as automatic watering with a nutrient solution and flower harvesting. To avoid influencing the thrips population with this agronomic practice, cut flowers were left close to the cutting site and in contact with the origin plant for at least two days. Pesticide applications were strictly avoided and, in order to initiate the experiment at very low or null densities of arthropods in the crops, we adopted preventive arthropod control measures and mass trapping by placing yellow sticky bands (0.3 x 6 m) among the plant rows. Traps were removed the day before the first artificial pest release.

A colony of *F. occidentalis* was initiated starting from individuals collected on protected rose crops and was maintained on caged rose plants in climatic chambers (24±1°C; 65% RH). The rose crop was infested homogeneously by *F. occidentalis* by releasing recently moulted thrips adults in the whole experimental crop. The artificial and homogenous infestation was performed through five weekly releases (from week 33 to 37) of 50 adults each (10 thrips per row) into the greenhouse compartment. Thrips were handled through a fine paint brush and gently released on young leaves.

Monitoring of thrips population dynamics was performed weekly on rose flowers using a standardized technique, the knock-down technique, which consists of tapping the rose flowers onto a white sheet of paper and counting larvae and adults [13]. Each week, 25 roses, located in the same positions as the temperature and humidity sensors, were sampled in the greenhouse using this non-destructive technique and the captured thrips were immediately released on the sampled plants. We recall that for marketable production only one rose was left per plant and that the greenhouse includes two double and one single row of plants. For double rows, the observations are the average of the two plants located around the climatic sensor sites. To sum up, there are 25 (= 5 x 2+ 5x2 + 5) observed roses corresponding to only15 (= 5+5+5) climatic sensor sites.

Statistical models

To investigate the relationship of the infestation levels of thrips on the climate conditions and to determine their preferences regarding temperature and humidity in the actual greenhouse conditions, we specifically developed two statistical models relating climatic variates to the larva and adult thrips locations inside the greenhouse.

For each sampled monitoring site i, i = 1, ...,15, and observation time t_j (in weeks) $t_j = j$; j = 1,..., 5., let H_{i,t_j} denotes the weekly mean of humidity, T_{i,t_j} the weekly mean of temperature, N^L_{i,t_j} the number of thrips larvae and N^A_{i,t_j} the number of thrips adults. First, we assumed that thrips populations in the different sites evolved independently (without specified spatial interactions) from each other. This did not exclude that the thrips individuals behaved similarly when moving from or settling in sites (i.e. mean behaviour but without individual interactions). Next, regarding the thrips dynamics, we sensibly assumed that it was a Markovian process in time where the present state randomly depended only on previous local climatic conditions and previous thrips population sizes. Moreover, in modelling the thrips dynamics we hypothesised the existence of optimal values (or ranges) for both temperature and humidity that favour the presence or development of thrips.

Our first model takes into consideration that the larvae resulted from egg laying events by adults that, in average, are known to occur 5.3 days before the L1 stage and 8.3 days before the L2 stage at 27°C [1]. Therefore, we considered that the density of *F. occidentalis* larvae at a given time *t* depends on temperature and humidity conditions where the adults were ovipositing (and feeding) at *t*-7 days. Climatic conditions at time *t*-7 has likely an influence on the egg laying activity of adults and subsequent location of the larvae at the time *t* (i.e., the propensity to successful develop from an egg), because larvae do not have wings and thus cannot move far.

In the same way, we took into consideration that the adults at time *t* were the result of egg laying events which took place 14 days earlier. We retained the climatic conditions (temperature and humidity) at *t*-14 because the required time to complete development of *F. occidentalis* at 25°C from oviposition to adult emergence is 14.3 days [25].

Besides, as thrips were artificially inoculated with merely two individuals, the adult and larva population sizes become noticeable only from the fourth week, and as a temporal Markovian assumption of one and two week lags was used, we only dealt with the last three observation dates regarding the first two as initial conditions and explaining covariates. Here are now the two statistical models, denoted M^L (L for larvae) and M^A (A for adults), devoted to the inference of some issues concerning how and to what extent larva and adult trips move and behave according to climate conditions.

Modelling of how larva counts statistically depend on adults and favourable climatic conditions of the week earlier?

Model M^L: For larva population dynamics, we assumed that at time t_j, conditionally to previous adult population size, i.e. N^A_{i,t_j-1} and climatic conditions H_{i,t_j-1} and T_{i,t_j-1}, the larva population size N^L_{i,t_j} is Poisson distributed with mean

$$\lambda^L_{i,t_j} = (e^{\alpha^L_0} + e^{\alpha^L_A} N^A_{i,t_j-1})(1 + e^{\{-\alpha^L_T (T_{i,t_j-1} - m^L_T)^2 - \alpha^L_H (H_{i,t_j-1} - m^L_H)^2\}}) \quad \textbf{(1)}$$

The parameter α_0^L or rather $e^{\alpha_A^L}$, stands for a completely random contribution, (denoted hereafter CRC) in thrips larvae to account for potential immigration, larvae and adults previously present but not observed, etc. ..., The component $e^{\alpha_A^L} N_{i,t_j-1}^A$ says that an adult observed the previous week is expected to engender $e^{\alpha_A^L}$ larvae during the next week. The last term refers, if ever, to the climatic component. It says that the expected number of larvae increases under optimal/favourable climatic conditions (optimal temperature value m_T^L and optimal humidity value m_H^L). It also says that temperature and humidity act independently on thrips preference and that these contributions decrease with Gaussian rates α_T^L and α_H^L as temperature and humidity departed from these optimal values. In this statistical model, the vector of the unknown parameters $\theta^L = (\alpha_0^L, \alpha_A^L, \alpha_H^L, m_H^L, \alpha_T^L, m_T^L)$ has length 6.

Modelling of how adult counts statistically depend on larvae present a week before and on climatic conditions present two weeks earlier?

Model M^A is similar to Model M^L. We assumed in this case that adult population depends on larvae population N_{i,t_j-1}^L of the previous week but, contrarily to the previous model, it depends on the attractiveness of climatic conditions present two weeks before

H_{i,t_j-2} and T_{i,t_j-2}. So, the Poisson distributions of counts N_{i,t_j}^A of adult thrips observed at time t_j has the following conditional expectation:

$$\lambda_{i,t_j}^A = \left(e^{\alpha_0^A} + e^{\alpha_L^A} N_{i,t_j-1}^L\right)\left(1 + e^{-\alpha_T^A (T_{i,t_j-2} - m_T^A)^2 - \alpha_H^A (H_{i,t_j-2} - m_H^A)^2}\right) \quad \textbf{(2)}$$

Similarly, we can recover in this formula 3 components: the completely random contribution, the larva contribution and the climatic component. The corresponding parameters have the same interpretation as in the first model but refer to adult thrips. Here, the vector of parameters is denoted $\theta^A = (\alpha_0^A, \alpha_L^A, \alpha_H^A, m_H^A, \alpha_T^A, m_T^A)$ and has the same length 6.

Actually, we tested other models of dynamics which revealed to be statistically "unstable" due to the relatively low size of the experimental data set (in time and in space). The two models proposed here for larva and adult populations of thrips revealed numerically stable proving that enough informative data was presently provided in order to test some relevant hypotheses.

Parameter Estimation and hypothesis testing.

The assumption that local thrips populations were independent from plant to plant and behaved according to a Markovian dynamics with the previous form of conditional Poisson distributions makes it possible to express the likelihood of the observed data. For simplicity, we only write down the log-likelihood of Model M^L (otherwise replace L by A for model M^A):

$$\ln(\theta^L) = \sum_{i=1}^{15} \sum_{j=3}^{5} \left(-\lambda_{i,t_j}^L + N_{i,t_j}^L \log(\lambda_{i,t_j}^L)\right) \tag{3}$$

First, for the estimation issue, the maximum likelihood estimators $\hat{\theta}^L$ and $\hat{\theta}^A$ in Table 1 were computed using library "bbmle" of R Software (R Development Core Team 2005). Note that the same subroutine also yields the approximate standard deviations of the parameters.

Table 1. Parameter estimates in full models for larva and adult presence in relation to climatic and other covariates.

Full Model M^L (Larvae) $-2\ln(\hat{\theta}^L) = -298.8826$			Full Model M^A (Adults) $-2\ln(\hat{\theta}^A) = -187.908$		
Parameter	*Estimate*	*Std. Error*	*Parameter*	*Estimate*	*Std. Error*
α_0^L	-0.155	0.210	α_0^A	1.264	0.087
α_A^L	0.039	0.114	α_L^A	-0.665	0.363
m_T^L	22.004	0.119	m_T^A	27.010	0.104
α_T^L	1.893	4.748	α_T^A	5.298	9.44e-06
m_H^L	86.380	0.745	m_H^A	62.790	0.215
α_H^L	0.045	0.023	α_H^A	1.24	5.78e-05

Next, concerning hypothesis testing, if we let the sign * stand for L (larvae) or A (adults), we tested several specific hypotheses, say "$H_{0X} : \theta^* \in \Theta_{0X}^*$" (corresponding to a sub-model M_{0X}^*) vs. the alternative $\theta^* \notin \Theta_{0X}^*$ (within the global model M^*), using the classical asymptotic log likelihood ratio test statistic $\Delta_{0X}^* = -2(\ln(\hat{\theta}_{M_{0X}^*}) - \ln(\hat{\theta}_{M^*}))$. Statistical theory asserts that under the assumptions defining sub-model M_{0X}^*, the statistic Δ_{0X}^* is asymptotically $\chi^2(k)$ distributed, k being the difference of the dimensions of parameter spaces under M^* and M_{0X}^* [34]. The significance of all our tests referred to the level of 1%. Actually, we gave the values of Δ_{0X}^* and the quantiles $P_{\chi^2(k)}^{-1}(0.99)$ of the significance levels to better see the departure from the corresponding p-values.

In this paper, we tested for both models three types of hypotheses. (1) We tested the non-effectiveness of the completely random contribution (CSC), that is M^*_{0R} "$\alpha^*_0 \approx -\infty$" or rather more practically $\alpha^*_0 = -12$ so that $e^{\alpha^*_0} \leq 10^{-5}$. (2) We also tested the non-effectiveness of larva (resp. adult) contribution to adult presence (resp. larva presence) that is M^A_{0L} : "$\alpha^A_{0L} \approx -\infty$", or rather $\alpha^A_{0L} = -12$ so that $e^{\alpha^A_{0L}} \leq 10^{-5}$ (resp. M^L_{0A} : "$\alpha^L_{0A} \approx -\infty$", or rather $\alpha^L_{0A} = -12$ so that $e^{\alpha^L_{0A}} \leq 10^{-5}$). The corresponding test statistic Δ^*_{0X} is expected to behave as a $\chi^2(1)$ distribution under the respective sub-models (1) and (2).

Finally, we tested (3) the global non-effectiveness of the climate component, via the sub-model M^*_{0C} assuming the hypothesis: "$e^{-\alpha^*_T (T_{i,t_j}-m^*_T)^2-\alpha^*_H (H_{i,t_j}-m^*_H)^2} = 0$". From a practical point of view, this is performed by assuming: $\alpha^*_T = \alpha^*_H = 50, \quad m^*_T = 10, \quad m^*_H = 50$ so that $e^{-\alpha^*_T (T_{i,t_j}-m^*_T)^2-\alpha^*_H (H_{i,t_j}-m^*_H)^2} \approx 0$ whatever are the values taken by T_{i,t_j}, H_{i,t_j} and in this case, Δ^*_{0C} should behave as a $\chi^2(4)$ under hypotheses defining M^*_{0C}.

RESULTS

Experimental results

The thrips populations began to increase slowly after the start of the experiment in calendar week 33. From week 37 on, there was a sudden and persistent increase of the number of thrips throughout the experiment until its end in week 39, then the number of thrips larvae gradually increased from week 37 to 39 (respectively 3, 75 and 161 larvae) together with the number of adults: 44, 119 and 182 adults (Fig 2). During weeks 38 and 39, the number of adult thrips and larvae increased 4 and 7 fold, respectively. At the end of week 39 the mean number of thrips per flower was equal to 13.7 and the experiment was ended, because this level was higher than the admitted economic threshold [13]. No thrips natural enemies or other pests were found on the experimental crop during the whole experiment.

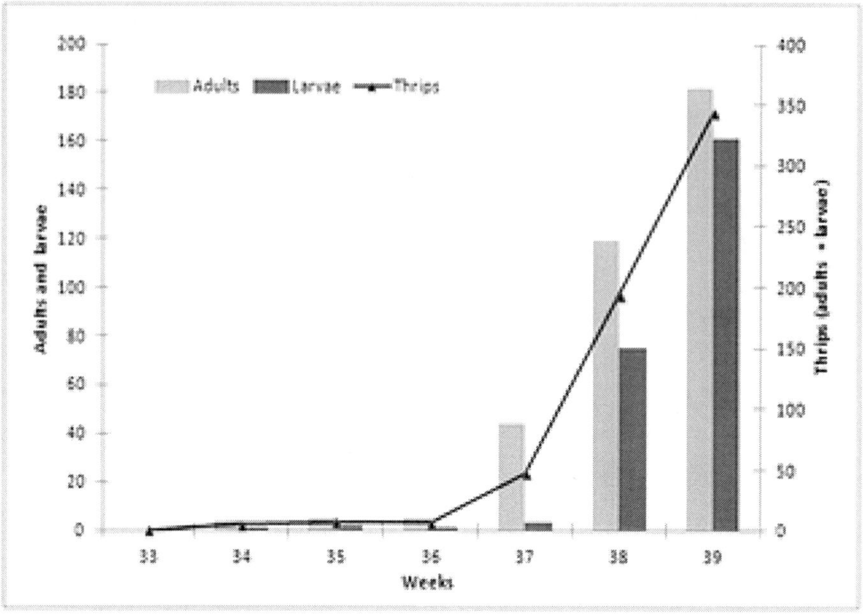

Fig 2. Time evolution of larvae and adults population of thrips in the experimental crop.

The thrips population was not distributed homogeneously within the crop cover (Fig 3A). It was higher in the south and in the north and lower in the central part of the compartment. The climatic conditions (diurnal temperature and humidity respectively) inside this greenhouse compartment measured in a horizontal plane located at 1.3 m above ground were unevenly distributed (Fig 3B and 3C). The diurnal temperature varied between 25 and 29°C (Fig 3A). Air humidity was also heterogeneous inside the greenhouse with the highest values observed in the central part. The lowest humidity occurred along the north-western part of the greenhouse.

Mapping larva and adult densities with respect to climatic conditions

The density map of thrips larvae with respect to air temperature and humidity (Fig 4) showed that the number of thrips larvae was high in spatial areas whose temperature and humidity are around 22°C and 86%, and decreased outside these zones. Similarly, when comparing the distribution pattern of thrips adults with respect to temperature and humidity (Fig 5), the highest population density of thrips adults was recorded at 27°C for temperature and 63% for humidity

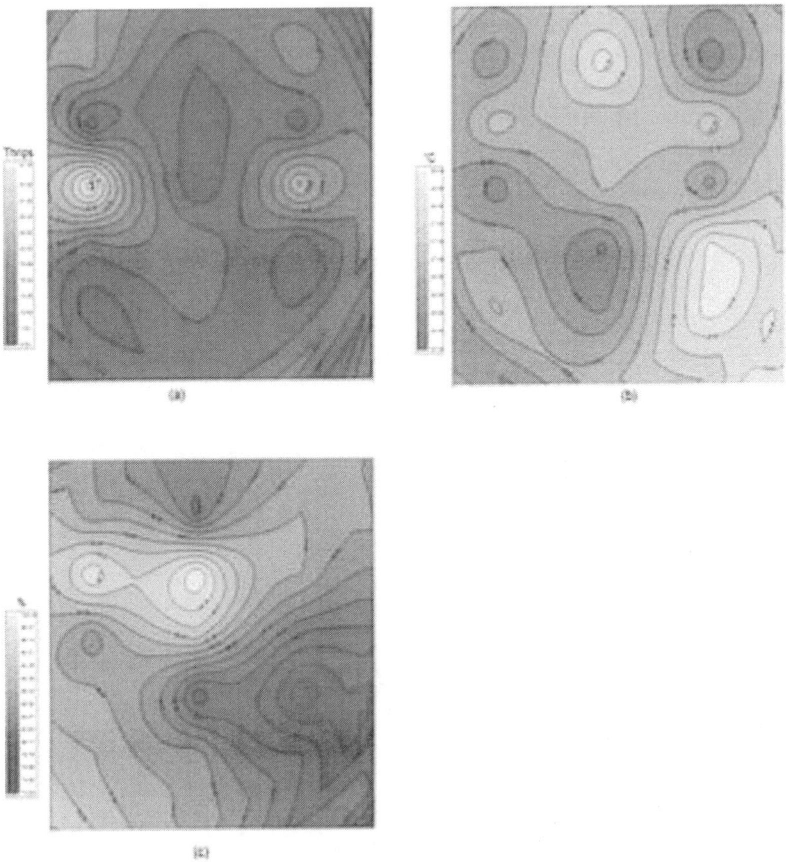

Fig 3. Contour maps of the (a) thrips population size (b) weekly mean diurnal temperature and (c) weekly mean air humidity within the experimental crop.

The size of circles corresponds to the number of larvae observed over the last three weeks. The * sign corresponds to estimated preferred climatic conditions.

The size of circles corresponds to the number of adults observed over the last three weeks. The * sign corresponds to estimated preferred climatic conditions.

Statistical results

Recall first that, the significance of all our tests referred to a level of 1% to which corresponds a quantile $P^{-1}_{\chi^2(k)}(0.99)$. The statistical results concerning hypothesis testing (or equivalently of submodels) are summed up in Tables 2 and 3. So, for both models, the effectiveness of a complete random contribution is highly significant. More specifically, for larva presence, we obtained:

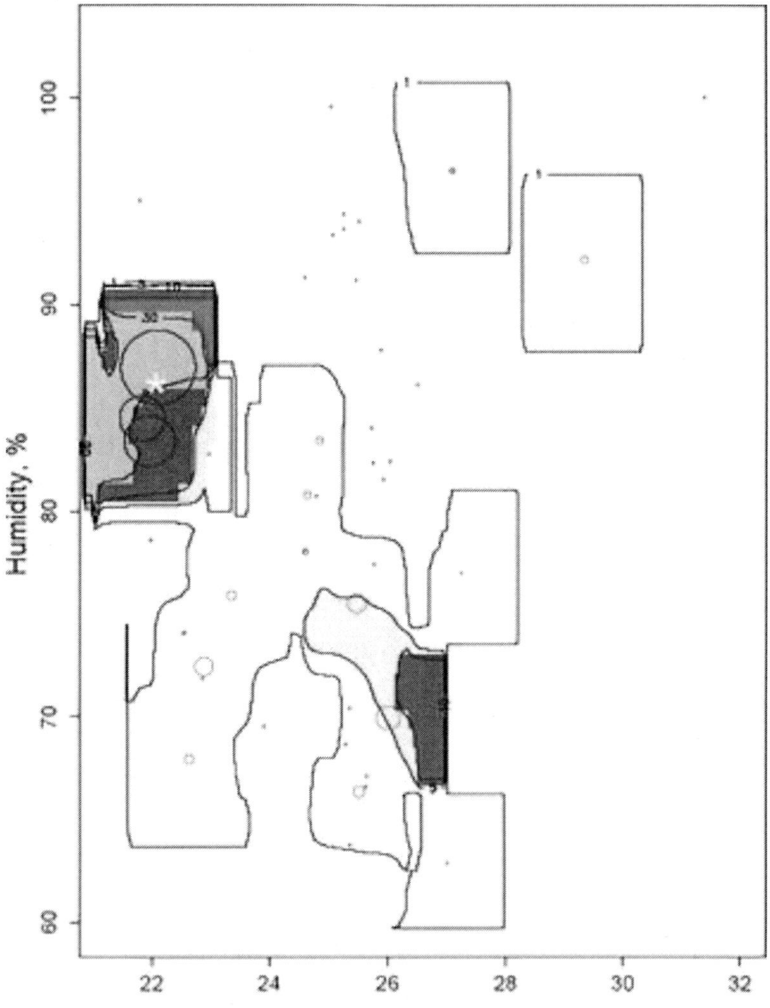

Fig 4. Within-crop distribution of thrips larvae in relation to the temperature and humidity.

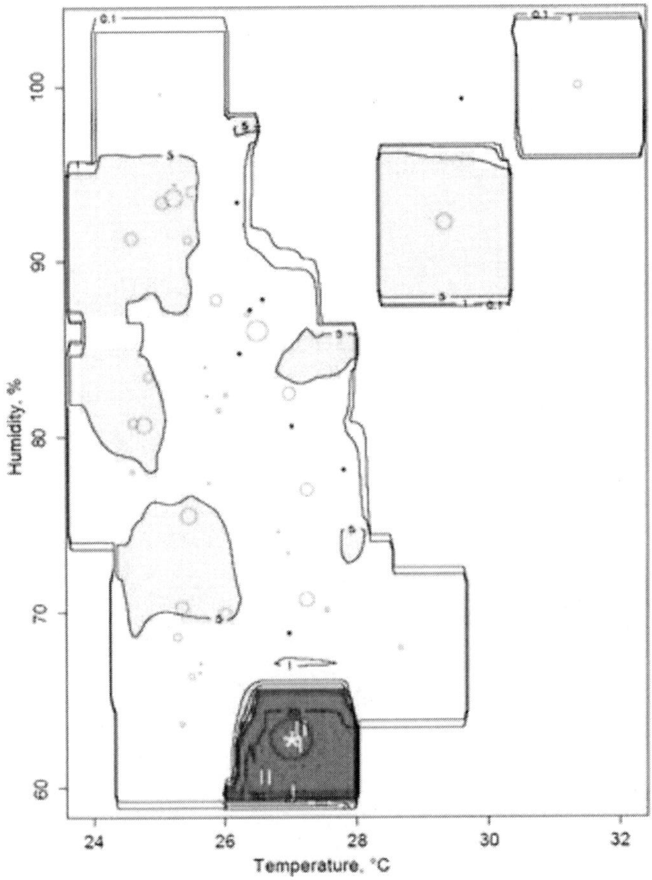

Fig 5. Within-crop distribution of thrips adults in relation to the temperature and humidity.

$\Delta_{0R}^{L} = -2(\ln(\hat{\theta}_{0R}^{L}) - \ln(\hat{\theta}^{L})) = 369.46 \gg 6.64 = P_{\chi^2(1)}^{-1}(0.99),\ (p_{\text{value}} \approx 0)$ and for adult presence we obtained an even greater significance $\Delta_{0R}^{A} = -2(\ln(\hat{\theta}_{0R}^{A}) - \ln(\hat{\theta}^{A})) = 2023.5 \gg 6.64 = P_{\chi^2(1)}^{-1}(0.99),\ (p_{\text{value}} \approx 0)$.

Next, concerning the contribution of the adults present a week before on the larva presence of the current weak revealed also highly significant: $\Delta_{0A}^{L} = -2(\ln(\hat{\theta}_{0A}^{L}) - \ln(\hat{\theta}^{L})) = 126.7 \gg 6.64 = {}_{P}^{\chi^2(1)-1}(0.99)(\text{i.e. } p_{\text{value}} \approx 0)$. Similarly, the contribution of larvae present a week before on adult presence of the current weak, even if lesser, was also highly

significant:
$\Delta^{A}_{0L} = -2(\ln(\hat{\theta}^{A}_{0L}) - \ln(\hat{\theta}^{A})) = 34 \gg 6.64 = P^{-1}_{\chi^2(1)}(0.99), \quad (\text{i.e. } p_{value} \approx 0)$

Table 2. Larva presence: Hypothesis testing on covariate effect.

Sub Models M^{L}_{0X}	$-2\ln(\hat{\theta}^{L}_{0X})$	$-2(\ln(\hat{\theta}^{L}_{0X}) - \ln(\hat{\theta}^{L})) \sim \chi^2(k)$	P-value
$H^{L}_{0R}: \alpha^{L}_{0} = -\infty$ "no CR contribution"	70.58	$369.46 \sim \chi^2$ (1)	0
$H^{L}_{0A}: \alpha^{L}_{A} = -\infty$ "no adult contribution"	-172.19	$126.69 \sim \chi^2$ (1)	0
H^{L}_{OC}: "no climatic effect"	**-238.566**	$60.3 \sim \chi^2$ (4)	0

Table 3. Adult presence: hypothesis testing on covariate effect.

Sub Models M^{A}_{0X}	$-2\ln(\hat{\theta}^{A}_{0X})$	$-2(\ln(\hat{\theta}^{A}_{0X}) - \ln(\hat{\theta}^{A})) \sim \chi^2(k)$	P_value
$H^{A}_{0R}: \alpha^{A}_{0} = -\infty$ "no CR contribution"	2015.58	$2023.5 \sim \chi^2$ (1)	0
$H^{A}_{0A}: \alpha^{A}_{L} = -\infty$ «no larva contribution »	-153.06	$34 \sim \chi^2$ (1)	0
H^{A}_{OC}: "no climate effect"	-161.45	$26.46 \sim \chi^2$ (4)	10^{-5}

Finally, for the effectiveness of climate effects, the tests revealed also very significant. When testing the contribution in model M^L we got $\Delta^{L}_{0C} = -2(\ln(\hat{\theta}^{L}_{0C}) - \ln(\hat{\theta}^{L})) = 60.32 \sim \chi(4), \quad (\text{i.e. } p_{value} \approx 0)$ and in model M^A $\Delta^{A}_{0C} = -2(\ln(\hat{\theta}^{A}_{0C}) - \ln(\hat{\theta}^{A})) = 26.45 \sim \chi(4), \quad (\text{i.e. } p_{value} \leq 10^{-4})$.

Since the full models M^L and M^A proved to be statistically justified regarding the effectiveness of the three distinct components, we turn now to the quantification of these contributions, that is, to parameter estimation issues (Table 1). So, for larva presence (model M^L), the completely random contribution on larva presence was about exp(-0.155) = 0.86 larvae per site and per week. Concerning the contribution of one single adult previously present, we obtained exp(0.039) = 1.04 larvae per adult. As for the optimal conditions, we got 22.00°C for temperature and 86.38% for humidity. With these most favourable climatic conditions, the contributions were twice. The values of the Gaussian decreasing rates in M^L were estimated to 1.89 for temperature and to 0.045 for humidity. Note that even if we took account of the difference between temperature and humidity ranges, the Gaussian decrease rate was more abrupt for temperature than for humidity.

For adult presence (model M^A), there were in average a completely random contribution of exp(1.26) = 3.54 adults per site per week and that one previously observed larva yielded exp(-0.665) = 0.51 adults in mean for the current week. As for the optimal climatic conditions for adults, these were

27.01°C for temperature and 62.79% for humidity. Under these optimal conditions, the number of adults doubled. Similarly, note the higher value of the Gaussian rate for temperature 5.298 compared to the Gaussian rate for humidity 1.24. Note also the high values of these Gaussian rates in adult model when compared to those of larva model. The higher the values of the Gaussian decreasing rates, the smaller are the favourable climatic zones.

DISCUSSION

Fine scale climate measurements and thrips distributions within a greenhouse rose crop were monitored to investigate the dependence of thrips infestation on climate conditions. For that purpose, a specific statistical model relating to space-time locations of thrips inside greenhouse was used to define the favourable climate conditions for thrips adults and larvae. On the one hand, data showed that air temperature and air humidity were very heterogeneously distributed inside the greenhouse. This heterogeneity was likely due to the difference of the transpiration activity of plants and air circulation inside the greenhouse [31]. On the other hand, our results highlighted that diurnal temperature and humidity exert a direct effect on the distribution of thrips (*F. occidentalis*) within the greenhouse crop. More precisely, we found that the non-homogeneous distribution of thrips infestation inside the greenhouse was partly due to the research of favourable climatic zones (Tables 2 & 3).

The dependence of *F. occidentalis* from temperature regimes has been demonstrated under well-controlled laboratory conditions by Robb [35], Robb & Parella [28] and Brodsgaard [36]. They showed how thrips gain an optimum development between 25°C and 27°C and that the female laying inflates from 133.6 eggs at 25°C to 228.6 eggs at 27.2°C. The increase of the *F. occidentalis* population density can also be due to the conjunction of a greater longevity at 27.2°C almost three times more than at 30°C according to Robb and Parella [28], a shorter development period (see Fig 6A) and a lower mortality rate (see Fig 6B) [35]. These studies are in line with our results relating to the temperature preference of adult thrips around 27°C (Fig 5). Within this favorable temperature, the rapid development of *F. occidentalis* and the high fertility of females, associated with a low mortality rate, had probably a significant effect on the rate of the pest population increase. In addition, our results yielded the mean favourable temperature for larvae.

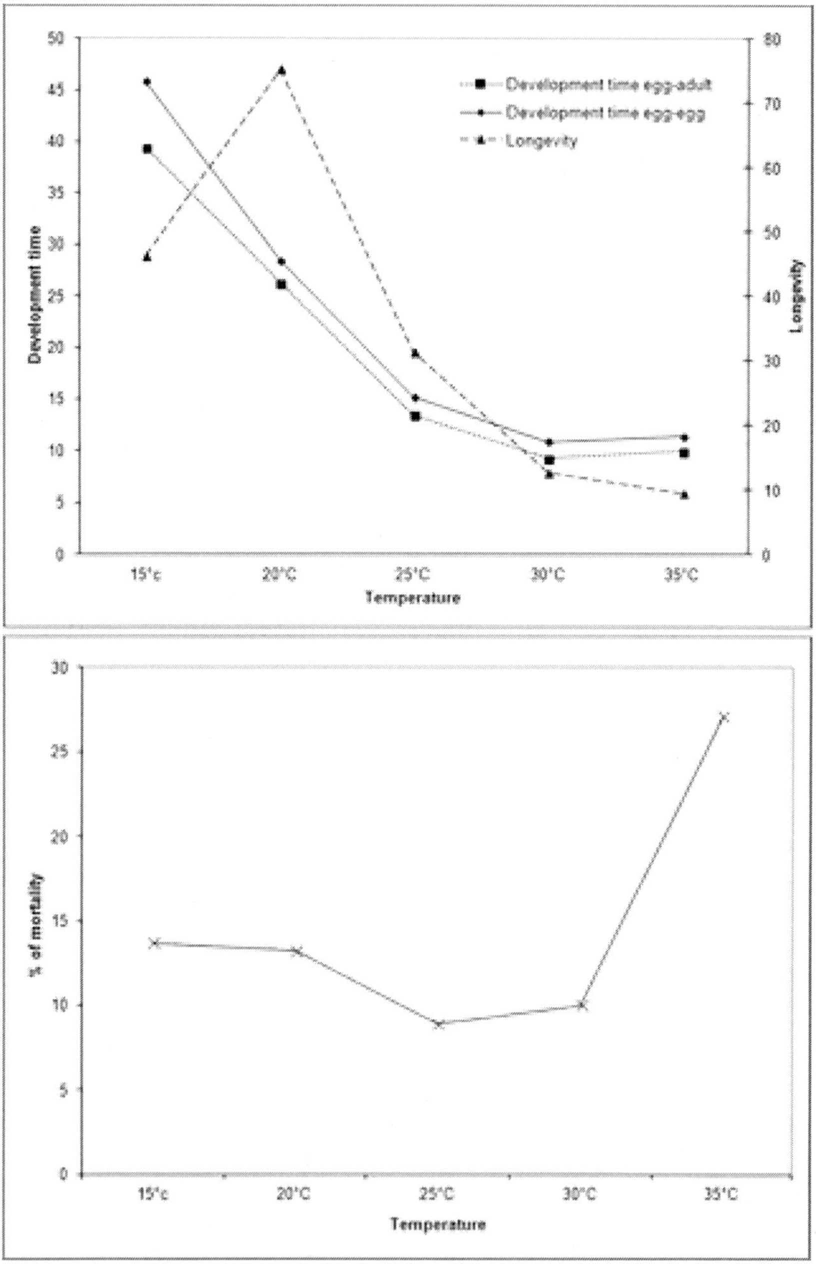

Fig 6. Dependence of development time, longevity (a) and mortality (b) of *F. occidentalis* on temperature feeding on chrysanthemum (modified from Robb, 1989).

Concerning the influence of humidity, Steiner et al. [37] and Shipp & Gillespie [38] showed that all stages of *F. occidentalis* are sensitive to humidity. The humidity range between 65% and 85% was unfavourable to the larvae and more favourable to the pupae and adults [37]. In this context, our statistical results were more precise in the sense that they pinpointed the optimal humidity values of 86% for larvae and 63% for the adults.

The higher concentration areas of thrips (adults or larvae) found in the greenhouse (Fig 3), may be due to the concomitance of two phenomena acting at two distinct scales: (i) locally, the higher population reproduction rate induced by optimal environmental conditions, and (ii) at a larger scale, the thrips displacements and in particular to areas with more favourable conditions. However, although the dispersal ability of *F. occidentalis* females has been estimated to be 0.18 to 0.29 m per day [39] corresponding to a maximum of 2 m per week in our study time scale, further specific tests on the *F. occidentalis* dispersal capacity on rose crop should be carried out to support this hypothesis. Indeed, the dispersal ability of insects can largely vary with the attributes and availability of their hosts [40, 41]. For this lack of information, these dispersal capacities were simply integrated in our models in what was defined as the completely random contributions. Although the estimated values were very different for larvae (0.86 larvae per site per week) and adult (3.54 adults per site per week), these random contributions revealed both very significant.

In addition, our results allowed the prediction of the weekly contributions of present adults (resp. larvae) to larva (resp. adult) presence for the next week. For example, in our study, with an average of 7 adults per site per week, the contribution was about 7x1.04 = 7.28 larvae. This value is very high compared to the random contribution of 0.86 larvae. Conversely, with an average of 5.3 larvae per site per week, the contribution was about 5.3x0.51 = 2.7 adults which is lower than the random contribution of 3.54 adults. These figures seemed to simply translate the action of the two mentioned scales, local climatic preferences and adult dispersal, as well as their respective importance.

Concerning the attractiveness rates toward the favorable climatic conditions, we first observed that, even if took account of the difference between the temperature and humidity ranges, the temperature contribution was more rapidly decreasing (three times) than the humidity contribution when departing from the favorable conditions in the two models. Second, we also observed that these decrease rates are very high for adults that for larvae suggesting that this can be due to the dispersal capacity of adults in their search of favorable conditions.

Besides validating, in real crop conditions, the life-table data generated in the laboratory [35], our findings could be put to good use to propose optimal greenhouse microclimate conditions combined with agricultural practices which consider both thrips limitations as well as plant health. This fully fits with the main principles of Integrated Pest Management (IPM) which combines prophylactic, biological and physical methods against bioagressors of crops, while minimizing the use of pesticides [42,43,18]. Indeed, cultural (agronomic and physical) control techniques developed from crop management or mechanical practices can be adapted to disadvantage pest population growth while having limited adverse effects on crop productivity. Our findings also suggest that once the climatic preferences of thrips are identified, the infestation can be effectively minimised by releasing natural enemies in small areas surrounding the *risk zones*, saving time for the infestation monitoring and amount of natural enemies to be released. In contrast, before giving overall conclusions on optimal microclimatic conditions to achieve within protected rose cultivations, further specific tests should be carried out to assess the effects on the thrips dynamics of other physical factors, e.g. the luminosity, airflows, ... Similar studies should be carried out focusing on other rose pests and diseases as well as on their interactions [44] according to various microclimatic regimes.

CONCLUSION

The favourable climatic conditions for thrips (*F. occidentalis*) adults and larvae were assessed from the comparison of the distribution of temperature, humidity and thrips within the greenhouse. Our results highlighted that diurnal temperature and humidity exert a direct effect on the distribution of thrips (*F. occidentalis*) within the greenhouse space.

These findings could be used to define a climate control strategies to fight against this insect by modifying temperature and humidity inside the greenhouse to make it unfavourable for this pest. It may also be used to identify the risk areas, and thus saving time and money for growers, by releasing natural enemies in small areas surrounding these risk zones. To be complete on the topic, photoperiod effect must be also investigated and the Computational Fluid Dynamics (CFD) simulation tool could be will be used to draw a risk maps inside greenhouse according to the climatic parameters. CFD could also help to define how to change the local climate at the crop level to turn them unfavourable for pests.

AUTHOR CONTRIBUTIONS

Conceived and designed the experiments: HF JP ND AB. Performed the experiments: HF JP AB. Analyzed the data: HF RS. Contributed reagents/materials/analysis tools: HF JP RS AB ND CP TB. Wrote the paper: HF JP RS AB ND CP TB.

REFERENCES

1. Lewis T. Thrips as Crop PestsCAB International, London; 1997.
2. Mound LA. Thysanoptera: diversity and interactions. Annu Rev Entomol. 2005; 50: 247–269. pmid:15355240
3. Morse JC, Hoddle MS. Invasion Biology of Thrips. Annu Rev Entomol. 2006; 5: 67–89.
4. Kirk WDJ, Terry LI. The spread of the western flower thrips Frankliniella occidentalis (Pergande). Agric Forest Entomol. 2003; 5: 301–310.
5. Reitz SR. Biology and Ecology of the western flower thrips (Thysanoptera: Thripidae): The making of a pest. Fla Entomol. 2009; 92: 7–13. pmid:19335835
6. Nyasani JO, Meyhöfer R, Subramanian S, Poehling HM. Seasonal abundance of western flower thrips and its natural enemies in different French bean agroecosystems in Kenya. J Pest Sci. 2013;86: 515–523.
7. Mound LA. Biological diversity, Chapter 6. In Lewis T. (ed.), Thrips as Crop Pests. CAB International, Wallingford. 1997.
8. Bournier A, Bournier JP. L'introduction en France d'un nouveau ravageur: Frankliniella occidentalis. Phytoma. 1987; 388: 14–17.
9. Steiner MY, Goodwin S. Methods for collecting and rearing thrips (Thysanoptera) and their natural enemies. Austral J Entomol. 1998; 37: 101–106.
10. Mehle N, Trdan S. Traditional and modern methods for the identification of thrips (Thysanoptera) species. J Pest Sci. 2012; 85: 179–190.
11. Bergh JC, LeBlanc JPR. Performance of Western flower thrips (Thysanoptera: Thripidae) on cultivars of miniature rose. J Econ Entomol. 1997; 90: 679–688.
12. Pizzol J, Poncet C, Hector S, Ziegler JP. Mise en place d'une protection biologique intégrée préventive contre les ravageurs des cultures de rosiers sous serre dans le sud de la France. IOBC/WPRS Bull. 2006; 29: 31–36.
13. Pizzol J, Nammour D, Hervouet P, Bout A, Desneux N, Mailleret L. Comparison of two methods of monitoring thrips populations in a greenhouse rose crop. J Pest Sci. 2010; 83:191–196
14. Bielza P, Quinto V, Fernandez E, Grávalos C, Contreras J. Genetics of spinosad resistance in Frankliniella occidentalis (Thysanoptera: Thripidae). J Econ Entomol. 2007; 100: 916–20. pmid:17598556
15. Immaraju JA, Paine TD, Bethke JA, Robb KL, Newman JP. Western Flower Thrips (Thysanoptera, Thripidae) resistance to insecticides in coastal California greenhouses. J Econ Entomol. 1992; 85: 9–14.
16. Rahman T, Broughton S, Spafford H. Can spinosad-resistant Frankliniella occidentalis (Pergande) (Thysanoptera: Thripidae) be managed with spinosad and predatory mites (Acari)? Cro Prot. 2012; 42: 281–288.
17. Weisenburger DD. Human health effects of agrichemicals use. Hum Pathol. 1993; 24: 571–576. pmid:8505035
18. Desneux N, Decourtye A, Delpuech JM. The sub lethal effects of pesticides on beneficial arthropods. Ann Rev Entomol. 2007; 52: 81–106. pmid:16842032

19. Biondi A, Mommaerts V, Smagghe G, Vinuela E, Zappalà L, Desneux N. The non-target impact of spinosyns on beneficial arthropods. Pest Manag Sci. 2012; 68: 1523–1536. pmid:23109262
20. Lu YH, Wu KM, Jiang YY, Guo YY, Desneux N. Widespread adoption of Bt cotton and insecticide decrease promotes biocontrol services. Nature. 2012; 487: 362–365. pmid:22722864
21. Egger B, Koschier EH. Behavioural responses of Frankliniella occidentalis Pergande larvae to methyl jasmonate and cis-jasmone. J Pest Sci. 2014; 87: 53–59.
22. Berndt O, Meyhofer R, Poehling H M. The edaphic phase in the ontogenesis of Frankliniella occidentalis and comparison of Hypoaspis miles and Hypoaspis acuilefer as predators of soil-dwelling thrips stages. Biol Control. 2004; 30: 17–24.
23. Navarro-Campos C, Pekas A, Aguilar A, Garcia-Marí F. Factors influencing citrus fruit scarring caused by Pezothrips kellyanus. J Pest Sci. 2013; 86:459–467.
24. Aguilar-Fenollosa E, Jacas JA. Effect of ground cover management on Thysanoptera (thrips) in clementine mandarin orchards. J Pest Sci. 2013; 86: 469–481.
25. Van Rijn P, Mollena C, Steenhuis-Broers G. Comparative life story studies of Frankliniella occidentalis and Thrips tabaci (Thysanoptera: Tripidae) on cucumber. Bull Entomol Res. 1995; 85: 285–297.
26. McDonald J, Bale J, Walters K. Effect of temperature on development of the Western Flower Thrips, Frankliniella occidentalis (Thysanoptera: Tripidae). Eur J Entomol. 1998; 94: 301–306.
27. Kumm S, Moritz G. Life-cycle variation, including female production by virgin females in Frankliniella occidentalis (Thysanoptera: Thripidae). J Appl Entomol. 2010; 134: 491–497.
28. Robb KL, Parrella MP. Western flower thrips, a serious pest of floricultural crops, pp. 343–358 In B. L. Parker, M. Skinner and T. Lewis [eds.], Towards Understanding the Thysanoptera. General Technical Report NE-147. US Department of Agriculture, Forest Service, Radnor, PA; 1991.
29. Stavrinides MC, Daane KM, Lampinen BD, Mills NJ. Plant water stress, leaf temperature, and spider mite (Acari: Tetranychidae) outbreaks in California vineyards. Environ Entomol. 2010; 39:1232–1241. pmid:22127173
30. Fatnassi H, Boulard T, Bouirden L. Simulation of climatic conditions in full-scale greenhouse fitted with insect-proof screens. Agric For Meteorol. 2003; 118: 97–111.
31. Fatnassi H, Boulard T, Poncet C, Chave M. Optimisation of greenhouse insect screening with computational fluid dynamics. Biosyst Eng. 2006; 93: 301–312.
32. Bartzanas T, Boulard T, Kittas C. Numerical Simulation of the airflow and temperature distribution in a tunnel greenhouse equipped with insect-proof screen in the openings. Comput Electron Agr. 2002; 34: 207–221.
33. Molina-Aiz FD, Fatnassi H, Boulard T, Roy JC, Valera DL. Comparison of finite element and finite volume methods for simulation of natural ventilation in greenhouses. Computers and Electronics in Agriculture. 2010; 72: 69–86.
34. Cox DR, Hinkley DV. Theoretical Statistics, Chapman and Hall; 1974.
35. Robb KL. Analysis of Frankliniella occidentalis (Pergande) as a pest of floricultural crops on California greenhouses. Ph.D. dissertation, Univ.Calif.,Riverside, 135pp; 1989.
36. Brodsgaard HF. Effect of photoperiod on the binomics of Frankliniella occidentalis (Pergande) (Thysanoptera, Thripidae). J Appl Entomol. 1994; 117: 498–507.
37. Steiner MY, Spohr LJ, Goodwin S. Relative humidity controls pupation success and dropping behaviour of western flower thrips, Frankliniella occidentalis (Pergande) (Thysanoptera:Thripidae). Austral J Entomol. 2011; 50: 179–186.
38. Shipp JL, Gillespie TJ. Influence of temperature and water vapor pressure deficit on survival of Frankliniella occidentalis (Thysanoptera: Thripidae). Environ Entomol. 1993; 22: 726–732.
39. Rhainds M, Shipp L. Dispersal of adult western flower thrips (Thysanoptera: Thripidae) in greenhouse crops. Canad Entomol. 2004; 136: 241–254.

40. Kareiva PM. Local movement in herbivorous insects: applying a passive diffusion model to mark–recapture field experiments. Oecol. 1983; 57: 322–327.
41. Zappalà L, Campolo O, Grande S, Saraceno F, Biondi A, Siscaro G et al. Dispersal of Aphytis melinus (Hymenoptera: Aphelinidae) after augmentative releases in citrus orchards. Eur J Entomol. 2012; 109: 561–568.
42. Kogan M. Integrated pest management: Historical perspectives and contemporary developments. Ann Rev Entomol. 198; 43: 243–270. pmid:9444752
43. van Lenteren JC. A greenhouse without pesticides: fact or fantasy?. Crop Pro. 2000; 19: 375–384.
44. Mouttet R, Bearez P, Thomas C, Desneux N. Phytophagous arthropods and a pathogen sharing a host plant: evidence for indirect plant-mediated interactions. PLoS ONE. 2011; 6(5): e18840. pmid:21611161

CHAPTER 3

Evolution of Growth Habit, Inflorescence Architecture, Flower Size, and Fruit Type in Rubiaceae: Its Ecological and Evolutionary Implications

Sylvain G. Razafimandimbison[1] *, Stefan Ekman[2] , Timothy D. McDowell[3] , Birgitta Bremer[1]

[1] Bergius Foundation, The Royal Swedish Academy of Sciences and Botany Department, Stockholm University, Stockholm, Sweden,

[2] Museum of Evolution, Uppsala University, Uppsala, Sweden, 3 Department of Biological Sciences, East Tennessee State University, Johnson City, Tennessee, United States of America

ABSTRACT

During angiosperm evolution, innovations in vegetative and reproductive organs have resulted in tremendous morphological diversity, which has played a crucial role in the ecological success of flowering plants. Morindeae (Rubiaceae) display considerable diversity in growth form, inflorescence architecture, flower size, and fruit type. Lianescent habit, head inflorescence, small flower, and multiple fruit are the predominant states, but arborescent habit, non-headed inflorescence, large flower, and simple fruit states occur in various genera. This makes Morindeae an ideal model for exploring the evolutionary appearances and transitions between the states of these characters. We reconstructed ancestral states for these four traits using a Bayesian approach and combined nuclear/chloroplast data for 61 Morindeae species. The aim was to test three hypotheses: 1) self-supporting habit is generally ancestral in clades comprising both lianescent and arborescent species; 2) changes from lianescent to arborescent habit are uncommon due to "a high degree of specialization and developmental burden"; 3) head inflorescences and multiple fruits in Morindeae evolved from non-headed inflorescences and simple fruits, respectively. Lianescent habit, head inflorescence, large flower, and multiple fruit are inferred for Morindeae, making arborescent habit, non-headed inflorescence, small flower, and simple fruit derived within the tribe. The rate of change from lianescent to arborescent habit is much higher than the reverse change. Therefore, evolutionary changes between lianescent and arborescent forms can be reversible, and their frequency and trends vary between groups. Moreover, these changes are partly attributed to a scarcity of host trees for climbing plants in more open habitats. Changes from large to small flowers might have been driven by shifts to pollinators with progressively shorter proboscis, which are associated with shifts in breeding systems towards dioecy. A single origin of dioecy from hermaphroditism is supported. Finally, we report evolutionary changes from headed to non-headed inflorescences and multiple to simple fruits.

INTRODUCTION

During angiosperm evolution, changes in vegetative and reproductive organs have resulted in remarkable morphological diversity, which has played an important role in the ecological success of flowering plants [1]. The fusion of clustered fruits into multiple fruits (or syncarps) has occurred repeatedly in different lineages [1]–[2]. Some multiple fruits are important food sources for a wide range of animals, and the evolution of this type of compound fruit has been hypothesized as a result of selection by large animals [3]. This is based on the fact that multiple fruits are generally favored and their seeds are effectively dispersed by large frugivorous dispersers [4]. Flowering plants from different groups produce edible multiple fruits that are economically important. Examples include jackfruits and breadfruits (Moraceae), pineapples (Bromeliaceae), and noni fruits (Rubiaceae). Despite their crucial roles in different ecosystems and for the human society, little is known about the evolution of multiple fruits. This is partly due to the lack of robust phylogenies for the lineages that contain species producing multiple fruits and species bearing simple fruits. Molecular-based phylogenies are essential for placing patterns of any heritable trait in an evolutionary context [5].

In the coffee family (Rubiaceae), most taxa with multiple fruits are members of the tribes Naucleeae [6]–[7] and Morindeae [8]–[10]. Fruits in Morindeae, belonging to the Psychotrieae alliance in the subfamily Rubioideae, are predominantly multiple fruits composed of two to many fully to basally fused drupaceous (fleshy) fruits (Figure 1C–D, H), which are derived from ovaries of the adjacent flowers. This type of compound fruit is found in three (*Coelospermum* Blume, *Gynochthodes* Blume (Figure 1H), and *Morinda* L. (Figure 1C)) of the five genera currently recognized in the tribe, and is absent in the other two genera (*Appunia* Hook.f. and *Siphonandrium* K.Schum.) the infructescences (fruiting stage of inflorescences) of which are formed by clusters of simple, drupaceous fruits (Figure 1B). A few members of *Coelospermum* are characterized by branched or headed infructescences bearing pedicellate (stalked), drupaceous fruits (Figure 1F), while some *Gynochthodes* species have infructescences composed of pedicellate, drupaceous fruits grouped in umbels (flat-topped or rounded flower/fruit clusters with the pedicels arising from more or less the same point) or fascicles (tight bundles). It has been postulated by McClatchey [11] that multiple fruits of the broadly circumscribed *Morinda* (*Morinda* sensu lato), which included all lianescent and arborescent *Morinda* species with multiple fruits recently transferred to *Gynochthodes* [8], evolved from an ancestor with umbels and simple fruits by suppression of the pedicels and fusion of the ovaries of the adjacent flowers. This would imply that multiple fruits of *Morinda* and *Gynochthodes* (both sensu Razafimandimbison et al. [8]–[9]) are derived in Morindeae.

Besides its fruit diversity, Morindeae are also diverse in growth form, inflorescence architecture, and flower size. Lianescent (climbing) habit, headed inflorescence, and small flower are the predominant states, and occur in different genera of the tribe. This makes Morindeae an attractive model for exploring the evolutionary appearances and transitions between the major states of the four major characters (i.e., growth habit, inflorescence architecture, flower size, and fruit type) from a phylogenetic perspective. The recently published molecular phylogeny of Morindeae [9] provides a solid basis for such a study. These traits have been revealed to be evolutionarily labile in Morindeae [9], however a proper ancestral state reconstruction (hereafter called ASR) is essential in order to better understand their evolution. Moreover, a combination of these four characters has been used for circumscribing the five recognized genera of the tribe (Table 1).

Figure 1. Characteristics and morphological variation of the tribe Morindeae (for details see text).

A–B: *Appunia debilis*; C: *Morinda citrifolia*; D: *Morinda pacifica*; E: *Coelospermum fragrans*; F: *Coelospermum balansanum*; G: *Gynochthodes kanalensis*; and H: *Gynochthodes retusa* (A–C by T. D. McDowell; D by F. Tronquet; E–G by J. T. Johansson; and H by K. Kainulainen).

Table 1. Morphological characteristics and other important information of the five recognized genera of the tribe Morindeae.

	Appunia Hook.f.	*Coelospermum* Blume	*Gynochthodes* sensu lato Blume	*Morinda* sensu stricto L.[1]	*Siphonandrium* K.Schum.[2]
Geographic distribution	Neotropics	Tropical Asia and Australasia	Tropical Asia, Australasia, and Madagascar	Pantropical	New Guinea
Number of species	Ca. 12	Ca. 11	Ca. 95	Ca. 40	1
Growth habit	Mostly frutescent	Mostly woody lianescent	Mostly woody lianescent,	Mostly arborescent and frutescent	Lianescent
Inflorescence architecture	Head inflorescences	Mostly non-headed inflorescences	Mostly head inflorescences	Head inflorescences	Head inflorescences
Flower size	Large (corolla tubes 5–10 (23–24) mm long > corolla lobes 0.5–7 (15–17) mm) long	Small (corolla tubes 3–7 (11) mm long < corolla lobes 4.5–16 mm) long	Small (corolla tubes 0.7–5.5 mm long < corolla lobes 1.5–11 mm) long	Large (corolla tubes 5–40 (80) mm long > corolla lobes 1–14 (22) mm) long	Small (corolla tubes ca. 3 mm long > corolla lobes ca. 5 mm long)
Breeding systems	Hermaphroditic	Androdioecious or dioecious or functionally dioecious	Androdioecious or dioecious or functionally dioecious	Hermaphroditic	Dioecious
Fruit type	Simple, drupaceous fruits	Mostly simple, drupaceous fruits	Mostly multiple fruits	Multiple fruits	Simple, drupaceous fruits

[1]All lianescent, dioecious species of *Morinda* with small flowers have recently been transferred to *Gynochthodes* [8].
[2]Filaments of *Siphonandrium* are tightly fused and its anthers are glued together, all forming a staminal tube. This feature is unique within Morindeae.

Morindeae comprise ca. 100 woody climbing species (ca. 62%), with only ca. 54 arborescent (tree or tree-like) and frutescent (shrubby) species (ca. 28%), and two suffrutescent (shrubby plants having woody stems only at the base) species. Frutescent plants are common in *Appunia*. Both arborescent and frutescent plants are predominant in *Morinda* but are extremely rare in *Coelospermum* (the arborescent *C. reticulatum* (F.Muell.) Benth.) and *Gynochthodes* (frutescent *G. decipiens* (Schltr.) Razafim. & B.Bremer and the arborescent *G. trimera* (Hillebr.) Razafim. & B.Bremer). Conversely, lianescent plants (woody vines or climbers) in the tribe are found mostly in *Gynochthodes* [8]–[9], with only one species in *Appunia* (*A. megalantha* C.M.Taylor & Lorence, [12]), two species in *Morinda* (*M. longiflora* G.Don and *M. morindoides* (Baker) Milne-Redh., [9]), and 10 species in *Coelospermum* [13]. There are at least two different (but not mutually exclusive) hypotheses regarding the evolutionary transitions between lianescent and arborescent growth forms. The first hypothesis considers self-supporting (arborescent or frutescent) habit to be the common ancestral condition for clades containing both lianescent and arborescent/frutescent species [14]–[15]. The second hypothesis states that the evolutionary change from lianescent to arborescent/frutescent habit is uncommon, because "the evolution of lianescence can carry a high degree of specialization and developmental burden that might limit evolution back to self-supporting growth forms" [16].

Head inflorescences (or heads) occur when two or more flowers are borne on a common receptacle (the end of the inflorescence stalk upon which the floral organs are borne). In Morindeae, these heads are composed of two to 50 flowers that are clustered tightly on the receptacles (Figure 1A, C–D, G).

These inflorescences may contain a single head or two to several heads, which are in turn arranged into various branching forms: umbels, corymbs (flat-topped or round-topped, racemose inflorescences with the lower pedicels longer than the upper), or panicles (branched clusters of flowers). Within a head the ovaries of the adjacent flowers may be fused or free. The majority of Morindeae species with heads have fused ovaries, but in *Appunia* (Figure 1A) and *Siphonandrium* the ovaries are free. It is important to note that the degree of the fusion of ovaries prior to fruit development (pre-genital fusion) varies greatly among species, from only basally to completely fused. Only 27 Morindeae species (nine from *Coelospermum* and 18 from *Gynochthodes*) bear non-headed inflorescences, which are arranged in umbels, or compound umbels, or fascicles (in the latter genus) and panicles or corymbs (in the former genus). We postulated that heads evolved from non-headed inflorescences in Morindeae. In addition, inflorescences in Morindeae are mostly terminal on the shoot, sometimes leaf-opposed ("pseudo-terminal") and axillary. Axillary inflorescences are found in *Siphonandrium*, at least three species of *Appunia*, and ca. 18 species of *Gynochthodes*. Leaf-opposed inflorescences distinguish the mostly Asian, arborescent *Morinda* clade (including the Neotropical *M. royoc* L., the pantropical *M. citrifolia* L., and the African *M. chrysorhiza* (Thonn.) DC. and *M. lucida* Benth.) from the remaining *Morinda* [9]; this type of inflorescence is also known to occur in *Appunia* (e.g., *A. surinamensis* (Bremek.) Steyerm.).

In Morindeae, flowers in the same inflorescence or head appear to open successively over days or weeks [17] (Figure 1A, C, E, G). The flowers vary greatly in size (Table 1); large flowers (corolla tube length/corolla lobe length >1) are found in *Appunia* and *Morinda* (Figure 1A, C) and presumably pollinated by larger insects, such as long-proboscis moths [18]. Plants with small flowers (corolla tube length/corolla lobe length <1) are restricted to *Coelospermum*, *Gynochthodes*, and *Siphonandrium* (Figure 1E, G), and are most likely to be pollinated by small insects (e.g., short-proboscis moths or small bees or flies) [19]–[20]. Overall, the species of *Morinda* have larger flowers than the species of *Appunia*, with the exception of the lianescent *A. megalantha* (corolla tubes of 23–24 mm long > corolla lobes of 15–17 mm long, [12]). Within *Morinda*, seven African species have much larger flowers than the remaining species of the genus. The flowers of *Coelospermum* are larger than those of the species of *Gynochthodes*[9]. Furthermore, Morindeae species vary in their breeding systems, and flowers are either bisexual or unisexual or functionally unisexual. Only the hermaphroditic condition has been reported in *Appunia* and *Morinda* [21], while the androdioecious (male and hermaphroditic individuals) [21]–[22], strict or functional dioecious [23]–[24], and hermaphroditic [21], [24] conditions are all known in *Gynochthodes* and *Coelospermum* [21]–[24]. The New Guinean *Siphonandrium* is dioecious (Table 1).

We reconstructed the evolution of fruit type, inflorescence architecture, flower size, and growth form across a phylogeny of the tribe Morindeae. We were particularly interested in testing the following hypotheses: 1) self-supporting habit is generally plesiomorphic in clades comprising both lianescent and arborescent species [14]; 2) evolutionary changes from lianescent to arborescent/frutescent habit are less frequent than the reverse change, from arborescent/frutescent to lianescent habit [16]; 3) and Head inflorescences and multiple fruits in Morindeae evolved from non-headed inflorescences and simple fruits, respectively. The ecological and evolutionary implications of the findings of this study are discussed.

RESULTS

The Bayesian majority rule consensus tree generated from the combined nrETS/nrITS/*trn*T-F data and shown in Figure 2 was fully resolved. Its overall topology is almost identical to that of the Bayesian majority rule consensus tree published in Razafimandimbison et al. [9].

Values above nodes are the posterior probabilities. Capital letters A–G denote selected nodes whose state probabilities were estimated for the states of the four characters (1–4). Data shown across the tips are growth habit (character 1: 0= lianescent, 1= arborescent, 2= herbaceous), inflorescence architecture (character 2: 0= headed inflorescences (heads); 1= non-headed inflorescences), flower size (character 3: 0= large, 1= small), and fruit type (character 4: 0= simple fruits, 1= fused or multiple fruits). SF and LF stand for small and large fruits, respectively.

Ancestral State Reconstructions of Growth form, Inflorescence Architecture, Flower Size, and Fruit Type in Morindeae

Two types of ASRs were performed (one with the outgroup *Damnacanthus indicus* C.F.Gaertn. and *Mitchella repens* L. (tribe Mitchelleae), hereafter called ASR with outgroup, and the other without the outgroup, hereafter called ASR without outgroup), to infer the ancestral states of growth form, inflorescence architecture, flower size, and fruit type (characters 1–4, respectively) at seven important nodes of Morindeae (nodes A–G, Figure 2). The results of these ASRs are summarized in Tables 2,3. For growth form (character 1) and flower size (character 3) the ratios q_{01}/q_{10} from the ASRs with outgroup, respectively, were 4.820 and 2.691 for Morindeae (Table 2). This indicates that the rates of changes from lianescent to arborescent habit and from large to small flower were higher than the rates of the reverse directions, from arborescent to lianescent habit and from small to large flower. In contrast, for inflorescence architecture (character 2) and fruit type (character 4), the ratios q_{01}/q_{10}, respectively, were 0.676 and 0.677 for Morindeae (Table 2); this means that the rates of changes from non-headed to headed inflorescence and from multiple to simple fruit are higher than the rates of the reverse changes, from headed to non-headed inflorescence and from simple to multiple fruit.

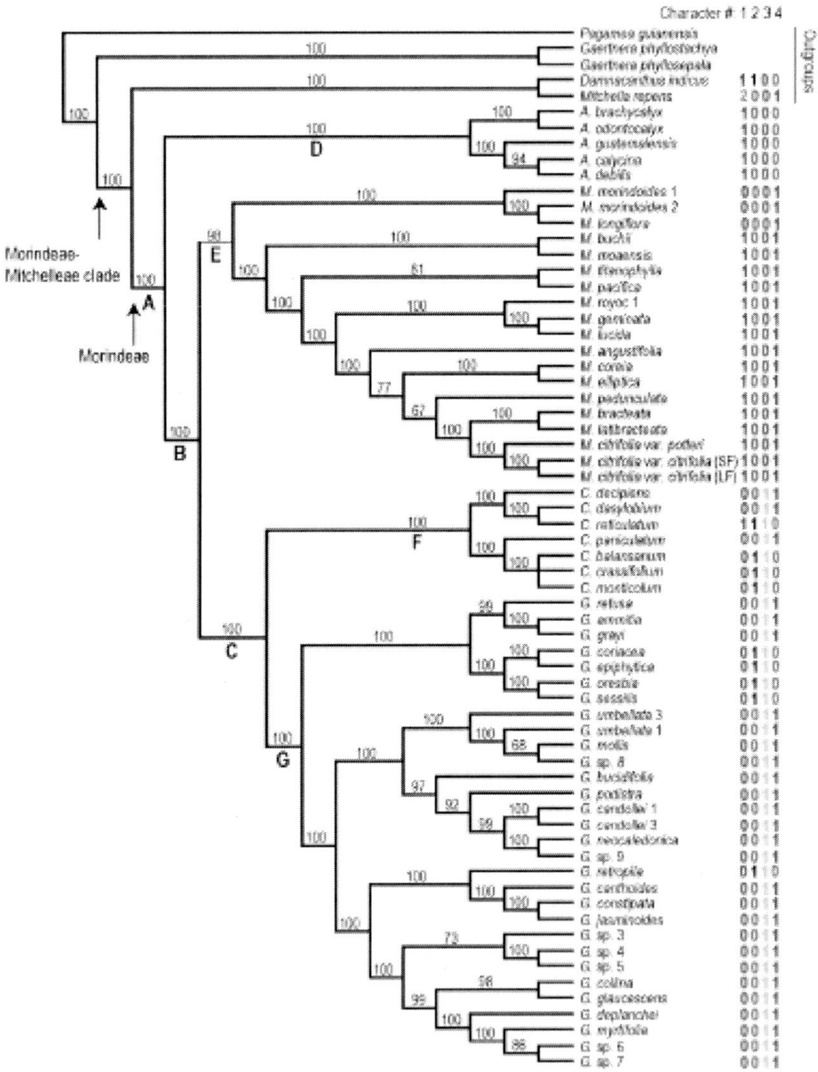

Figure 2. Bayesian majority rule consensus tree from the combined nrETS/nrITS/*trn*T-F data of 61 Morindeae taxa.

Table 2. Bayesian reconstruction of ancestral states in the four characters (1–4) at seven nodes (A–G) across a posterior sample of trees including Morindeae but no outgroup.

	Character*			
	1 (growth form)	2 (inflorescence architecture)	3 (flower size)	4 (fruit type)
Coded character states	lianescent = 0 and arborescent = 1	headed = 0 and non-headed = 1	large = 0 and small = 1	simple = 0 and fused = 1
q_{01}/q_{10}	4.820	0.676	2.691	0.677
κ	0.961 (0.363–1.508)	0.711 (0.111–1.315)	1.124 (0.239–1.950)	0.787 (0.142–1.332)
Node A (Morindeae)	0.861	0.584	0.980	0.370
Node B (the *Morinda-Coelospermum-Gynochthodes* clade)	0.992	0.586	0.878	0.181
Node C (the *Coelospermum-Gynochthodes*)	0.998	0.584	0.000	0.185
Node D (*Appunia*)	0.007	0.909	1.000	0.989
Node E (*Morinda*)	0.860	0.910	1.000	0.025
Node F (*Coelospermum*)	0.998	0.565	0.000	0.214
Node G (*Gynochthodes*)	1.000	0.583	0.000	0.186

*For each character 1–4, the following information is provided: the ratio of the average rate q_{01} to the average rate q_{10}, the average and 95% highest posterior density (HPD) of κ, and the marginal posterior probabilities of having state 0 in each of the seven nodes (A–G). As all characters are binary, the marginal posterior probability of having state 1 is one minus the probability of state 0. The 95% HPD of κ excludes 0 in all cases, which is a strong indication that branch lengths carry information about the amount of change in the morphological characters.

Table 3. Bayesian reconstruction of ancestral states in the four characters (1–4) at seven nodes (A–G) across a posterior sample of trees including Morindeae as well as the outgroup taxa *Damnacanthus indicus* and *Mitchella repens* (Mitchelleae).

	Character			
	1 (growth form)	2 (inflorescence architecture)	3 (flower size)	4 (fruit type)
Coded character states	lianescent (0), arborescent (1), and herbaceous (2)	headed (0) and non-headed (1)	large (0) and small (1)	simple (0) and fused (1)
κ	**0.981 (0.394–1.516)**	**0.671 (0.061–1.259)**	**1.144 (0.221–1.996)**	**0.784 (0.128–1.312)**
Node of Morindeae-Mitchelleae clade	0.258, 0.216, 0.524	0.513	0.996	0.359
Node A (Morindeae)	0.733, 0.199, 0.067	0.566	0.994	0.371
Node B (the *Morinda-Coelospermum-Gynochthodes* clade)	0.970, 0.011, 0.019	0.568	0.886	0.166
Node C (the *Coelospermum-Gynochthodes*)	0.991, 0.001, 0.008	0.565	0.000	0.171
Node D (*Appunia*)	0.007, 0.980, 0.013	0.916	1.000	0.987
Node E (*Morinda*)	0.733, 0.202, 0.065	0.917	1.000	0.023
Node F (*Coelospermum*)	0.990, 0.001, 0.009	0.540	0.000	0.207
Node G (*Gynochthodes*)	0.998, 0.000, 0.002	0.564	0.000	0.172

*For each character 1–4, the average and 95% highest posterior density of κ is provided. For character 1, we also provide the marginal posterior probabilities of having state 0, 1, and 2, respectively, in each of the eight selected nodes (the Mitchelleae-Morindeae root node and nodes A–G). For character 2–4, we provide the marginal posterior probabilities of having state 0 for the same eight nodes. The 95% HPD of κ excludes 0 in all cases, which is a strong indication that branch lengths carry information about the amount of change in the morphological characters.

The node of the Morindeae-Mitchelleae clade (= the Mitchelleae-Morindeae common ancestor) (Figure 2) was inferred with strong and moderate support, respectively, as large flower and multiple fruit in the ASR with outgroup; however, the results of this analysis were inconclusive for growth habit and inflorescence architecture (Table 3). The outcomes of the ASRs with and without outgroup were very similar for the seven important nodes of Morindeae (nodes A–G, Figure 2) (Tables 2,3). At node A (the Morindeae common ancestor) the lianescent habit, head inflorescence, large flower, and multiple fruit states were inferred; however the support was weak for head inflorescence, moderate for lianescent habit and multiple fruit, and strong for large flower. Within Morindeae lianescent habit was strongly inferred at node B (the *Morinda-Coelospermum-Gynochthodes* clade), node C (*Coelospermum-Gynochthodes* clade), node E (*Morinda*), node F (*Coelospermum*), and node G (*Gynochthodes*), while arborescent habit was resolved at node D (*Appunia*). For inflorescence architecture (character 2) head inflorescence was inferred at nodes B-G; the support was moderate for nodes B–C and F–G but strong for nodes D–E. For flower size (character 3) large flower was strongly inferred at nodes B and D–E, whereas nodes C and F–G were unambiguously resolved as small flower. Finally, for fruit type (character 4) fused fruit (multiple fruit) was strongly inferred at nodes B–C and nodes E–G, however simple fruit was highly resolved at node D (Tables 2,3).

DISCUSSION

We performed ASRs with and without outgroup in order to assess the influence of the outgroup taxa on the outcomes. The fact that the results of these two ASRs are very similar for the seven nodes of Morindeae (nodes A–G, Figure 2, Tables 2,3) suggests that the inclusion of the outgroup taxa (*Damnacanthus indicus* and *Mitchella repens*) has almost no effect on the analyses. The ASR with outgroup infers large flower with strong support and multiple fruit with moderate support at the node of the Morindeae-Mitchelleae clade. Lianescent habit, head inflorescence, large flower, and multiple fruit are inferred at node A (Morindeae) (Tables 2,3). If these inferences are correct, these states are interpreted as plesiomorphic for Morindeae and are plesiomorphic within Morindeae, with respect to the later changes (i.e., arborescent habit, non-headed inflorescence, small flower, and simple fruit) in the group.

Evolution of Growth form in Morindeae and its Ecological and Evolutionary Implications

Two-thirds of the species in Morindeae are represented by the lianescent species of *Gynochthodes*, while only two *Morinda* and one *Appunia* species are lianas. Conversely, a single species of *Coelospermum* (*C. reticulatum*) and two *Gynochthodes* (*G. decipiens* and *G. trimera*, not investigated in this study) species are arborescent. The fact that lianescent habit is inferred at node A (Morindeae) means that this state is plesiomorphic at node B (the *Morinda-Coelospermum-Gynochthodes* clade), node C (the *Coelospermum-Gynochthodes* clade), and nodes E–G (*Morinda*, *Coelospermum*, and *Gynochthodes*, respectively). Arborescent habit is inferred as apomorphic at node D (*Appunia*), and seems to have arisen at least three times within Morindeae: *Appunia*, *Morinda* (node E: *M. butchii* Urb. to *M. citrifolia* L., Figure 2), and *Coelospermum* (node F: *C. reticulatum*, Figure 2). Our findings provide no support for the reported prevalence of a plesiomorphic arborescent habit in lineages containing both lianescent and arborescent plants [14]. In fact, a plesiomorphic lianescent habit and multiple independent origins of arborescent from lianescent habit have recently been inferred for the primarily lianescent subfamily Secamonoideae in the family Apocynaceae [16] and the family Menispermaceae [25]. Moreover, the rate of change from lianescent to arborescent habit in Morindeae is significantly higher than the reverse change, from arborescent to lianescent habit ($q_{01}/q_{10}=4.820>1$, Table 2); this is inconsistent with Lahaye et al.'s [16] claim that "the evolution of lianescence can carry a high degree of specialization and developmental burden that might limit evolution back to self-supporting growth forms". Based on the evidence presented above we argue that evolutionary changes between arborescent and lianescent habits can be reversible, and that their frequency and trends seem to vary between groups. In addition, the weak-stem condition of shrubs and treelets in *Appunia* (observations by T. D. McDowell) and the scandent- or vining-branch condition of shrubs or treelets in the Neotropical *Morinda royoc* [26] may be viewed as a reflection of their origins from lianescent forms. We find no evidence of any reversal from arborescent to lianescent habit in the 61 Morindeae species included in this study. On the other hand, the sole lianescent species (*A. megalantha*) in the otherwise arborescent *Appunia*, not investigated in this study due to lack of material for sequencing, may represent a unique case of an evolutionary reversal from arborescent to lianescent habit in the tribe.

Furthermore, the acquisition of arborescent habit in *Appunia* seems to have coincided with the diversification of the genus in the Neotropics. Consequently, the evolutionary changes from lianescent to arborescent habit within Morindeae may in part be attributed to reduced competition for open ground and a scarcity of host trees for climbing plants in more open habitats [15]–[16]. This could explain the abundance of some species of the Asian, arborescent *Morinda* in sparse forests on hill slopes or open disturbed forests and the common occurrence of many Asian, lianescent *Gynochthodes* in forests or thickets on mountains [27]. The pantropical, arborescent *Morinda citrifolia* L. is also commonly found on seashores and sparse forests throughout its geographic ranges [26]–[27]. Similarly, the Neotropical *M. royoc* is common in pine savannas and coastal strands [26]. Finally, five *Appunia* species of the Neotropical Guianas region are shrubs, which frequently occur at forest edges, in clearings along riverbanks, and in disturbed, opened sites (observations by T. D. McDowell).

Evolution of Inflorescence Architecture and its Evolutionary Implications

The majority of species in Morindeae with head inflorescence belong to *Gynochthodes*, although they occur in all five recognized genera (Table 1). Conversely, nine of the 11 *Coelospermum* species and 18 *Gynochthodes* species have non-headed inflorescences. If the weakly inferred head inflorescence at node A (Morindeae) is correct, this state is interpreted as plesiomorphic within Morindeae (for nodes B–G); this is inconsistent with our hypothesis of a derived head inflorescence within the tribe. The inferred plesiomorphic head inflorescence for nodes B–G, although weakly supported for nodes B–C and F–G, is consistent with highly to moderately supported plesiomorphic multiple fruits in nodes B–C and F–G. Multiple fruits can only be produced by taxa with head inflorescences, although plants with headed inflorescences can also produce simple fruits (e.g., *Appunia* (node D), Figure 2). The evolutionary changes from headed to non-headed inflorescence occurred at least four times within Morindeae: twice each in *Coelospermum* (the Australian *C. reticulatum* and the New Caledonian *C. balansanum* group, Figure 2) and *Gynochthodes* (the *G. coriacea* group and the Australian *G. retropila* (Halford & A.J.Ford) Razafim. & B.Bremer, Figure 2). This is, to our current knowledge, the first report of evolutionary changes from headed to non-headed inflorescences in Rubiaceae.

The findings of this study raise new interesting questions. We do not know if the formation of non-headed from head inflorescences passes through the development of pedicels (umbels) followed by the formation of inflorescence branches in the umbellate forms to produce elongated, branched inflorescences. Alternatively, the non-headed inflorescence could be derived from a branched inflorescence of many heads if flower number was reduced to leave only one flower per receptacle. Unfortunately, discrete state ASR cannot tell us anything about the intermediate evolutionary changes leading to the formation of non-headed inflorescences in *Coelospermum* and *Gynochthodes*. Detailed comparative morphological and developmental studies combined with phylogeny are essential in order to elucidate the underlying developmental basis between the states of inflorescence architecture in Morindeae [28]–[29].

Evolution of Flower Size in Morindeae and its Ecological and Evolutionary Implications

Almost all species of Morindeae with large flowers belong to the arborescent *Appunia* and *Morinda*, with the exception of the sole lianescent *Appunia* species, *A. megalantha* [12], and the two lianescent *Morinda* species, *M. longiflora* and *M. morindoides*. Conversely, Morindeae plants with small flowers are mostly the lianescent species of *Coelospermum*, *Gynochthodes*, and *Siphonandrium*, except the two arborescent *Gynochthodes* species (*G. decipiens* and *G. trimera*) and the lianescent *G. sublanceolata* Miq. Our ASRs strongly infer large flowers at the Morindeae-Mitchellae root node as well as node A (Morindeae), meaning that this state is plesiomorphic for the tribe Morindeae, the *Morinda-Coelospermum-Gynochthodes* clade (node B), *Appunia* (node D), and *Morinda* (node E). Small flowers are derived for the *Coelospemum-Gynochthodes* clade (node C), *Coelospermum* (node F), and *Gynochthodes* (node G). In other words, small flowers seem to have evolved only once from the large flowers within Morindeae (Figure 2). It is worth noting that *G. sublanceolata* and *G. decipiens*, with large flowers but not included in this study, may represent one or two cases of reversals from small to large flowers.

The *Coelospemum-Gynochthodes* clade (node C) contains over 60% of the species in the tribe, and produce small flowers with inconspicuous colors that are most likely to be pollinated by small insects (e.g., short-proboscis moths or small bees or flies). Pollinators with progressively shorter proboscis may have been driving the transition from large to small flowers and an accompanying increase in speciation rate. Furthermore, change from large to small flowers in the *Coelospemum-Gynochthodes* clade appears to have been associated with a gender dimorphism transition. Androgynoecious (male and hermarphroditic) and dioecious conditions are only known from the lianescent species of *Gynochthodes* and *Coelospermum*[19]–[24] (Table 1). Thus, the high incidence of dioecy in the *Coelospermum-Gynochthodes* clade is correlated with woody, climbing growth habit, small flowers pollinated probably by unspecialized pollinators, and fleshy fruits. This pattern is consistent with those that have been reported from island habitats and various tropical forests [30]–[37]. Therefore, this study presents further support for the importance of these traits in the evolution of dioecy. On the other hand, it is important to note that all hermaphroditic members of *Appunia* and *Morinda* with large flowers also have the woody (but arborescent or frutescent) habit and fleshy fruits. This suggests that woodiness and fruit fleshiness alone cannot fully predict dioecy in the tribe Morindeae. In sum, the members of the *Coelospemum-Gynochthodes* clade display island syndrome characteristics, which are consistent with the fact that many of their species are indeed island endemics [13], [24].

In contrast to the *Coelospermum-Gynochthodes* clade, the large, mostly white flowers of *Appunia* and *Morinda* may be pollinated by larger insects, such as long-proboscis moths. This is consistent with the report on the Asian, arborescent *Morinda coreia* Buch.-Ham. being pollinated by hawkmoths in India [18]. The fact that the species of *Appunia* and *Morinda* are hermarphroditic suggests a single origin of dioecy in the *Coelospermum-Gynochthodes* clade from hermaphroditism. Members of *Appunia* and *Morinda* are predominantly distributed in continental areas (Africa mainland, continental Asia, and South and Central America), and show characteristics of the mainland pollinations and floral traits [30], [33]–[34], [36]–[38].

Evolution of Multiple Fruits in Morindeae and its Ecological and Evolutionary Implications

Most Morindeae, about 90% of the species, bear multiple fruits. The majority of these species belong to *Gynochthodes* and *Morinda*, with only three species (two investigated in this study, Figure 2) in *Coelospermum*. The infructescences of *Appunia* and *Siphonandrium* are composed of simple, drupaceous fruits. Our ASRs with moderate certainty infer multiple fruits at node Morindeae-Mitchelleae and node A (Morindeae). If correct, this state is plesiomorphic for Morindeae, the *Coelospermum-Gynochthodes* clade (node C), *Morinda* (node E), *Coelospermum* (node F), and *Gynochthodes* (node G) (Tables 2,3). This is inconsistent with the hypothesis of a derived multiple fruit for the broadly delimited *Morinda* (including the lianescent *Morinda* species transferred to *Gynochthodes* sensu Razafimandimbison et al. [8]–[9]), as postulated by McClatchey [11]. Simple, drupaceous fruits are derived for *Appunia* (node D) and seem to have arisen at least five times within Morindeae: once in *Appunia*, twice each in *Coelospermum* (the Australian *C. reticulatum* and the New Caledonian *C. balansanum* group), and *Gynochthodes* (the *G. coriacea* group and the Australian *G. retrophila*) (Figure 2). This is, to our knowledge, the first report of an evolutionary transition from multiple to simple fruits in Rubiaceae. Within the *Coelospermum-Gynochthodes* clade (node C) the evolutionary change from multiple to simple fruits coincides with that of from headed to non-headed inflorescences. However, it is interesting that *Appunia* (node D) seems to have retained the plesiomorphic headed inflorescences but acquire simple fruits.

Like the acquisition of arborescent habit, the derivation of simple fruits in *Appunia* seems to have coincided with the divergence of the *Appunia* lineage in the Neotropics. The change from multiple to simple fruits in this genus is in part attributed to shifts in seed dispersal vectors. Seeds of the simple, drupaceous fruits of *Appunia* species are presumably dispersed by birds, whereas seeds of the larger multiple fruits are dispersed effectively by large frugivorous animals [3]. The same mechanism seems to underlie the evolutionary change from multiple to simple fruits within the *Coelospemum-Gynochthodes* clade.

The degree of ovary fusion prior to fruit development in head inflorescences varies greatly from only a basal, partial fusion to completely fused ovaries among *Morinda* and *Gynochthodes*. This variation, which is rarely mentioned by Rubiaceae systematists [26], merits consideration for its ecological implications. Clusters of simple fruits of *Appunia* are likely to be dispersed individually by frugivorous birds. Multiple fruits composed of partly to fully fused ovaries are presumably dispersed as single units, while those formed by basally fused ovaries could well be dispersed individually by frugivorous birds or as single units by larger frugivorous dispersers. Furthermore, we suspect that in many members of *Gynochthodes* and *Morinda* ovaries of the adjacent flowers are basally fused prior to and during maturation of the anthers, and that ovary fusion extends midway during fructification. This type of ovary fusion was reported for *Breonia richardsonii* Razafim. in the tribe Naucleeae of the subfamily Cinchonoideae (Rubiaceae) by Razafimandimbison [39].

Future Perspectives

The Bayesian phylogenetic approach used here provides a sound framework for examining the evolution of distinctive vegetative (growth habit) and reproductive traits (flower, inflorescence, and fruit structures), which have broad ecological importance and potential impact on our understanding of speciation and diversity. Methods, which rely upon mapping discrete character states across a phylogeny, inevitably reduce the complexity of character variation among a diverse group of species. Thus, the arborescent habit includes all non-liana woody shrubs and trees (large trunked trees (e.g., noni, *Morinda citrifolia*), suffrutescent plants (e.g., *M. buchii* Urb.), shrubs or treelets with scandent- or vining-branches (e.g., *Morinda royoc*), and weakly branching treelets (e.g., *Appunia debilis* Sandwith)). Similarly, the character states "large flower" and "small flower" and their diagnosis based upon corolla tube/lobe ratio summarize diverse flower sizes. The presence or absence of head inflorescences involves the complication of comparing much-branched inflorescences with unbranched inflorescences: either may have flowers in heads or not. Fruit fusion, though variable in degree, is summarized in the character states as simple or multiple. Despite the simplification of diverse characteristics into discrete character states, the essential outcomes of these analyses are clearly evident across the phylogenetic span of this inquiry: repeated shifts have occurred in the evolution of the growth habit, inflorescence architecture, flower size, and fruit across the species of the Morindeae. Moreover, the direction of these evolutionary changes has often been unexpected and at odds with currently accepted hypotheses. Finally, the findings of this study provide a new context for viewing patterns of character evolution and examining their ecological and developmental basis.

MATERIALS AND METHODS

Taxon Sampling and Data Collection

The sampling used for this study coincided with the molecular phylogenetic study of Morindeae by Razafimandimbison et al. [9], on the basis of which new generic limits of the tribe were established. This latter study resulted in the transfer of all lianescent, dioecious *Morinda* species to *Gynochthodes* and all species of *Sarcopygme* Setch & Christoph. to *Morinda* [8]–[9]. Accordingly, the newly combined names of *Morinda* and *Gynochthodes*, respectively, were utilized in this study to replace the names of the sampled *Sarcopygme* and lianescent *Morinda* used in Razafimandimbison et al. [9]. Five Morindeae taxa (*Appunia tenuiflora* (Benth.) Jacks & Hook.f., *Morinda royoc* L. 2, and *Gynochthodes candollei* Montrouz. 2, 4, and 5) with incomplete sequences were excluded from this study to decrease the percentage of missing information in the combined nrETS/nrITS/*trn*T-F matrix and obtain a well-resolved phylogeny of Morindeae for basing our ASRs. We investigated a total of 66 taxa, and all information about the voucher specimens and sequences used in the study is published in Razafimandimbison et al. [9].

All morphological characteristics of the five genera of Morindeae summarized in Table 1 were based on data from field notes made by SGR (for *Coelospermum*, *Gynochthodes*, and the paleotropical *Morinda*) and by TDM (for *Appunia* and the Neotropical *Morinda*). This was coupled with data compiled by SGR from herbarium specimens on loan from many herbaria (BR, K, L, MO, P, S, TAN, TEF, UPS, [40]) and the literature [8]–[9], [12]–[13], [21]–[22], [24], [26]–[27].

Laboratory Work and Phylogenetic Analyses

The protocols used for DNA extraction, amplification, and sequencing are outlined in Razafimandimbison et al. [9]. The alignment of the combined nrETS/nrITS/*trn*T-F data was re-adjusted after the removal of *A. tenuiflora*, *M. royoc* 2, and *G. candollei* 2, 4, and 5. We treated each of the three gene regions as a separate partition and selected likelihood models following Razafimandimbison et al. [9]. As a consequence, we applied separately parameterized GTR+ G models to the *trn*T-F and nrITS partitions and a separately parameterized HKY+ G model to the nrETS partition. The gamma distributed rate heterogeneity across sites was approximated with four discrete categories. Flat Dirichlet priors were applied to the state frequencies and to the substitution rates of the GTR model, whereas a flat beta distribution was used as prior for the transition-to-transversion rate. A uniform prior on the interval (0.1, 50) was applied to the gamma curve shape parameter α. The prior on branch lengths was an exponential distribution with mean 0.1. Rate heterogeneity across partitions was modeled according to a proportional model with a flat Dirichlet prior. Tree topologies were treated as a priori equally likely. Three runs of Metropolis-coupled MCMC was run for 25×10^6 generations, each run starting from a random tree with initial branch lengths set to 0.1. Each run included four chains, three of which were incrementally heated to a temperature of 0.15 to ensure swap rates between adjacent chains between 10 and 70%. Every 1000th generation of the cold MCMC chain was sampled. Stationarity and convergence of runs, as well as the correlation of split frequencies between the runs were checked using the program AWTY [41]. We checked the effective sample size (ESS) of parameters using the program Tracer v.1.5.0 [42]. Trees sampled from the first 12.5×10^6 generations were discarded as burn-in. All saved trees (after excluding burn-ins) from the three independent runs were pooled for a consensus tree.

Reconstruction of Ancestral States

A variety of comparative phylogenetic methods have been used for reconstructing ancestral states of characters and mapping character changes across lineages: maximum parsimony [43]–[44], maximum likelihood [45], Bayesian inference [46]–[48], and stochastic character mapping [49]. The influences of method choice in reconstructing ancestral states of characters are well documented [2], [50]; it has recently been demonstrated that homoplasious characters are sensitive to choice of method [2].

The Bayesian approach implemented in the computer program BayesTraits v. 1.0 [48] appears to preserve the highest amount of uncertainty in ASR of discrete characters [2], [50]. It takes into account both phylogenetic uncertainty and branch length, and also permits one to explore a variety of models for character transition and to investigate nodes of interest [2], [50]. We performed ASRs of the four characters of the tribe

Morindeae (growth habit, inflorescence architecture, flower size, and fruit type) using the software BayesTraits as described by [48] and on two posterior tree samples, one in which all outgroup taxa (Figure 2) had been pruned and one in which we kept the outgroup taxa *Mitchella repens* and *Damnacanthus indicus* of the tribe Mitchelleae, known from previous studies to be the closest relatives of the Morindeae [51]. *Pagamea guianensis* Aubl. and *Gaertnera phyllostachya* Baker were pruned from the analyses, because they represent the poorly sampled tribe Gaertnereae and appear on long branches in the phylogeny. Before proceeding with the ASR, we checked trees for the node-density artifact [52] using the on-line implementation at http://www.evolution.reading.ac.uk/pe/index.html. The following four discrete characters were reconstructed for the ingroup taxa: growth form lianescent (0), arborescent (including frutescent and suffrutescent plants of *Morinda*, the weak-stemmed shrubs or treelets of *Appunia*, and the scandent- or vining branched shrubs or treelets of *Morinda royoc* L., i.e., all non-liana woody shrubs and trees) (1), and herbaceous (only relevant for the outgroup taxon *M. repens*) (2); inflorescence headed (0) or non-headed (1); flowers large (corolla tube length/corolla lobe length >1) (0) or small (corolla tube length/corolla lobe length <1) (1); and fruits simple (0) or fused (1). State probabilities were estimated for the following seven selected nodes in the Bayesian majority rule consensus tree (Figure 2): Morindeae (node A), the *Morinda-Coelospermum-Gynochthodes* clade (node B), the *Coelospermum-Gynochthodes* clade (node C), *Appunia* (node D), *Morinda* (node E), *Coelospermum* (node F), and *Gynochthodes* (node G). Node A corresponds to the root of the tree when the outgroup taxa had been excluded. In addition, we reconstructed the root node (joining the Mitchelleae and Morindeae) in the analyses involving the two outgroup taxa of Mitchelleae. Reversible-jump MCMC was used to integrate over models. For single binary characters, there are four possible models, one two-rate model in which forward (q_{01}) and backward (q_{10}) rates are free, one single-rate model in which q_{01} and q_{10} are constrained to be equal, two single-rate models in which either q_{01} or q_{10} is estimated, and the reverse rate is fixed to zero. Ratios of q_{01} to q_{10} deviating from 1 indicate that the rate of change in one direction is higher than in the opposite direction.

We used a uniform prior on the models and an exponential prior on rates, the mean of which was seeded by a uniform hyperprior on the interval (0, 10). By applying an exponential prior on rates we say that moderate rates are a priori more likely than high rates and that strong evidence from the data is required to accept high-rate estimates. We also included the branch-length transformation parameter κ in the model [53]. This parameter raises original branch lengths to the κ power. If $\kappa = 0$, all branches are equally long, i.e., change is independent of branch lengths. If $\kappa = 1$, branch lengths are not modified and change is perfectly proportional to the original branch lengths. κ >1 indicates that change accelerates with increasing branch length and 0< κ <1 indicates that change decelerates with decreasing branch lengths. The prior on κ is a uniform distribution on the interval (0, 5) (A. Meade, pers. com.). The MCMC was run for 220×10^6 generations, the first 20×10^6 of which were discarded as burnin. A sample was saved from the posterior every 1000th generation. The rate deviation of the normal distribution was set to obtain an MCMC acceptance rate between 20% and 40%. Each analysis was conducted three times to check that similar harmonic mean likelihoods were obtained across runs.

ACKNOWLEDGMENTS

We thank the following herbaria for allowing access to their collections: BR, CAY, K, L, MO, P, S, TAN, TEF, UPS, and US; the DGF (Direction Générale des Forêts) and MNP (Madagascar National Parks) in Madagascar for issuing collecting permits for SGR; Missouri Botanical Garden, Madagascar Program for logistical support; Parc Botanique et Zoologique de Tsimbazaza and Missouri Botanical Garden, Madagascar Program (Lalao Andriamahefarivo and Faranirina Lantoarisoa) for arranging collecting permits for SGR; Kent Kainulainen for technical assistance with Figure 1; and Andrew J. Ford, Johan T. Johansson, Frédéric Tronquet, and Kent Kainulainen for kindly providing photos of Morindeae taxa.

AUTHOR CONTRIBUTIONS

Conceived and designed the experiments: SGR. Performed the experiments: SGR SE. Analyzed the data: SGR SE. Contributed reagents/materials /analysis tools: BB. Wrote the paper: SGR SE TDM BB.

REFERENCES

1. Endress PK (1994) Diversity and evolutionary biology of tropical flowers. Cambridge: University Press. 511p p.PK Endress1994Diversity and evolutionary biology of tropical flowers.Cambridge: University Press511p
2. Xiang QY, Thomas DT (2008) Tracking character evolution and biogeographic history through time in Cornaceae - Does choice of methods matter? J Syst Evol 46: 349–374.QY XiangDT Thomas2008Tracking character evolution and biogeographic history through time in Cornaceae - Does choice of methods matter?J Syst Evol46349374

3. Eyde RH (1985) The case for monkey-mediated evolution in big-bracted dogwoods. Arnodia 45: 2–9.RH Eyde1985The case for monkey-mediated evolution in big-bracted dogwoods.Arnodia4529
4. Corlett RT (1998) Frugivory and seed dispersal by vertebrates in the Oriental (Indomalayan) Region. Biol Rev Cambridge Phil Soc 73: 413–448.RT Corlett1998Frugivory and seed dispersal by vertebrates in the Oriental (Indomalayan) Region.Biol Rev Cambridge Phil Soc73413448
5. Rowe N, Speck T (2005) Plant growth forms: an ecological and evolutionary perspective. New Phytol 166: 61–72.N. RoweT. Speck2005Plant growth forms: an ecological and evolutionary perspective.New Phytol1666172
6. Razafimandimbison SG, Bremer B (2001) Tribal delimitation of Naucleeae (Rubiaceae): inference from molecular and morphological data. Syst Geogr Pl 71: 515–538.SG RazafimandimbisonB. Bremer2001Tribal delimitation of Naucleeae (Rubiaceae): inference from molecular and morphological data.Syst Geogr Pl71515538
7. Razafimandimbison SG, Bremer B (2002) Phylogeny and classification of Naucleeae (Rubiaceae) inferred from molecular (nrITS, rbcL, and trnT-F) and morphological data. Am J Bot 89: 1027–1041.SG RazafimandimbisonB. Bremer2002Phylogeny and classification of Naucleeae (Rubiaceae) inferred from molecular (nrITS, rbcL, and trnT-F) and morphological data.Am J Bot8910271041
8. Razafimandimbison SG, Bremer B (2011) Nomenclatural changes and taxonomic notes in the tribe Morindeae (Rubiaceae). Adansonia 33: 281–307.SG RazafimandimbisonB. Bremer2011Nomenclatural changes and taxonomic notes in the tribe Morindeae (Rubiaceae).Adansonia33281307
9. Razafimandimbison SG, McDowell TD, Halford DA, Bremer B (2009) Molecular phylogenetics and generic assessment in the tribe Morindeae (Rubiaceae-Rubioideae): how to circumscribe Morinda L. to be monophyletic? Mol Phylogenet Evol 52: 879–886.SG RazafimandimbisonTD McDowellDA HalfordB. Bremer2009Molecular phylogenetics and generic assessment in the tribe Morindeae (Rubiaceae-Rubioideae): how to circumscribe Morinda L. to be monophyletic?Mol Phylogenet Evol52879886
10. Razafimandimbison SG, McDowell TD, Halford DA, Bremer B (2010) Origin of the pantropical and nutriceutical Morinda citrifolia L. (Rubiaceae): comments on its distribution range and circumscription. J. Biogeogr 37: 520–529.SG RazafimandimbisonTD McDowellDA HalfordB. Bremer2010Origin of the pantropical and nutriceutical Morinda citrifolia L. (Rubiaceae): comments on its distribution range and circumscription.J. Biogeogr37520529
11. McClatchey WC (2003) Diversity of growth forms, and uses in the Morinda citrifolia L. complex. In: Nelson SC, editor. pp. 5–10.WC McClatchey2003Diversity of growth forms, and uses in the Morinda citrifolia L. complex.SC Nelson510editor.Proceeding of the 2002 Hawai'i Noni Conference.Honolulu: University of Hawaii at Manoa. Honolulu: University of Hawaii at Manoa.
12. Taylor CM, Lorence D (2010) Rubiacearum Americanum Magna Hama Pars XXII: Notable new species of South American Coutarea, Morinda, Patima, and Rosenbergiodebndron. Novon 95–105. CM TaylorD. Lorence2010Rubiacearum Americanum Magna Hama Pars XXII: Notable new species of South American Coutarea, Morinda, Patima, and Rosenbergiodebndron.Novon 95–105
13. Johansson JT (1988) Revision of Caelospermum Blume (Rubiaceae, Rubioideae, Morindeae). Blumea 33: 265–297.JT Johansson1988Revision of Caelospermum Blume (Rubiaceae, Rubioideae, Morindeae).Blumea33265297
14. Speck T, Rowe NP, Civeyrel L, Classen-Bockhoff R, Neinhuis C, et al. (2003) The potential of plant biomechanics in functional biology and systematics. In: Stuessy TF, Mayer V, Hörandl E, editors. pp. 241–271.T. SpeckNP RoweL. CiveyrelR. Classen-BockhoffC. Neinhuis2003The potential of plant biomechanics in functional biology and systematics.TF StuessyV. MayerE. Hörandl241271editors.Deep morphology:

toward a renaissance of morphology in plant systematics.Lichtenstein: ARG Ganter Verlag. Lichtenstein: ARG Ganter Verlag.

15. Whitlock BA, Hale AM (2011) The phylogeny of Ayenia, Byttneria, and Rayleya (Malvaceae s.l.) and its implication for the evolution of growth forms. Syst Bot 36: 129–136.BA WhitlockAM Hale2011The phylogeny of Ayenia, Byttneria, and Rayleya (Malvaceae s.l.) and its implication for the evolution of growth forms.Syst Bot36129136
16. Lahaye R, Giveyrel L, Speck T, Rowe NP (2005) Evolution of shrub-like growth forms in the lianoid subfamily Secamonoideae (Apocynaceae s.l.) of Madagascar: phylogeny, biomechanics, and development. Amer J Bot 92: 1381–1396.R. LahayeL. GiveyrelT. SpeckNP Rowe2005Evolution of shrub-like growth forms in the lianoid subfamily Secamonoideae (Apocynaceae s.l.) of Madagascar: phylogeny, biomechanics, and development.Amer J Bot9213811396
17. Robbrecht E (1988) Tropical woody Rubiaceae. Opera Bot Belg 1: 1–271.E. Robbrecht1988Tropical woody Rubiaceae.Opera Bot Belg11271
18. Raju AJS, Rao SP, Ezradaman V, Zafar R, Kalpana PR, et al. (2004) The hawkmoth Macroglossum gyrans and its interaction with some plant species at Visakhapatnam. Zoos' Print J 19: 1595–1598.AJS RajuSP RaoV. EzradamanR. ZafarPR Kalpana2004The hawkmoth Macroglossum gyrans and its interaction with some plant species at Visakhapatnam.Zoos' Print J1915951598
19. Halford DA, Ford AJ (2009) Two species of Morinda L. (Rubiaceae) from north-east Queensland. Austrobaileya 8: 81–90.DA HalfordAJ Ford2009Two species of Morinda L. (Rubiaceae) from north-east Queensland.Austrobaileya88190
20. Halford DA, Ford AJ (2009) Coelospermum purpureum Halford & A.J. Ford (Rubiaceae), a new species from north-east Queensland. Austrobaileya 8: 69–76.DA HalfordAJ Ford2009Coelospermum purpureum Halford & A.J. Ford (Rubiaceae), a new species from north-east Queensland.Austrobaileya86976
21. Burck MW (1883) Sur l'organisation florale chez quelques Rubiacées. Suite. Ann Jard Bot Buitenzorg 3: 109.MW Burck1883Sur l'organisation florale chez quelques Rubiacées. Suite.Ann Jard Bot Buitenzorg3109
22. Johansson JT (1994) The genus Morinda (Morindeae, Rubioideae, Rubiaceae) in New Caledonia. Taxonomy and phylogeny. Opera Bot 122: 5–67.JT Johansson1994The genus Morinda (Morindeae, Rubioideae, Rubiaceae) in New Caledonia. Taxonomy and phylogeny.Opera Bot122567
23. Liu Y, Luo Z, Wu X, Bai X, Zhang D (2012) Pollinators with progressively shorter proboscis may have been driving the transition from large to small flowers and an accompanying increase in speciation rate. Pl Syst Evol 298: 775–785.Y. LiuZ. LuoX. WuX. BaiD. Zhang2012Pollinators with progressively shorter proboscis may have been driving the transition from large to small flowers and an accompanying increase in speciation rate.Pl Syst Evol298775785
24. Wong KM (1984) A synopsis of Morinda (Rubiaceae) in the Malay Peninsula, with two new species. Malayan Nat J 38: 89–98.KM Wong1984A synopsis of Morinda (Rubiaceae) in the Malay Peninsula, with two new species.Malayan Nat J388998
25. Ortiz RD, Kellogg CEA, Werff HVD (2007) Molecular phylogeny of the mooseed family (Menispermaceae): implications for morphological diversification. Am J Bot 94: 1425–1438.RD OrtizCEA KelloggHVD Werff2007Molecular phylogeny of the mooseed family (Menispermaceae): implications for morphological diversification.Am J Bot9414251438
26. Burger W, Taylor CM Burger W, editor. (1993) Family # 202 Rubiaceae. 33: 1–333.W. BurgerCM Taylor1993Family # 202 Rubiaceae.W. Burger331333editor.Flora Costaricensis.Fieldiana Bot. Fieldiana Bot.
27. Tao C, Taylor CM (2011) Rubiaceae. Fl China 19: 220–229.C. TaoCM Taylor2011Rubiaceae.Fl China19220229
28. Endress PK (2010) Disentangling confusions in inflorescence morphology: patterns and diversity of reproductive shoot ramification in angiosperms. J Syst Evol 48: 225–239.PK Endress2010Disentangling confusions in inflorescence morphology:

patterns and diversity of reproductive shoot ramification in angiosperms.J Syst Evol48225239
29. Feng CM, Xiang QY, Franks RG (2011) Phylogeny-based developmental analyses illuminate evolution of inflorescence architectures in dogwoods (Cornus s.l., Cornaceae). New Phytol 191: 850–869.CM FengQY XiangRG Franks2011Phylogeny-based developmental analyses illuminate evolution of inflorescence architectures in dogwoods (Cornus s.l., Cornaceae).New Phytol191850869
30. Bawa KS (1980) Evolution of dioecy in flowering plants. Ann Rev Ecol Syst 11: 15–39.KS Bawa1980Evolution of dioecy in flowering plants.Ann Rev Ecol Syst111539
31. Bawa KS (1982) Outcrossing and the incidence of dioecism in island floras. Am Nat 119: 866–871.KS Bawa1982Outcrossing and the incidence of dioecism in island floras.Am Nat119866871
32. Bawa KS (1994) Pollination of tropical dioecious angiosperms: a reassessment? No, not yet. Am J Bot 81: 456–460.KS Bawa1994Pollination of tropical dioecious angiosperms: a reassessment? No, not yet.Am J Bot81456460
33. Bawa KS, Bullock SH, Perry DR, Coville RE, Grayum MH (1985) Reproductive biology of tropical lowland rain forest trees. II. Pollination systems. Am J Bot 72: 346–356.KS BawaSH BullockDR PerryRE CovilleMH Grayum1985Reproductive biology of tropical lowland rain forest trees. II. Pollination systems.Am J Bot72346356
34. Bawa KS, Opler PA (1975) Dioecism in tropical forest trees. Evolution 29: 167–179.KS BawaPA Opler1975Dioecism in tropical forest trees.Evolution29167179
35. Muenchov GE (1987) Is dioecy associated with fleshy fruit? Am J Bot 74: 287–293.GE Muenchov1987Is dioecy associated with fleshy fruit?Am J Bot74287293
36. Renner S, Ricklefs RE (1995) Dioecy and its correlates in the flowering plants. Am J Bot 82: 596–606.S. RennerRE Ricklefs1995Dioecy and its correlates in the flowering plants.Am J Bot82596606
37. Weller SG, Sakai AK (1999) Using phylogenetic approaches for the analysis of plant breeding system evolution. Ann Rev Ecol Syst 30: 167–199.SG WellerAK Sakai1999Using phylogenetic approaches for the analysis of plant breeding system evolution.Ann Rev Ecol Syst30167199
38. Tetsuto A (2006) Threatened pollination systems in native floras of the Ogasawara (Bonin) Islands. Ann Bot 98: 317–334.A. Tetsuto2006Threatened pollination systems in native floras of the Ogasawara (Bonin) Islands.Ann Bot98317334
39. Razafimandimbison SG (2002) A systematic revision of Breonia (Rubiaceae-Naucleeae). Ann Missouri Bot Gard 89: 1–37.SG Razafimandimbison2002A systematic revision of Breonia (Rubiaceae-Naucleeae).Ann Missouri Bot Gard89137
40. Holmgren PK, Holmgren NH, Barnett LC (1990) Index herbarium. Part I: the herbaria of the world, 8th edition. New York: New York Botanical Garden. 693 p. PK HolmgrenNH HolmgrenLC Barnett1990Index herbarium. Part I: the herbaria of the world, 8th edition. New York: New York Botanical Garden.693 p
41. Nylander JAA, Wilgenbusch JC, Warren DL, Swofford DL (2008) AWTY (Are We There Yet?): A system for graphical exploration of MCMC convergence in Bayesian phylogenetics. Bioinformatics 24: 581–583.JAA NylanderJC WilgenbuschDL WarrenDL Swofford2008AWTY (Are We There Yet?): A system for graphical exploration of MCMC convergence in Bayesian phylogenetics. Bioinformatics24581583
42. Rambaut A, Drummond AJ (2009) Tracer version 1.5. Edinburgh: University of Edinburgh. Available from. A. RambautAJ Drummond2009Tracer version 1.5. Edinburgh: University of Edinburgh.Available fromhttp://tree.bio.ed.ac.uk /software/tracer/. http://tree.bio.ed.ac.uk/software/tracer/.
43. Maddison DR, Maddison WP (1992) MacClade: Analysis of phylogeny and character evolution. Version 3.0. Sunderland: Sinauer Associates. DR MaddisonWP Maddison1992MacClade: Analysis of phylogeny and character evolution. Version 3.0.Sunderland: Sinauer Associates
44. Maddison D, Maddison WP (2007) Mesquite: a modular system for evolutionary analysis (online). Version 2.01. Available from. D. MaddisonWP

Maddison2007Mesquite: a modular system for evolutionary analysis (online). Version 2.01.Available fromhttp://mesquiteproject.org. http://mesquiteproject.org.
45. Pagel M (1999) The maximum likelihood approach to reconstructing ancestral character states of discrete on phylogenies. Syst Biol 48: 612–622.M. Pagel1999The maximum likelihood approach to reconstructing ancestral character states of discrete on phylogenies.Syst Biol48612622
46. Huelsenbeck JP, Ronquist F, Nielsen R, Bollback JP (2001) Bayesian inference of phylogeny and its impact on evolutionary biology. Science 294: 2310–2314.JP HuelsenbeckF. RonquistR. NielsenJP Bollback2001Bayesian inference of phylogeny and its impact on evolutionary biology.Science29423102314
47. Ronquist F (2004) Bayesian inference of character evolution. Trends Ecol Evol 9: 475–481.F. Ronquist2004Bayesian inference of character evolution.Trends Ecol Evol9475481
48. Pagel M, Meade A, Barker D (2004) Bayesian estimation of ancestral states on phylogenies. Syst Biol 53: 673–684.M. PagelA. MeadeD. Barker2004Bayesian estimation of ancestral states on phylogenies.Syst Biol53673684
49. Huelsenbeck J, Bollback JP (2001) Empirical and hierarchical Bayesian estimation of ancestral states. Syst Biol 50: 351–366.J. HuelsenbeckJP Bollback2001Empirical and hierarchical Bayesian estimation of ancestral states.Syst Biol50351366
50. Ekman S, Andersen HL Wedin M (2008) The limitations of ancestral state reconstruction and the evolution of the ascus in the Lecanorales (Lichenized Ascomycota). Syst Biol 57: 141–156.S. EkmanM. Andersen HL Wedin2008The limitations of ancestral state reconstruction and the evolution of the ascus in the Lecanorales (Lichenized Ascomycota).Syst Biol57141156
51. Razafimandimbison SG, Rydin C, Bremer B (2008) Evolution and trends in the Psychotrieae alliance (Rubiaceae): A rarely reported evolutionary change from one-seeded carpels to many-seeded carpels. Mol Phylogenet Evol 48: 207–223.SG RazafimandimbisonC. RydinB. Bremer2008Evolution and trends in the Psychotrieae alliance (Rubiaceae): A rarely reported evolutionary change from one-seeded carpels to many-seeded carpels.Mol Phylogenet Evol48207223
52. Venditti C, Meade A, Pagel M (2006) Detecting the node-density artifact in phylogeny reconstruction. Syst Biol 55: 637–643.C. VendittiA. MeadeM. Pagel2006Detecting the node-density artifact in phylogeny reconstruction.Syst Biol55637643
53. Pagel M (1994) Detecting correlated evolution on phylogenies: a general method for the comparative analysis of discrete characters. Proc Royal Soc London, ser B 255: 37–45.M. Pagel1994Detecting correlated evolution on phylogenies: a general method for the comparative analysis of discrete characters.Proc Royal Soc London, ser B2553745

CHAPTER 4

Antimicrobial Activity of Fruit Latexes from Ten Laticiferous Plants

Ravi Kant Upadhyay

D. D. U. Gorakhpur University, Gorakhpur, India

ABSTRACT

In the present investigation antibacterial activity of latexes from ten Indian plant species Spondias dulcis (Amra), Diospyros melanoxylon (Tendu), Terminalia bellirica (Wahera), Ficus glomerata (Gular), Phyllanthus emblica (Awla), Thevetia neriifolia (Kaner), Carica papaya (Papita), Calotropis procera (Ak), Ficus benghalensis (Bargad), Artocarpus heterophyllus (Kathal) collected from Gorakhpur, North India were determined in various in vitro systems. MIC (minimum inhibitory concentration) and MBC (minimum bactericidal concentration) values were determined in growth inhibitory bioassays by using different increasing concentrations of various latex extracts. Latex samples were diluted by using serial micro dilution method up to 10^{-10} method with Luria broth culture medium. These values were obtained significantly 2 - 3 times lower than that of broad spectrum antibiotic drugs. Besides this, inhibition zone diameters were measured in agar disc diffusion assay. A known volume i.e. 0.1 - 20 µg/µl of each latex were coated on separate sterile filter paper discs (Whatman No. 1) measuring 6 mm in size. Latex fractions registered significantly higher growth inhibition than that of broad spectrum antimicrobial drugs. Present study indicates the potential use of shows that both latex and its components and latex as are valuable source of medicinal products/active principles that can be used for treatment of life threatening infectious diseases. Because of higher inhibitory and cidal potential obtained in latexes than the synthetic drugs these that could lead to become efficient phytomedicines mainly to have and develop as complete drug formulations against to control infectious microbes.

Keywords: Plant Latex, Carica papaya, Calotropis procera, Ficus benghalensis, Artocarpus heterophyllus, Growth Inhibition, Antimicrobial Susceptibility

1. INTRODUCTION

Latex is milky fluid secreted by the ducts of laticiferous tissue found in roots, stems, leaves and fruits of flower-

ing plants [1] . It is a mixture of many pharmaceutically active substances which showed very high anti-bacterial [2] , antiviral [3] , antifungal [4] anti-protozoan activity. Latex from several plant species such as A. ochroleuca [2] Calotropis procera, Plumeria rubra, Carica candamarcensis, Euphorbia tirucalli (EtLP) [5] , Erythrina variegata [6] , Jatropha curcas, Hancornia speciosa Gomes (Apocynaceae) [7] , C. procera [8] , and Euphorbia lathyris [9] have shown very high antimicrobial activity. The leaves of J. curcas contain apigenin, vitexin and isovitexin, which along with other factors showed strong antibacterial activity against Staphylococcus aureus and Escherichia coli [10] . Hevea bransilensis showed antifungal activity against Trichosporon cutaneum and Cryptococcus neoformans [5] . It also exhibit good activity against multiple drug resistant Staphylococcus aureus (NCTC 11994) and Salmonella typhimurium (ATCC 1255) and Candida albicans (ATCC 10231) [11] . The latex of fig fruit (Ficus carica) is used in traditional medicine for the treatment of skin infections. Its hexanic and hexane-ethyl acetate (v/v) extracts inhibit virus replication and work against herpes simplex type 1 (HSV-1), echovirus type 11 (ECV-11) and adenovirus (ADV) infections [12] . Latex from Phyllanthus sp. showed antiviral effects [13] .

Plant latex is a natural source of antiprotozoal drugs [14] [15] which show schizonticidal activity [16] . Poly-o- acylated jatrophene diterpenes found in latex of Pedilanthus tithymaloides showed antiplasmodial activity [17] . Similarly Xylopia aromatica, Casearia sylvestris, Cupania vernalis and Aspidosperma macrocarpon [18] Casearia sylvestris var. lingua latex showed anti-protozoan activity against Trpanosoma cruzi [19] . C. scoparioides latex shows antimalarial, leishmanicidal and trypanocidal activity in vitro [20] . Latex from few plant species such as Himatonthus sucouba [21] showed Leishmanicidal activity while lectins isolated from

Synadenium carinatum latex exhibit anti-protozoan activity against Leishmania, amazonensis promastigotes/amastigotes [22] . Moreover, Croton lechleri latex contains crofelemer which shows anti-secretory antidiarrheal activity [23] .

In present time drug resistance in microbes is a very serious problem. With the time new chemical drugs have been discovered for control of these pathogens. But due to indiscriminate use of these synthetic drugs, most of the pathogens have developed resistance. Though in the beginning, these synthetic antibiotics have shown very good fighting potential against pathogens but later on these become ineffective because microbes have developed resistance to them. Besides, synthetic drugs generate wider side effects and heavy cross reactivity in the body tissues and cells. Therefore, for the treatment of infectious diseases, many complementary and alternative medicines are to be discovered for safe therapeutic purposes. In present time for effective control of microbial infections, plant products considered as best solutions because of their low side effects. Until the date, very few active components have been isolated from plants and still their antimicrobial susceptibility is yet to check. In spite of the fact lot of work has been done on antimicrobial plant product but no such work is reported on plant latex. Hence, in the present investigation antimicrobial susceptibility of plant latex was screened for effective control of infectious pathogens. Until the date, very few active components have been isolated from plants and still their antimicrobial susceptibility is yet to check. In spite of the fact, lot of work has been done on antimicrobial plant product but no such work is reported on plant latex. Hence, in the present investigation antimicrobial susceptibility of plant latex was screened for effective control of infectious pathogens. The main objective of the present study was to establish the antibacterial activity of fruit latex from four plant species Spondias dulcis (Amra), Diospyros melanoxylon (Tendu), Terminalia bellirica (Wahera), Ficus glomerata (Gular), Phyllanthus emblica (Awla), Thevetia neriifolia (Kaner), Carica papaya (Papita), Calotropis procera (Ak), Ficus benghalensis (Bargad), Artocarpus heterophyllus (Kathal) (Figure 1) pathogenic bacterial strains i.e. Klebsiella pneumoniae, E. coli, Bacillus cereus. Staphylococcus aureus, Lactobacillus acidophilus, Micrococcus luteus and Streptococcus pneumoniae.

2. EXPERIMENTAL

2.1. Plant Material

Fruit latexes were collected from unripe fresh fruits in a aseptic glass vessel by making sharp incisions at fixed time interval [24] . For draining maximum amount fruits were cut open from its top then slightly squeezed to collect the latex in sterile tube then stored at −20°C until further use. Fruit latexes from eight plant species Spondias dulcis (Amra), Diospyros melanoxylon (Tendu), Terminalia bellirica (Wahera), Ficus glomerata (Gular), Phyllanthus emblica (Awla), Thevetia neriifolia (Kaner), Carica papaya (Papita), Calotropis procera (Ak), Ficus benghalensis (Bargad), Artocarpus heterophyllus (Kathal) (Figure 1) were used for various anti-microbial bioassays.

1

2

3

4

5

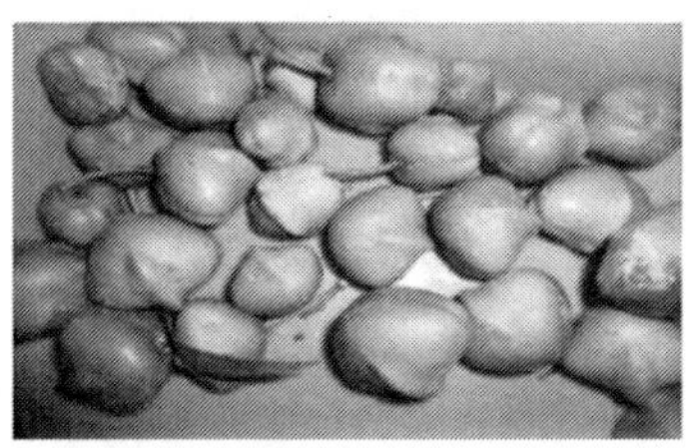

6

7a

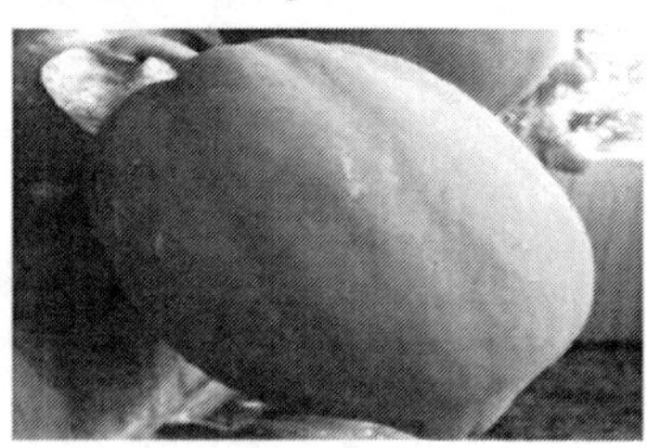

7b

Figure 1. Showing various latex bearing fruits from plant species i.e. 1. Spondias dulcis (Amra), 2. Diospyros melanoxylon (Tendu), 3. Terminalia bcllirica (Wahera), 4. Ficus glomerata (Gular), 5. Phyllanthus emblica (Awla), 6. Thevetia neriifolia (Kaner), 7(a & b) Carica papaya (Papita), 8. Calotropis procera (Ak), 9. Ficus benghalensis (Bargad), 10 (a & b) Artocarpus heterophyllus (Kathal).

2.2. Isolation and Extraction of Plant Latex

Collected plant latex was lyophilized and powdered in vacuum in cold. It was extracted with different solvents. For better fractionation active fractions from the latex were portioned between different solvents on the basis of their polarity from low to high and non polar. Mostly portioning was done between hexane and aqueous methanol, petroleum ether and chloroform. Further a portion of dried latex was extracted with distilled water, 1.5% acetic acid, 1.5% Sodium bicarbonate and 1.5% sulfuric acid and diethyl ether to separate various fractions by following the method of Steven McCay and Paul Mahlberg, 1973 [25] . For testing antimicrobial activity a known quantity of vacuum dried plant latex was added to fresh solvents. Latex samples were stored in cold at 4°C.

2.3. Maintenance of Bacterial Culture

Cultures of seven pathogenic bacterial strains each of Escherichia coli (ATCC 25922), Bacillus cereus (ATCC 11778), Lactobacillus acidophilus (ATCC 53103), Micrococcs luteus (ATCC 9341), Staphylococcus aureus (ATCC 25923), Klebsiella pneumoniae, (ATCC 15380) and Streptococcus pneumoniae (ATCC 12755) was maintained in the laboratory in Luria Broth (2% w/v) regularly for five days at 37°C before use in experiments. For experiments, a portion (100 μl) of the overnight bacterial culture was mixed in the tests and control for inoculation. For activity anti-microbial testing bacterial cultures were sub cultured after every 8th day in solid agar plates and stored at 4°C.

2.4. Determination of Antimicrobial Activity: MIC and MBC Values

For determination of antimicrobial activity bacterial growth inhibition was accessed in presence of different increasing concentrations of plant latex fractions. For this purpose, latex samples were diluted by using serial micro dilution method with Luria Broth culture medium. In tests latex was added to fresh suspension after making the serial dilutions up to 10^{-10}. Each sample of latex was assayed in triplicate. Before conducting experiments all the conditions for in vitro anti-microbial activity were standardized to determine MIC and MBC values according to Karamon et al., 2003 [26] . The MIC values were determined in culture flask after 24 h of incubation at 37°C. The turbidity in the culture flasks was considered as visible growth of microorganisms. Further, it was standardized in terms of absorbance at 600 nm in a visible spectrophotometer. For determination of minimum bactericidal concentration (MBC) growth inhibitory assays were also performed. For this purpose, inoculums size was adjusted to prepare a final colony number as 10^8 colony forming units (CFU/ml) in sterile agar plates. The incubation of test and control cultures was performed at 37°C for 24 h. For comparison, both negative and positive controls were set and bacterial colony number was counted in all test and control Petri plates. The least concentration at which no visible growth obtained in agar plates was considered as MBC value. For evaluation of inhibition two parallel controls were set for each test sample.

2.5. Determination of Inhibition Zone Diameter

Antimicrobial activity of latex was evaluated by agar disc diffusion method and inhibition zone diameters were measured according to Bauer et al., 1966 [27] in presence of each latex sample separately. Agar-agar was used as media for the test microorganisms. For antimicrobial activity testing plant latex samples were diluted by adding equal volume of solvent and preceded further two to obtain two fold serial micro-dilution. From this, a known volume i.e. 2 - 32 µl of each latex coated on sterile filter paper discs (Whatman No. 1) measuring 6 mm in size. These latex impregnated discs were dried under laminar flow cabinet. Before experiment inoculum size was determined and adjusted to prepare a final colony number as 10^8 colony forming units (CFU/ml) in sterile agar plates. Bacterial inoculum was evenly distributed on to the surface of agar plates by using sterile rubber pad spreader. Now latex coated discs were positioned on the inoculated agar surface. Each latex sample was assayed in triplicate. Sterile distilled water was used as negative control while different antibiotics i.e. tetracycline, ampicillin and ciprofloxacin were used as standard (positive control) to compare the bacterial growth. All treated and untreated plates were incubated for 24 h at 37°C. Inhibition zone diameter surrounding the filter paper discs were measured in each test and control Petri plate.

3. RESULTS

3.1. Determination of MIC and MBC Values

The antimicrobial activity of solvent and aqueous extracts from fruit latexes of eight plant species were tested against seven pathological strains of bacteria. The MIC values obtained in vitro suspension culture of different plant extracts are shown in Table 1. The MIC value of all the fruit latex extracts was found to be very low. The MIC value for Spondias dulcis (Amra), hexanoic extract was obtained lowest against S. pneumoniae (0.12 µg/mL), while methanolic extract was found more susceptible to Klibesella pneumoniae as the MIC value was lowest i.e. 0.28 µg/mL. Its petroleum extract and chloroform extracts have shown higher activity against Klibesella pneumoniae as MIC values obtained were 0.64 and 0.40 µg/mL. Its water extract has shown higher activity against E. coli, B. cereus and Lactobacillus acidophilus as the MIC value obtained was 0.72 µg/mL (Table 1). Diospyros melanoxylon (Tendu), methanolic extract has shown higher MIC value 0.070 µg/mL against K. pneumoniae, while hexanoic extract i.e. 0.48 µg/mL against S. pneumoniae. Its water extract has shown MIC value 0.72 µg/mL against Klebsiella pneumoniae, and Streptococcus pneumoniae (Table 1). Terminalia bellirica (Wahera), methanolic extract has shown higher MIC value against E. coli and M. luteus 0.14, and 0.24 µg/mL while water extract has shown lowest MIC value 0.72 E. coli and Lactobacillus acidophilus (Table 1). Ficus glomerata (Gular) methanolic extract has shown higher MIC value against Klebsiella pneumoniae, M luteus 0.14

µg/mL and 0.24 µg/mL against Lactobacillus acidophilus (Table 1). Phyllanthus emblica (Awla methanolic extract has shown higher MIC value against Klebsiella pneumoniae and Streptococcus pneumoniae 0.14 µg/mL (Table 1). Its water extract has shown lowest MIC against 0.36 0.14 µg/mL against Staphylococcus aureus and Lactobacillus acidophilus. Methanolic extract of Thevetia neriifolia (Kaner), has shown lower MIC value 0.14 µg/mL and 0.24 µg/mL against E. coli and S. aureus. Its water extract was found active against three bacterial strains as the MIC values obtained are very low E. coli, B. cereus Lactobacillus acidophilus 0.72 µg/mL. Calotropis procera has shown very high antmicrobial susceptibility as the MIC value obtained was least in all cases.

Table 1. Determination of MIC and MBC (µg/ml) values of aqueous and solvent extracts of different plant latexes against certain bacterial strains.

Fractions	Conc. range in (µg/ml)	*K. pneumoniae*	*E. coli*	*M. luteus*	*S. pneumoniae*	*S. aureus*	*B. cereus*	*L. acidophillus*
				Spondias dulcis (Amra)				
Hexanoic	0.030 - 30.72	0.96 (2.76)	1.92 (3.84)	0.48 (0.96)	0.12 (0.48)	0.96 (1.92)	0.24 (0.96)	0.12 (0.24)
Methanolic	0.035 - 35.84	0.28 (1.12)	0.56 (1.12)	0.24 (0.96)	0.24 (0.48)	0.48 (0.96)	1.92 (7.68)	0.96 (1.92)
Pet. Ether	0.040 - 40.96	0.64 (2.56)	0.64 (1.28)	2.56 (5.12)	0.64 (1.28)	1.28 (5.12)	1.28 (5.12)	2.56 (5.12)
Chloroform	0.025 - 51.20	0.40 (1.60)	1.60 (3.20)	1.60 (3.20)	0. 80 (1.60)	0.80 (1.60)	1.60 (3.20)	3.20 (6.40)
Water	0.045 - 46.08	1.44 (2.88)	0.72 (2.88)	1.44 (2.88)	1.44 (2.88)	2.88 (5.76)	0.72 (2.88)	0.72 (2.88)
				Diospyros melanoxylon (*Tendu*)				
Hexanoic	0.030 - 30.72	0.96 (1.92)	1.92 (3.84)	0.96 (1.92)	0.48 (0.96)	1.92 (3.84)	1.92 (3.84)	0.96 (3.84)
Methanolic	0.035 - 35.84	0.070 (0.28)	0.14 (0.56)	0.28 (0.56)	0.56 (2.24)	0.28 (0.56)	0.28 (1.12)	0.056 (1.12)
Pet. Ether	0.040 - 40.96	0.64 (2.56)	1.28 (2.56)	1.28 (2.56)	2.56 (5.12)	0.80 (0.16)	0.28 (1.12)	0.32 (0.64)
Chloroform	0.025 - 51.20	0.20 (0.80)	3.20 (6.40)	1.60 (3.20)	0.40 (1.60)	1.60 (3.20)	0. 80 (1.60)	0.80 (1.60)
Water	0.045 - 46.08	0.72 (2.88)	1.44 (2.88)	2.88 (5.76)	0.72 (2.88)	0.72 (2.88)	2.88 (5.76)	1.44 (2.88)
				Terminalia bellirica (*Wahera*)				
Hexanoic	0.030 - 30.72	0.48 (0.96)	0.96 (1.92)	1.92 (3.84)	0.48 (0.96)	3.84 (7.68)	3.84 (7.68)	0.48 (1.92)
Methanolic	0.035 - 35.84	0.56 (1.12)	0.14 (0.24)	0.24 (0.48)	0.24 (1.12)	0.56 (1.12)	1.12 (2.24)	0.56 (1.12)
Pet. Ether	0.040 - 40.96	0.64 (2.56)	0.64 (2.56)	1.28 (2.56)	1.28 (5.12)	1.28 (2.56)	0.64 (2.56)	0.64 (1.28)
Chloroform	0.025 - 51.20	0.40 (1.60)	1.60 (3.20)	0.80 (1.60)	0.40 (1.60)	0.40 (1.60)	0. 80 (1.60)	0.40 (1.60)
Water	0.045 - 46.08	1.44 (2.88)	0.72 (2.88)	1.44 (2.88)	1.44 (2.88)	2.88 (5.76)	1.44 (2.88)	0.72 (2.88)
				Ficus glomerata (*Gular*)				
Hexanoic	0.030 - 30.72	0.96 (1.92)	1.92 (3.84)	0.96 (3.84)	0.48 (1.92)	0.48 (1.92)	1.92 (3.84)	0.48 (0.96)
Methanolic	0.035 - 35.84	0.14 (0.28)	0.24 (0.48)	0.14 (0.28)	0.56 (1.12)	0.56 (1.12)	0.56 (1.12)	0.28 (0.56)
Pet. Ether	0.040 - 40.96	1.28 (2.56)	0.64 (2.56)	1.28 (2.56)	2.56 (5.12)	1.28 (2.56)	0.64 (1.28)	0.64 (1.28)
Chloroform	0.025 - 51.20	1.60 (3.20)	0. 80 (1.60)	0.40 (1.60)	1.60 (3.20)	1.60 (3.20)	0. 80 (1.60)	0.40 (1.60)
Water	0.045 - 46.08	0.36 (1.44)	1.44 (2.88)	2.88 (5.76)	0.72 (2.88)	0.72 (2.88)	1.44 (2.88)	1.44 (2.88)

				Phyllanthus emblica (*Awla*)				
Hexanoic	0.030 - 30.72	1.92 (3.84)	1.92 (7.68)	3.84 (6.78)	1.92 (3.28)	0.48 (0.96)	0.48 (0.96)	0.96 (1.92)
Methanolic	0.035 - 35.84	0.14 (0.56)	0.28 (0.56)	0.14 (1.12)	0.14 (0.56)	1.12 (2.24)	1.12 (2.24)	0.56 (1.12)
Pet. Ether	0.040 - 40.96	0.64 (1.28)	1.28 (2.56)	0.64 (2.56)	1.28 (2.56)	2.56 (5.12)	0.64 (1.28	0.64 (2.56)
Chloroform	0.025 - 51.20	0. 80 (1.60)	0.40 (1.60)	1.60 (3.20)	0.40 (1.60)	0. 80 (1.60)	0.80 (1.60)	0.40 (1.60)
Water	0.045 - 46.08	0.72 (2.88)	0.72 (2.88)	0.72 (2.88)	0.36 (1.44)	0.36 (1.44)	0.72 (2.88)	0.36 (1.44)
				Thevetia neriifolia (*Kaner*)				
Hexanoic	0.030 - 30.72	0.96 (1.92)	0.24 (0.96)	0.96 (1.92)	0.48 (1.92)	0.96 (4.84)	1.92 (3.84)	0.96 (1.92)
Methanolic	0.035 - 35.84	0.28 (1.12)	0.14 (1.12)	0.28 (2.24	0.56 (1.12)	0.14 (0.42)	0.28 (2.24)	0.28 (2.24)
Pet. Ether	0.040 - 40.96	0.64 (5.12)	0.32 (0.64)	0.64 (2.56)	0.64 (2.56)	2.56 (5.12)	2.56 (5.12)	0.64 (0.128)
Chloroform	0.025 - 51.20	0.40 (1.60)	1.60 (3.20)	0. 80 (1.60)	0.40 (1.60)	0. 80 (1.60)	1.60 (3.20)	0.40 (1.60)
Water	0.045 - 46.08	1.44 (2.88)	0.72 (2.88)	1.44 (2.88)	1.44 (2.88)	1.44 (2.88)	0.72 (2.88)	0.72 (1.44)
				Carica papaya (*Papita*)				
Hexanoic	0.030 - 30.72	0.24 (0.48)	0.48 (0.96)	0.48 (1.92)	2.92 (3.84)	0.96 (1.92)	0.48 (1.92)	0.96 (1.92)
Methanolic	0.035 - 35.84	0.28 (0.56)	0.14 (0.56)	0.28 (0.56)	0.56 (1.12)	0.28 (0.56)	0.28 (0.56)	0.14 (0.56)
Pet. Ether	0.040 - 40.96	0.32 (0.64)	0.64 (1.28)	0.64 (1.28)	1.28 (2.56)	1.28 (2.56)	2.56 (1.28)	0.64 (1.28)
Chloroform	0.025 - 51.20	0.80 (1.60)	0.40 (0.80)	0.80 (1.60)	0.80 (1.60)	0.20 (0.40)	0.80 (1.60)	0. 40 (0.80)
Water	0.045 - 46.08	0.36 (1.44)	0.72 (1.44)	1.44 (2.88)	0.44 (2.88)	0.72 (2.88)	0.36 (0.72)	0.72 (1.44)
				Calotropis procera (*Ak*)				
Hexanoic	0.030 - 30.72	0.48 (0.96)	0.96 (1.92)	0.48 (1.92)	1.92 (3.84)	0.48 (1.92)	0.96 (1.92)	0.48 (1.92)
Methanolic	0.035 - 35.84	0.14 (0.56)	0.28 (0.56)	0.56 (1.12)	0.28 (1.12)	0.14 (0.24)	0.28 (0.56)	0.56 (1.12)
Pet. Ether	0.040 - 40.96	0.64 (1.28)	0.32 (0.64)	0.32 (0.64)	1.28 (2.56)	2.56 (5.12)	2.56 (1.24)	0.64 (1.28)
Chloroform	0.025 - 51.20	0.40 (1.60)	0.80 (1.60)	0.40 (1.60)	0.40 (1.60)	0.80 (1.60)	0.80 (1.60)	0.80 (1.60)
Water	0.045 - 46.08	0.72 (1.44)	0.36 (1.44)	0.72 (2.88)	1.44 (2.88)	1.44 (2.88)	0.72 (1.44)	0.36 (0.72)
				Ficus benghalensis (Bargad)				
Hexanoic	0.030 - 30.72	0.24 (0.48)	0.48 (1.92)	0.48 (1.92)	0.96 (3.84)	0.24 (1.92)	0.48 (1.92)	0.96 (1.92)
Methanolic	0.035 - 35.84	0.28 (0.56)	0.14 (0.56)	0.56 (1.12)	0.56 (1.12)	0.56 (0.24)	0.28 (0.56)	0.28 (0.56)
Pet. Ether	0.040 - 40.96	0.32 (1.28)	0.64 (1.28)	0.32 (0.64)	0.64 (1.28)	1.28 (2.56)	2.56 (1.24)	0.32 (0.64)
Chloroform	0.025 - 51.20	0.80 (1.60)	0.40 (0.80)	0.40 (1.60)	0.80 (1.60)	1.60 (3.20)	0.40 (0.80)	0. 80 (1.60)
Water	0.045 - 46.08	0.72 (1.44)	0.72 (1.44)	1.44 (2.88)	1.44 (2.88)	0.72 (2.88)	0.36 (0.72)	0.36 (0.72)
				Artocarpus heterophyllus (*Kathal*)				
Hexanoic	0.030 - 30.72	0.96 (1.92)	0.48 (1.92)	1.92 (3.84)	1.92 (3.84)	0.96 (1.92)	1.92 (3.84)	0.96 (3.84)
Methanolic	0.035 - 35.84	0.28 (1.12)	0.28 (1.12)	0.56 (1.12)	0.56 (1.12)	1.12 (2.24)	0.14 (0.28)	1.12 (2.24)
Pet. Ether	0.040 - 40.96	0.64 (1.28)	0.64 (2.56)	0.64 (2.56)	0.32 (0.64)	0.32 (0.64)	0.64 (1.28)	1.12 (2.24)
Chloroform	0.025 - 51.20	0.80 (1.60)	0.40 (1.60)	0.80 (1.60)	1.60 (3.20)	0.80 (1.60)	1.60 (3.20)	1.60 (3.20)
Water	0.045 - 46.08	1.44 (2.88)	0.72 (1.44)	0.72 (1.44)	1.44 (2.88)	1.44 (2.88)	0.72 (1.44)	0.72 (1.44)
				Antibiotics				
Tetracycline	0.15 - 38.4 μg/ml	4.8 (9.6)	0.8 (1.6)	1.60 (3.2)	1.60 (3.20)	4.8 (9.6)	1.60 (3.2)	0.80 (1.6)
Ampicillin	0.15 - 38.4 μg/ml	3.2 (6.4)	1.60 (3.20)	1.60 (3.2)	1.60 (3.2)	6.40 (12.8)	1.60 (3.2)	1.60 (3.2)
Ciprofloxacin	0.15 - 38.4 μg/ml	1.6 (3.20)	3.2 (6.4)	0.8 (1.6)	0.8 (1.6)	1.60 (3.20	0.8 (1.6)	0.8 (1.6)

*Plant latexes were diluted by serial micro-dilution method with Luria broth culture media. **Antibiotic solutions were prepared by same method and concentration range was 0.15 - 38.4 μg/ml.

It was ranging from 0.14 - 0.56 µg/mL against different bacterial strains. Its water extract has shown good antimicrobial activity as the MIC value was ranging from 0.36 - 1.44 µg/mL (Table 1). Methanolic extract of Artocarpus heterophyllus has shown lowest MIC value 0.14 - 0.56 µg/mL against Klebsiella pneumoniae, E. coli, M. luteus, Streptococcus pneumoniae and B. cereus (Table 1). MIC values obtained in case of broad spectrum antibiotic drugs tetracycline, ampicillin and ciprofloxacin were higher than the latex samples. These were ranging from 0.8 - 6.40 µg/mL against different bacterial strains (Table 1).

Besides MIC values MBC values of latex extracts were also found lower than the broad spectrum antibiotics assayed for comparison. These are displayed in Table 1. Spondias dulcis (Amra) methanolic extract has shown lowest MBC value 0.48 µg/mL against S. pneumoniae. It was quite higher in water extract as 2.88 µg/mL in each case. Similar MBC value was also obtained in Diospyros melanoxylon (Tendu), water extract. Diospyros melanoxylon methanolic extract has shown lower MBC value in K. pneumoniae, E. coli, M luteus and auerus i.e. 0.56 µg/mL (Table 1). Terminalia bellirica (Wahera), methanolic extract has shown MBC value against E. coli and M luteus 0.24 and 0.48 µg/mL while water extract has shown lowest MIC value 2.88 µg/mL against all bacterial strains (Table 1). Ficus glomerata (Gular) methanolic extract has shown lower MBC value 0.28 - 0.56 µg/mL against Klebsiella pneumoniae, E coli, and Lactobacillus acidophilus (Table 1). Its water extract has shown MBC value 1.44 µg/mL K pneumioniae. Phyllanthus emblica (Awla) methanolic extract has shown MBC value 0.56 µg/mL against Klebsiella pneumoniae, E. coli and Streptococcus pneumoniae 0.14 µg/mL (Table 1). Its water extract has shown lowest MBC S value against Streptococcus pneumoniae, Staphylococcus aureus and L. acidophilus 1.44 µg/mL. Methanolic extract of Thevetia neriifolia (Kaner), Calotropis procera (Ak) and Artocarpus heterophyllus (katahal) showed lower MBC values ranging from 0.56 - 2.44 µg/mL while water extract from these plants have shown MBC S values between 0.72 - 2.88 µg/mL (Table 1). A higher MBC values were obtained in broad spectrum antibiotics which depict low susceptibility in of these drugs in comparison to latex extracts. The MBC values were noted in a range of 1.6 - 9.6 µg/mL (Table 1). This suggests that fruit latex extracts, possess strong bactericidal agents that can be, used as prevailing drugs for the control of bacterial growth in vivo.

3.2. Determination of Inhibition Zone Diameters

Besides determination of MIC and MBC values the effectiveness of all fruit latex extracts was confirmed by filter paper disc diffusion bioassay. After 24 hrs of in vitro Kirby Bauer test growth inhibition were measured in tests and controls. However, IZD obtained in presence of each latex extract was considered as final bactericidal marker. These are displayed in Table 2. On the basis of size of growth inhibition zone pathogens were divided in to three categories i.e. resistant (>7), intermediate (>12) and susceptible >18 mm respectively The zone of inhibition above 7 mm was considered as positive results. Methanolic extract of Spondias dulcis (Amra) methanolic extract has shown highest inhibition zone diameter 28.13 against K. pneumoniae followed by B. cereus 25.51, M. luteus 23.2, L. acidophilus 21.2. In its hexane extract highest inhibition zone diameter 20.36 mm against S. aureus, Petroleum ether extract against M. luteus 19.76, chloroform against B cereus 16.10. Water extract has shown inhibition zone diameter 15.32 against E coli (Table 2), Similarly methanolic extract of D melanoxylon (Tendu) has shown inhibition zone diameter between 19.65 - 25.48 mm against all different pathogens subsequently. It has shown highest inhibition zone diameter 25.48 against L. acidophilus (Table 2). Its water extract has shown inhibition zone diameter 16.26 against E. coli. Terminalia bellirica (Wahera) methanolic extract has shown IZD 24.54 against K. pneumoniae. Its water extract has shown highest inhibition zone diameter 18.32 against L. acidophilus. Ficus glomerata (Gular) methanolic extract has shown highest inhibition zone diameter between 12.86 - 17.64 mm against all different bacterial strains Phyllanthus emblica (Amla). It has shown highest inhibition zone diameter against M. luteus 28.66 followed by K. pneumoniae 27.82 and 27.34 mm (Table 2). Its aqueous extract has shown IZD between 13.65 - 19.60 (Table 2). Thevatia nerifolia (kaner) hexanoic extract has shown highest IZD 24.36 mm against L. acidophilus. Methanolic extract has shown highest inhibition zone diameter against L acidophilus 30.74 mm. Petroleum ether extract has shown 17.60 mm against E. coli (Table 2). Its water extract has shown highest inhibition zone diameter against L. acidophilus 18.46 mm. Calotropis procera solvent extract has shown inhibition zone diameter between 13.80 - 28.72 mm. Its water extract has shown highest inhibition zone diameter between 13.54 - 17.60. Artocarpus heterphyllus methanolic extract has shown highest inhibition zone diameter 26.43 mm against B cereus. Its water extract has shown inhibition zone diameter against E. coli 21.24 mm. In all tests inhibition zone diameter was found higher in solvent extracts in comparison to water extract. Furthermore, all solvent and aqueous extracts of latexes from assayed from different plant species have shown significantly higher inhibition zone diameter than the broad spectrum antibiotics as these were noted 7.46 - 14.86 mm. From the experiments methanolic extract were found to be highly susceptible against all the bacterial strains. As their growth inhibition zone diameter was obtained between 17.6 - 31.63 mm in size (Table 2). This inhibition was observed as 4-µg concentration of each extract at 24 h in agar disc diffusion bioassay. It was also observed that

Table 2. The quantitations of microbial activity of different aqueous and solvent extracts of plant latexes were measured by agar diffusion method. The effectiveness of different extracts is demonstrated by the size of the micro-organisms growth inhibition zone around the filter paper disc, which is typically expressed as the diameter of the zone in mm.

Fractions	Conc. µg/disc	*K. pneumoniae*	*E. coli*	*M. luteus*	*S. pneumoniae*	*S. aureus*	*B. cereus*	*L. acidophillus*
				Spondias dulcis (Amra)				
Hexanoic	4.00	17.33	15.36	18.46	19.33	20.36	14.70	19.23
Methanolic	4.00	28.13	17.66	23.20	17.60	19.40	25.51	21.20
Pet. Ether	4.00	16.43	17.46	19.76	17.43	18.30	16.26	19.16
Chloroform	4.00	12.34	13.40	14.30	12.86	15.12	16.10	13.18
Water	4.00	14.30	15.32	13.38	14.64	12.80	14.50	16.80
				Diospyros melanoxylon (Tendu)				
Hexanoic	4.00	21.31	17.42	13.40	15.60	13.40	14.11	13.22
Methanolic	4.00	24.35	21.71	20.21	19.65	24.33	23.37	25.48
Pet. Ether	4.00	14.66	17.37	18.63	17.72	19.56	20.20	15.47
Chloroform	4.00	13.67	14.50	17.24	13.23	14.50	19.60	13.80
Water	4.00	13.45	16.26	13.82	17.60	14.20	14.50	12.76
				Terminalia bellirica (Wahera)				
Hexanoic	4.00	13.47	15.75	19.67	16.33	12.40	18.74	20.65
Methanolic	4.00	24.54	23.36	21.44	25.80	24.23	21.73	18.46
Pet. Ether	4.00	16.30	15.37	18.14	19.26	16.42	19.23	21.78
Chloroform	4.00	14.20	16.76	15.43	14.20	12.86	14.80	15.74
Water	4.00	13.62	18.80	13.20	13.40	14.25	13.72	18.32
				Ficus glomerata (Gular)				
Hexanoic	4.00	21.53	19.53	17.33	25.26	21.33	20.46	25.30
Methanolic	4.00	25.43	23.53	20.56	29.60	22.76	29.53	31.63
Pet. Ether	4.00	21.16	17.57	23.24	19.30	23.57	17.86	15.84
Chloroform	4.00	13.26	16.74	14.80	14.24	15.63	12.80	14.78
Water	4.00	14.22	17.64	15.67	13.44	13.84	13.66	12.86
				Phyllanthus emblica (Awla)				
Hexanoic	4.00	19.66	25.43	20.17	25.31	22.71	21.26	17.30
Methanolic	4.00	27.82	25.13	28.66	27.34	23.16	29.53	31.63
Pet. Ether	4.00	23.06	16.50	29.26	18.30	21.50	19.36	16.83
Chloroform	4.00	14.62	18.12	13.26	12.76	15.40	14.34	13.24
Water	4.00	13.65	19.6	13.56	13.82	13.20	14.10	14.64
				Thevetia neriifolia (Kaner)				
Hexanoic	4.00	20.17	17.23	19.67	24.28	22.14	19.61	24.36
Methanolic	4.00	23.87	22.13	21.82	27.45	22.12	28.36	30.74
Pet. Ether	4.00	22.10	15.56	28.66	17.42	20.23	18.46	17.56
Chloroform	4.00	14.22	14.70	13.74	13.80	14.76	13.28	13.66
Water	4.00	13.60	15.42	14.24	13.88	12.80	14.70	18.46

				Carica papaya (Papita)				
Hexanoic	4.00	17.24	19.33	15.65	21.26	23.46	19.67	25.23
Methanolic	4.00	23.67	21.18	20.64	29.34	25.34	23.20	20.24
Pet. Ether	4.00	17.54	16.21	19.33	18.66	21.27	18.56	19.30
Chloroform	4.00	15.46	16.80	14.10	15.72	13.77	14.22	15.33
Water	4.00	15.68	14.71	15.22	13023	15.24	13.38	16.26
				Calotropis procera (Ak)				
Hexanoic	4.00	20.43	18.68	16.42	23.46	22.13	21.65	26.71
Methanolic	4.00	24.16	22.68	21.77	28.72	24.54	28.86	23.53
Pet. Ether	4.00	22.66	15.74	21.22	17.68	22.57	19.36	17.10
Chloroform	4.00	16.24	15.20	13.80	14.60	17.23	13.80	13.67
Water	4.00	17.60	14.22	13.54	14.46	16.30	13.68	14.56
				Ficus benghalensis **(Bargad)**				
Hexanoic	4.00	21.20	17.24	15.33	21.20	19.20	14.56	16.87
Methanolic	4.00	23.78	21.22	19.76	24.34	21.44	27.23	21.54
Pet. Ether	4.00	17.62	14.36	19.56	15.35	21.27	18.76	16.35
Chloroform	4.00	14.54	14.46	14.22	15.78	16.87	14.82	14.71
Water	4.00	14.64	15.23	19.43	15.66	18.25	14.32	15.66
				Artocarpus heterophyllus (Kathal)				
Hexanoic	4.00	19.51	18.61	17.87	26.72	20.43	19.36	16.30
Methanolic	4.00	22.41	20.19	19.33	23.60	21.23	26.43	20.77
Pet. Ether	4.00	15.66	16.73	19.10	15.40	22.54	18.22	15.84
Chloroform	4.00	13.46	14.28	15.24	14.68	14.20	13.86	16.80
Water	4.00	14.22	21.24	13.82	14.24	15.62	16.22	14.74
				Antibiotics				
Tetracycline	10.0	7.46	12.13	13.24	14.86	7.22	10.61	12.33
Ampicillin	10.0	8.77	9.23	10.35	14.21	9.20	13.44	12.36
Ciprofloxacin	10.0	8.16	12.82	13.82	13.21	7.24	12.23	11.17

*Control no antibiotic, *strength of activity is presented as resistant (0.7 mm), intermediate (>12 mm) and susceptible (>18 mm).

solvent extracts have shown higher inhibition zone diameter in comparison to water extracts.

4. DISCUSSION

In present study various antimicrobial susceptibility bioassays were conducted in vitro and most of the plant latex fractions were found susceptible to both gram-positive and gram-negative bacteria. Bacterial pathogens including Gram-positive bacteria such as Lactobacillus acidophilus, Streptococcus pneumoniae, Staphylococcus aureus, Micrococcus luteus, Bacillus cereus and Gram-negative bacteria Klebsiella pneumoniae, Escherichia coli have shown susceptibility both in presence of solvent and aqueous extracts. In various bioassays solvent extracts have shown very low MIC value than broad spectrum antibiotic drugs that indicates higher susceptibility in latex samples. In all the latex extracts an overall MIC value was obtained that ranges from 0.14 - 2.88 μg/ml. Among all the latex extracts methanolic extract has shown higher susceptibility to bacterial strains followed by hexanoic, petroleum ether and cholrofrom chloroform extract. Methanolic extracts from all different plant species have shown lowest MIC value that was in a range of 0.122 - 0.245 μg/ml (Table 1). It was also noted that solvent extracts have shown higher susceptibility in comparison to water extracts as the MIC and MBC values obtained in solvent extracts were found significantly lesser than the water extracts. Diospyros melanoxylon (Tendu), methanolic extract has shown higher MIC value 0.070 μg/mL against K. pneumoniae, while hexanoic extract i.e. 0.48 μg/mL against S. pneumoniae. Its water extract has shown MIC value 0.72 μg/mL against Klebsiella pneumoniae, and Streptococcus pneumoniae (Table 1). Terminalia bellirica (Wahera), methanolic extract has shown higher MIC value against E. coli and M luteus 0.14, and 0.24 μg/mL while water extract has shown lowest MIC value 0.72 E. coli and Lactobacillus acidophilus (Table 1). Ficus glomerata (Gular) methanolic extract has shown higher MIC value in a range of 0.14 μg/mL - 0.24 μg/mL against Lactobacillus acidophilus (Table 1). Phyllanthus emblica (Awla) methanolic extract has shown higher MIC value against Klebsiella pneumoniae and Streptococcus pneumoniae 0.14 μg/mL (Table 1). Its water extract has shown lowest MIC against 0.36 0.14 μg/mL against Staphylococcus aureus and Lactobacillus acidophilus. Methanolic extract of Thevetia neriifolia (Kaner), has shown lower MIC value 0.14 μg/mL and 0.24 μg/mL against E. coli and S. aureus. Its water extract was found active against three bacterial strains as the MIC values obtained are very low E. coli, B. cereus Lactobacillus acidophilus 0.72 μg/mL. Calotropis procera has shown very high antimicrobial susceptibility as the MIC value obtained was least in all cases. It was ranging from 0.14 - 0.56 μg/mL against different bacterial strains. Its water extract has shown good antimicrobial activity as the MIC value was ranging from 0.36 - 1.44 μg/mL (Table 1). MIC values obtained in case of broad spectrum antibiotic drugs tetracycline, ampicillin and ciprofloxacin were higher than the latex samples. These were ranging from 0.8 - 6.40 μg/mL against different bacterial strains (Table 1). Presence of this antibacterial and antifungal activity against pathogenic bacteria and fungi in plant latexes may be due to presence of certain chemical components such as acetogenins, alkaloids, acids, terpenes, enzymes, protein, resins, tannins, sterols, sugars, oils, glycolipids and cystein proteases. The leaves of J. curcas contain apigenin, vitexin and isovitexin, which along with other factors show antibacterial activity against Staphylococcus aureus and Escherichia coli [10] .

Furthermore, MBC (minimum bactericidal concentration) values of solvent and aqueous extracts were also determined. Spondias dulcis (Amra) methanolic extract has shown lowest MBC value 0.48 μg/mL against S.

pneumoniae, It was quite higher in water extract as 2.88 μg/mL in each case. Similar MBC value was also obtained in Diospyros melanoxylon (Tendu), water extract. Diospyros melanoxylon methanolic extract has shown lower MBC value in K. pneumoniae, E. coli, M luteus and S. auerus i.e. 0.56 μg/mL (Table 1). Terminalia bellirica (Wahera), methanolic extract has shown MBC value against E. coli and M luteus 0.24 and 0.48 μg/mL while water extract has shown lowest MIC value 2.88 μg/mL against all bacterial strains (Table 1). Ficus glomerata (Gular) methanolic extract has shown lower MBC value 0.28 - 0.56 μg/mL against Klebsiella pneumoniae, E coli, and Lactobacillus acidophilus (Table 1). Its water extract has shown MBC value 1.44 μg/mL K pneumioniae. Phyllanthus emblica (Awla) methanolic extract has shown MBC value 0.56 μg/mL against Klebsiella pneumoniae, E. coli and Streptococcus pneumoniae 0.14 μg/mL (Table 1). Its water extract has shown lowest MBS value against Streptococcus pneumoniae, Staphylococcus aureus and L. acidophilus 1.44 μg/mL. Methanolic extract of Thevetia neriifolia (Kaner), Calotropis procera (Ak) and Artocarpus heterophyllus (katahal) showed lower MBC values ranging from 0.56 - 2.44 μg/mL while water extract from these plants have shown MBS MBC values between 0.72 - 2.88 μg/mL (Table 1). Aqueous extracts in all the cases have shown lower MBC value ranging from 0.10 - 0.66 μg/ml. It has shown very high susceptibility to all the bacterial strains tested in the antimicrobial bioassays. In vitro data obtained reveals that latex samples from ten plant species were highly susceptible to both bacteria Among eight plant species Calotropis procera (Ak), and Carica papaya (Papita), have shown highest susceptibility to different microbial strains as lowest MIC values was obtained in latex extracts. MIC values obtained in methanolic extracts were between 0.14 - 0.56 μg/ml. Similar MIC value was also obtained in Ficus benghalensis (Bargad) in vitro antimicrobial assays (Table 1).

Similar, antibacterial activity is reported in latexes of A. ochroleuca [2] , Hyaenanche globosa (Euphorbiaceae) [28] , Hancornia speciosa Gomes (Apocynaceae) against Klebsiella, Pantoea, Enterobacter and Burkholderia [7] . Hevea bransilensis latex shows antifungal activity against Trichosprum cutaneum and Cryptococcus neoformans [5] . It also exhibits a good activity against multiple drug resistant Staphylococcus aureus (NCTC 11994), Salmonella typhimurium (ATCC 1255) and Candida albicans (ATCC 10231) [11] . When compared MIC values obtained in broad spectrum antibiotics against all seven bacterial strains were approximately 3.0 times higher than the latex extracts. It indicates that bacterial strains have developed resistance against these antibiotics. It was further confirmed by 2.5 times higher MBCS obtained in these antibiotics in in vitro bioassays.

The crude latex of A. ochroleuca exhibited a potent antibacterial effect against Candida albicans, Staphylococcus aureusa, Escherichia coli, Salmonella typhi, and Bacillus subtilis [2] . Similarly, a South African plant;

Hyaenanche globosa (Euphorbiaceae) showed poisonous activity due to presence of some toxic compounds [28] . Latex of Aloe harlana Reynolds shows antimicrobial activity of aqueous and methanol extracts of Juniperus oxycedrum [26] . Methonolic extract of three Salvia species [29] have shown strong antimicrobial activity [30] . The ethanolic extract of the fruits of H. globosa (F.E) contain monophene which showed inhibitory and cytotoxic activity in HeLa cells' and other cancer cell lines [31] . Similarly, Tutin and hyenanchin isolated from H. globosa have shown significant ($P < 0.01$) inhibition on cell viability/proliferation [28] . Latex of Jatropha curcas is used externally to treat infection, piles and sores among the domestic livestock. The leaves of J. curcas contain apigenin, vitexin and isovitexin, which along with other factors show antibacterial activity against Staphylococcus aureus and Escherichia coli [10] . Similarly, latex of Argemone ochroleuca Sweet has shown antibacterial activity against Bacillus subtilis, Enterobacter aerogenes, Micrococcus luteus, Escherichia coli, and Staphylococcus aureus [2] . Latex from Hancornia speciosa Gomes (Apocynaceae) has shown activity against Klebsiella, Pantoea, Enterobacter and Burkholderia [7] while Hevea bransilensis latex shows antifungal activity against Trichosprum cutaneum and Cryptococcus neoformans [5] . It also exhibits a good activity against multiple drug resistant Staphylococcus aureus (NCTC 11994) and Salmonella typhimurium (ATCC 1255) and Candida albicans (ATCC 10231) [11] . Similarly, antibacterial activity is reported in C. procera by Jain et al., (1996) [8] . Hexane extract of Tunisian caprifig latex from the unripe fruit of Ficus carica showed strong anti microbial activity [32] .

Besides determination of MIC and MBC values the effectiveness of all fruit latex extracts was confirmed by filter paper disc diffusion bioassay. After 24 hrs of in vitro Kirby Bauer test growth inhibition were measured in tests and controls. However, IZD obtained in presence of each latex extract was considered as final bactericidal marker. These are displayed in Table 2. On the basis of size of growth inhibition zone pathogens were divided in to three categories i.e. resistant (>7), intermediate (>12) and susceptible > 18 mm respectively. The zone of inhibition above 7 mm was considered as positive results. In present study broad-spectrum antibiotics have shown marginal activity or intermediate effect while latex samples have shown IZD more than 13 mm in each case. It shows that latexes contain pharmaceutically important active principles which are solvent extractable. Methanolic extract of Spondias dulcis (Amra) methanolic extract has shown highest inhibition zone diameter 28.13 mm against K. pneumoniae followed by B cereus 25.51 mm, M luteus 23.2 mm, L acidophilus 21.2 mm. In its hexane extract highest inhibition zone diameter 20.36 mm against S. aureus, Petroleum ether extract against M luteus 19.76 mm, chloroform against B cereus 16.10 mm. Water extract has shown inhibition zone diameter 15.32 mm against E coli (Table 2). Almost similar activity was reported in methanolic extracts from all ten different species. An overall activity susceptibility was displayed from MIC and MBC and IZD in a order of methanolic < hexanoic < petroleum ether < chloroform < water. Similarly methanolic extract of D melanoxylon (Tendu) has shown inhibition zone diameter between 19.65 - 25.48 mm against all different pathogens subsequently. It has shown highest inhibition zone diameter 25.48 mm against L. acidophilus (Table 2). Its water extract has shown inhibition zone diameter 16.26 mm against E. coli. In all tests inhibition zone diameter obtained in presence of solvent extracts were higher in comparison to water extracts from all plant species screened. Furthermore, these extracts have shown significantly higher inhibition zone diameter 15 - 31.63 mm than the broad spectrum antibiotics as these were noted 7.46 - 14.86 mm in. From the experiments methanolic extract were found to be highly susceptible against all the bacterial strains. As their growth inhibition zone diameter was obtained between 17.6 - 31.63 mm in size (Table 2). This inhibition was observed as at 4-μg concentration of each extract at 24 h in agar disc diffusion bioassay. It was also observed that solvent extracts have shown higher inhibition zone diameter in comparison to water extracts. It may be due to higher diffusion of latex components coated on filter paper discs.

Further, in susceptibility bioassays growth inhibition was found to be dose dependent. Similarly, increase in radial growth was dose dependent. The results indicate that the highest radial growth inhibition was obtained in methanolic extract Ficus glomerata and Calatropis procera at 4 μg/mL in comparison to broad spectrum antibiotics which have shown intermediate radial growth inhibition zone diameter less than 13 mm (Table 2). It indicates that these antibiotics are worth less as the inhibition zone diameter was obtained near 7 mm which indicates no susceptibility to microbes or marginal activity. From results it is clear that latex components are potential inhibitors of bacteria growth in suspension and solid agar culture. Similar antimicrobial activity was found in Croton bonplandinum [33] [34] [10] , Juniperous oxycedrum [26] , Himatanthus articulatus [34] , Salvia species [29] , Argemone ochroleuca [2] , Carica papaya [5] , Calatropis procera [30] , Hevea brasilensis [35] and Ficus cariaca latex against resistant pathogens [3] .

This activity in plant latexes is due to presence of various active components such as Alpha-D-mannosidase and N-acetyl-beta-D-glucosaminidase [5] , dieterpenes [4] , certain proteases [36] and chitinases [37] . Similar activity against phyto-pathogens is reported in latexes of Calotropis procera (CpLP), Plumeria rubra (PrLP), Carica candamarcensis (P1G10) and Euphorbia tirucalli (EtLP) due to presence of antimicrobial proteins [6] .

Crofelmer isolated from Croton lecheri [23] and Cotinifolia from Euphorbia cotinifolia latex have shown high antimicrobial activity [38] . Hevea brasilensis latex shows antimicrobial effects against oral micro-organisms [39] . Latex from Jatropha curcas and Artocarpus heterophyllus were found active against all seven bacterial strains microbes due to presence active principles of Latex from Himatanthus succubae exhibited Leishmanicidal activity [21] . Fruit chitinases in banana latex have shown significantly higher antimicrobial susceptibility to the antibiotics [40] . Similar antimicrobial activity is reported in Calotropis procera latex due to presence of Osmotin [41] . Procerain, a stable cysteine protease isolated from the latex of Calotropis procera and anthrone and chromone shows strong antimicrobial activity [11] [42] . Latex proteins prevent septic shock due to lethal infection by Salmonella enterica serovar Typhimurium [43] . Hevea brasiliensis on oral microorganisms [39] , Saponins-rich fraction of Calotropis procera leaves elicit no antitrypanosomal activity [44] , while synergistic effect was shown in of Carica papaya latex sap and fluconazole on Candida albicans growth [5] . Himatanthus sucuuba latex was found active against Leishmania amazonensis. [21] . Latex from Apocynaceae species showed antiplasmodial activity [45] while Euphorbia species showed antileishmanial and antitrypanosomal [46] [47] . More often, plant latex contain many proteolytic enzymes and other proteins which lyse bacterial cell wall, therefore and latex act heavily upon human and veterinary pathogens.

It was cleared from bioassays that plant latex is more bactericidal to the gram-positive bacteria. This may be due to three important reasons i.e. first latex components are more diffusible and of low molecular weight and solvent and water soluble. Second absence of lipo-polysaccharide layer, that function as an effective barrier against any incoming bio-molecule [48] [49] . There might be another possibility that latex may successfully inhibit microbial respiration and increase the plasma membrane permeability, which results in to ion leakage from the cell [50] [51] . Higher diffusion also occurs due to hydrophilic nature of bacterial cell wall [52] that results an increase in antimicrobial properties of active components [3] . No doubt latexes show multiple biological activities such as cytotoxic [53] , antiproliferative and antiplasmodial [54] , anti-inflmmatory [55] , Fibrinolytic [56] , Anti-helmintic [57] , anti-viral [58] antitumor activity [59] and strong antifungal activity against mycelial fungi [35] [41] . It is due to presence of proteases, chitinases and Osmotin, diterpenes and saponins which play important role in plant defense [41] [60] . These components displayed higher antimicrobial potential that can work as strong antimicrobial agents for of chemotherapy [25] for multidrug resistant human pathogens [61] .

In the present study fruit latex fractions from ten latex secreting plant species registered significantly higher growth inhibition than that of broad spectrum antimicrobial drugs. The MIC and MBC values obtained in these fractions were significantly lower than that of antibiotics of broad spectrum activity. In addition, in presence of latex extracts IZD obtained were found more than 14 mm. Highest inhibition zone diameter against M. luteus (33.2 mm) followed by K. pneumoniae (29.26 mm) and L. acidophilus (28.43 mm) was obtained in presence of methanolic extract of Thevetia neriifolia. Though, no report is available on presence of volatile components in plant latexes but very high diffusion activity obtained in pure latex samples at room temperature, showed association of some volatile action that may be more active against bacteria in comparison to crude latex. Therefore, major anti-bacterial action in fungi seems to be diffusion action of plant latex or inhibition of metabolism of fungal strain. However, from previous studies no report is available on its volatile action in vitro. However, plant latex can be used as antimicrobial agent without any side effect [46] . So far studies have been done number of medicinal plants have been explored for their antimicrobial activity with fewer side effects and reduced toxicity. Therefore, it can be concluded that latex components if purified might prove potent therapeutic agents, which could become possible only after identification of major latex constituents. However, it could be concluded that latex induced inhibition and cidal activity against bacterial and fungal strains is due to presence of significant active principles. No doubt, if latex components are identified these could be used as therapeutic agents against various pathogenic bacterial and fungal strains.

5. CONCLUSION

Ravi Kant Upadhyay Latex extracts prepared from ten laticiferous Indian plant species have shown strong antibacterial and antifungal activity. It is due to presence various latex components such as proteins, proteases, chitinases, osmotin, alkaloids, glycosides, diterpenes and saponins which can act as potent therapeutic agents. Latex samples have shown higher growth inhibition and cidal activity against bacterial and fungal strains than the broad spectrum antibiotic drugs due to presence of significant active principles. No doubt, if latex components are identified these could be used as multiple therapeutic purposes and food preservation applications.

REFERENCES

1. Pickard, W.F. (2008) Laticifers and Secretory Ducts: Two Other Tube Systems in Plants. New Phytologist, 177, 877- 888.
2. Alamri, S.A. and Moustafa, M.F. Antibacterial Activity of the Latex of Argemone ochroleuca Sweet. Saudi Medicinal Journal, 31, 1207-1210. [Citation Time(s):6]
3. Aref Lazreg, H., Gaaliche, B., Fekih, A., Mars, M., Aouni, M., Pierre Chaumon, J. and Said, K. (2011) In Vitro Cytotoxic and Antiviral Activities of Ficus carica Latex Extracts. Natural Product Research, 25, 310-319.
4. Avila, L., Perez, M., Sanchez-Duffhues, G., Hernández-Galán, R., Muñoz, E., Cabezas, F., Quiñones, W., Torres, F. and Echeverri, F. (2010) Effects of Diterpenes from Latex of Euphorbia lactea and Euphorbia laurifolia on Human Immunodeficiency Virus Type 1 Reactivation. Phytochemistry, 71, 243-248.
5. Giordani, R., Cardinas, M.L., Moulin, T.J. and Regl, P. (1991) Antifungal Action of Carica papaya Latex Isolation of Fungal Cell Wall Hydrolyzing Enzymes. Mycoses, 34, 469-477.
6. Tanaka, H., Sato, M. and Fujiwara, S. (2002) Antibacterial Activity of Iso-Flavonoids Isolated from Erythrina variegata against Methicillin Resistant Staphylococcus aureus. Letters in Applied Microbiology, 35, 228-489.
7. Silva, T.F., Coelho, M.R., Vollú, R.E., de Vasconcelos Goulart, F.R., Alviano, D.S. and Alviano Csseldin, L. (2011) Bacterial Community Associated with the Trunk Latex of Hancornia speciosa Gomes (Apocynaceae) Grown in the Northeast of Brazil. Antonie Van Leeuwenhoek, 99, 523-532.
8. Jain, S.C., Sharma, R., Jain, R. and Sharma, R.A. (1996) Antimicrobial Activity of Calotropis procera. Fitoterapia, 67, 275-277.
9. Itokawa, H., Ichihara, Y., Watanabe, K. and Takeya, K. (1989) An Antitumor Principle from Euphorbia Lathyris. Planta Medica, 55, 271-272.
10. Thomas, R., Sah, N.K. and Sharma, P.B. (2008) Therapeutic Biology of Jatropha curcas: A Mini Review. Current Pharmaceutical Biotechnology, 9, 315-324.
11. Asamenew, G., Bisrat, D., Mazumder, A. and Asres, K. (2011) In Vitro Antimicrobial and Antioxidant Activities of Anthrone and Chromone from the Latex of Aloe harlana Reynolds. Phytotherapy Research, 25, 1756-1760.
12. Lazreg-Aref, H., Mars, M., Fekih, A., Aouni, M. and Said, K. (2012) Chemical Composition and Antibacterial Activity of a Hexane Extract of Tunisian Caprifig Latex from the Unripe Fruit of Ficus carica. Pharmaceutical Biology, 50, 407-412.
13. Unander, D.W., Webster, G.L. and Blumberg, B.S. (1995) Usage and Bioassays in Phyllanthus (Euphorbiaceae). IV. Clustering of Antiviral Uses and Other Effects. Journal of Ethnopharmacology, 45, 1-18.
14. Camacho, M.D.R., Phillipson, J.D., Croft, S.L., Kirby, G.C., Warhurst, D.C., Castaneda, P. and Mata, R. (2000) Antiprotozoal and Cytotoxic Activities of Carbocyclic Triterpenes and Biflavonoids from Celaenodendrum mexicanum. Planta Medica, 66, 463-468.
15. Frederich, M., Tits, M. and Angenot, L. (2008) Potential Antimalarial Activity of Indole Alkaloids. Transactions of the Royal Society of Tropical Medicine and Hygiene, 102, 11-19.
16. Sharma, J.D. and Sharma, P. (2000) In-Vitro Schizonticidal Screening of Calotropis procera. Fitoterapia, 71, 77-79.
17. Adzu, B. and Salawu, O.A. (2009) Screening Diospyros mespiliformis Extract for Antimalarial Potency. International. Journal. Biological Chemical Science, 3, 271-276.
18. de Mesquita, M.L., Grellier, P., Mambu, L., de Paula, J.E. and Espindola, L.S. (2007) In Vitro Antiplasmodial Activity of Brazilian Cerrado Plants Used as Traditional Remedies. Journal of Ethnopharmacology, 110, 165-170.

19. Espindola, L.S., Vasconcelos Júnior, J.R., de Mesquita, M.L., Marquié, P., de Paula, J.E., Mambu, L., et al. (2004) Trypanocidal Activity of a New Diterpene from Casearia sylvestris var. Lingua. Planta, 70, 1093-1095.
20. Weniger, B., Robledo, S., Arango, G.J., Deharo, E., Aragón, R., Muñoz, V., et al. (2001) Antiprotozoal Activities of Colombian Plants. Journal of Ethnopharmacology, 78, 193-200.
21. Soares, D.C., Andrade, A.L., Delorenzi, J.C., Silva, J.R., Freire-de-Lima, L., Falcão, C.A., Pinto, A.C., Rossi-Bergmann, B. and Saraiva, E.M. (2010) Leishmanicidal Activity of Himatanthus sucuuba Latex against Leishmania amazonensis. Parasitology International, 59, 173-177.
22. Afonso-Cardoso, S.R., Silva, C.V., Ferreira, M.S. and Souza, M.A. (2011) Effect of the Synadenium carinatum Latex Lectin (ScLL) on Leishmania (Leishmania) amazonensis Infection in Murine Macrophages. Experimental Parasitology, 128, 61-67.
23. Tradtrantip, L., Namkung, W. and Verkman, A.S. (2010) Crofelemer, an Antisecretory Antidiarrheal Proanthocyanidin Oligomer Extracted from Croton lechleri, Targets Two Distinct Intestinal Chloride Channels. Molecular Pharmacology, 77, 69-78.
24. Baranov, A.U. and Elmuradov, B.J. (2009) Extraction and Characterization of Latex and Natural Rubber-Bearing Plants. Journal of Agricultural and Food Chemistry, 58, 734-743.
25. McCay, S. and Mahlberg, P. (1973) Study of Antimicrobial Activity and Bacteriology of Latex and Asclepias syriaca L. Antimicrobial Agent and Chemotherapy, 3, 247-253.
26. Karaman, I., Sahin, F., Gullus, M., Ogutue, H., Sengul, M. and Adiguzel, A. (2003) Antimicrobial Activity of Aqueous and Methanol Extracts of Juniperus oxycedrum L. Journal of Ethnophamacology, 85, 231-235.
27. Bauer, A.W., Kirby, W.M., Sherris, J.C. and Turuk, M. (1966) Antibiotic Susceptibility Testing by a Standardized Single Disk Method. American Journal of Clinical Pathology, 45, 493-496.
28. Momtaz, S., Lall, N., Hussein, A., Ostad, S.N. and Abdollahi, M. (2010) Investigation of the Possible Biological Activities of a Poisonous South African Plant; Hyaenanche globosa (Euphorbiaceae). Pharmacognosy Magzine, 6, 34-41.
29. Bel Hadj Salah, K., Mahjoub, M.A., Chaumont, J.P., Michel, L., Millet-Clerc, J., Chraeif, I., Ammar, S., Mighri, Z. and Aouni, M. (2007) Chemical Composition and in Vitro Antifungal and Antioxidant Activity of the Essential Oil and Methanolic Extract of Teucrium sauvagei Le Houerou. Natural Product Research, 20, 1089-1097.
30. Larhsini, M., Bousad, M., Lazrek, H.B., Jana, M. and Amarouch, H. (1997) Evaluation of Antifungal and Molluscicidal Properties of Extracts of Calotropis procera. Fitoterapia, 68, 371-373.
31. Mahmoud, T.S., Marques, M.R., do Ó Pessoa, C., Lotufo, L.V.C., Magalhães, H.I.F., de Moraes, M.O., de Lima, D.P. and Tininis, A.G. (2011) In Vitro Cytotoxic Activity of Brazilian Middle West Plant Extracts. Revista Brasileira de Farmacognosia Brazilian Journal of Pharmacognosy, 21, 456-464.
32. Lazreg-Aref, H., Mars, M., Fekih, A., Aouni, M. and Said, K. (2012) Chemical Composition and Antibacterial Activity of a Hexane Extract of Tunisian caprifig Latex from the Unripe Fruit of Ficus carica. Pharmaceutical Biology, 50, 407-412.
33. Amrita, A., Mall, H.V., Dixit, K. and Gupta, S. (1989) Fungitoxic Properties of Latex of Plants with Special Reference to That of Croton bonplandianum Boill. Pharmaceutical Biology, 27, 25-28.
34. Sequeira, B.J., Vital, M.J.S., Pohlit, A.M., Pararos, I.C. and Cauper, G.S.B. (2009) Antibacterial and Antifungal Activity of Extracts and Exudates of the Amazonian Medicinal Tree Himatanthus articulates (Vahl) Woodson (Common neme: sucuba). Memórias do Instituto Oswaldo Cruz, 104, 34-39.
35. Daruliza, K.M., Yang, K.L., Lam, K.L., Priscilla, J.T., Sunderasan, E. and Ong, M.T. (2011) Anti-Candida albicans Activity and Brine Shrimp Lethality Test of Hevea

brasiliensis Latex B-Serum. European Review for Medical and Pharmacological Sciences, 15, 1163-1171.
36. Siritapetawee, J., Thammasirirak, S. and Samosornsuk, W. (2012) Antimicrobial Activity of a 48-kDa Protease (AMP48) from Artocarpus heterophyllus Latex. European Review for Medical and Pharmacological Sciences, 16, 132-137.
37. Taira, T., Ohdomari, A., Nakama, N., Shimoji, M. and Ishihara, M. (2005) Characterization and Antifungal Activity of Gazyumaru (Ficus microcarpa) Latex Chitinase: Both the Chitin-Binding and the Antifungal Activities of Class 1 Chitinase Are Reinforced with Increasing Ionic Strength. Bioscience Biotechnology Biochemistry, 69, 811-818.
38. Kumar, R., Singh, K.A., Tomar, R. and Jagannadham, M.V. (2011) Biochemical and Spectroscopic Characterization of a Novel Metalloprotease, Cotinifolin from an Antiviral Plant Shrub: Euphorbia cotinifolia. Plant Physioliology Biochemistry, 49, 721-728.
39. Kanokwiroon, K., Teanpaisan, R., Wititsuwannakul, D., Hooper, A.B. and Wititsuwannakul, R. (2008) Antimicrobial Activity of a Protein Purified from the Latex of Hevea brasiliensis on Oral Microorganisms. Mycoses, 51, 301-307.
40. Sanchez-Monge, R., Blanco, C., Díaz-Perales, A., Collada, C., Carrillo, T., Aragoncillo, C. and Salcedo, G. (1999) Isolation and Characterization of Major Banana Allergens: Identification as Fruit Class I Chitinases. Clinical Experimental Allergy, 29, 673-680.
41. de Freitas, C.D., Nogueira, F.C., Vasconcelos, I.M., Oliveira, J.T., Domont, G.B. and Ramos, M.V. (2011) Osmotin Purified from the Latex of Calotropis procera: Biochemical Characterization, Biological Activity and Role in Plant Defense. Plant Physiology Biochemistry, 49, 738-743.
42. Dubey, K. and Jagannadham, M.V. (2003) Procerain, a Stable Cysteine Protease from the Latex of Calotropis procera. Phytochemistry, 62, 1057-1071.
43. Lima-Filho, J.V., Patriota, J.M., Silva, A.F., Filho, N.T., Oliveira, R.S., Alencar, N.M. and Ramos, M.V. (2010) Proteins from Latex of Calotropis procera Prevent Septic Shock Due to Lethal Infection by Salmonella enterica Serovar Typhimurium. Journal of Ethnopharmacology, 129, 327-334.
44. Ibrahim, M.A., Aliyu, A.B., Meduteni, K. and Yunusa, I. (2013) Saponins-Rich Fraction of Calotropis procera Leaves Elicit No Antitrypanosomal Activity in a Rat Model. Asian Pacific Journal Tropical Biomedicine, 3, 569-572.
45. Wong, S.K., Lim, Y.Y., Abdullah, N.R. and Nordin, F.J. (2011) Assessment of Antiproliferative and Antiplasmodial Activities of Five Selected Apocynaceae Species. BMC Complement Alternative Medicine, 11, 3.
46. Mazoir, N., Benharref, A., Bailén, M., Reina, M., González-Coloma, A. and Martínez-Díaz, R.A. (2011) Antileishmanial and Antitrypanosomal Activity of Triterpene Derivatives from Latex of Two Euphorbia Species. Zeitschrift für Naturforschung C, 66, 360-366.
47. Hohmann, J. and Molnár, J. (2004) Euphorbiaceae Diterpenes: Plant Toxins or Promising Molecules for the Therapy? Acta Pharmaceutica Hungarica, 74, 149-157.
48. Inouye, S., Takizawa, T. and Yamaguchi, H. (2001) Antibacterial Activity of Essential Oils and Their Major Constituents against Respiratory Tract Pathogen by Gaseous Contact. Journal Antimicrobial. Chemotherapy, 47, 565-573.
49. Delaquis, P.J., Stewart, S., Cazaux, S. and Toivonen, P. (2002) Survival and Growth of Listeria monocytogenes and Escherichia coli O157: H7 in Ready-to-Eat Iceberg Lettuce Washed in Warm, Chlorinated Water. Journal Food Protection, 65, 459-464.
50. Lambert, R.J., Skandamis, P.N., Coote, P.J. and Nycas, G.J. (2001) A Study of the Minimum Inhibitory Concentration and Mode of Action of Oregano Essential Oil, Thymol and Carvacrol. Journal of Applied Microbiology, 91, 453-462.
51. Walsh, S.E., Maillard, J.Y., Russel, A.D., Catrenich, C.E., Charbonneau, D.L. and Bartolo, R.J. (2003) Activity and Mechanism of Action of Selected Biocidal Agents on Gram-Positive and -Negative Bacteria. Journal of Applied Microbiology, 94, 240-247.
52. Knobloch, K., Weigand, H., Weis, N., Schwar, H.M. and Vigenschow, H. (1986) Action of Terpenoids on Energy Metabolism. In: Gruyter, D., Ed., Progress in Essential Oil

Research: 16th International Symposium on Essential Oils, Germany, Berlin, pp:429-445.

53. Olapeju, O.A., Amao, A.H., Ernesto, F., Taglialatela-Scafati, O., Heinz, C.S. and Müller, W.E.G. (2011) Cytotoxic Activity of Crude Extracts as well as of Pure Components from Jatropha Species, Plants Used Extensively in African Traditional Medicine. Evidence-Based Complementary and Alternative Medicine, 2011, 1-7.
54. Wong, S.K., Lim, Y.Y., Abdullah, N.R. and Nordin, F.J. (2011) Assessment of Antiproliferative and Antiplasmodial Activities of Five Selected Apocynaceae Species. BMC Complement Alternative Medicine, 11, 3.
55. Kumar, V.L. and Roy, S. (2007) Calotropis procera Latex Extract Affords Protection against Inflammation and Oxidative Stress in Freund's Complete Adjuvant-Induced Monoarthritis in Rats. Mediators Inflammation, 2007, Article ID: 47523.
56. Rajesh, R., Raghavendra Gowda, C.D., Nataraju, A., Dhananjaya, B.L., Kemparaju, K. and Vishwanath, B.S. (2005) Procoagulant Activity of Calotropis gigantea Latex Associated with Fibrin(ogen)olytic Activity. Toxicon, 46, 84-92.
57. Nagaty, H.F., Rifatt, M.A. and Morsy, T.A. (1959) Trials of the Effect on Dog Ascaris in Vivo Produced by the Latex of Ficus carica and Papaya carica Growing in Cairo Gardens. Annals of Tropical Medicine and Parasitology, 53, 215- 219.
58. Sabry, Y.M., Sanaa, M.F., Gad-Rab, N.H., Ahmad, A. and Shoreit, M. (2010) Antiviral Activity of Latex from Ficus nitida against Plant Viruses. Global Journal of Biotechnology & Biochemistry, 5, 198-205.
59. Mousinho, K.C., Oliveira, C.D., Ferreira, J.R., Carvalho, A.A., Magalhães, H.I., Bezerra, D.P., Alves, A.P., Costa-Lotufo, L.V., Pessoa, C., de Matos, M.P., Ramos, M.V. and Moraes, M.O. (2011) Antitumor Effect of Laticifer Proteins of Himatanthus drasticus (Mart.) Plumel-Apocynaceae. Journal of Ethnopharmacol, 137, 421-426.
60. Lee, J.H. and Stein, B.D. (2011) Antimicrobial Activity of a Combination of Mume fructus, Schizandrae fructus, and Coptidis rhizoma on Enterohemorrhagic Escherichia coli O26, O111, and O157 and Its Effect on Shiga Toxin Releases. Foodborne Pathogens and Disease, 8, 643-646.
61. Ahmad, I. and Beg, A.G. (2001) Antimicrobial and Phytochemical Activity on 45 Indian Medicinal Plant against Multidrug Resistant Human Pathogens. Journal of Ethnopharmacology, 74, 113-123.

CHAPTER 5

Transcriptomic Analysis of Flower Development in Wintersweet (*Chimonanthus praecox*)

Daofeng Liu, Shunzhao Sui, Jing Ma, Zhineng Li, Yulong Guo, Dengpan Luo, Jianfeng Yang, Mingyang Li*

Chongqing Engineering Research Center for Floriculture, Key Laboratory of Horticulture Science for Southern Mountainous Regions, Ministry of Education, College of Horticulture and Landscape Architecture, Southwest University, Chongqing, China

ABSTRACT

Wintersweet (*Chimonanthus praecox*) is familiar as a garden plant and woody ornamental flower. On account of its unique flowering time and strong fragrance, it has a high ornamental and economic value. Despite a long history of human cultivation, our understanding of wintersweet genetics and molecular biology remains scant, reflecting a lack of basic genomic and transcriptomic data. In this study, we assembled three cDNA libraries, from three successive stages in flower development, designated as the flower bud with displayed petal, open flower and senescing flower stages. Using the Illumina RNA-Seq method, we obtained 21,412,928, 26,950,404, 24,912,954 qualified Illumina reads, respectively, for the three successive stages. The pooled reads from all three libraries were then assembled into 106,995 transcripts, 51,793 of which were annotated in the NCBI non-redundant protein database. Of these annotated sequences, 32,649 and 21,893 transcripts were assigned to gene ontology categories and clusters of orthologous groups, respectively. We could map 15,587 transcripts onto 312 pathways using the Kyoto Encyclopedia of Genes and Genomes pathway database. Based on these transcriptomic data, we obtained a large number of candidate genes that were differentially expressed at the open flower and senescing flower stages. An analysis of differentially expressed genes involved in plant hormone signal transduction pathways indicated that although flower opening and senescence may be independent of the ethylene signaling pathway in wintersweet, salicylic acid may be involved in the regulation of flower senescence. We also succeeded in isolating key genes of floral scent biosynthesis and proposed a biosynthetic pathway for monoterpenes and sesquiterpenes in wintersweet flowers, based on the annotated sequences. This comprehensive transcriptomic analysis presents fundamental information on the genes and pathways which are involved in flower development in wintersweet. And our data provided a useful database for further research of wintersweet and other Calycanthaceae family plants.

INTRODUCTION

The small, evolutionarily ancient Calycanthaceae family comprises four genera, namely *Calycanthus* L., in North America, *Idiospermum* Blake, in Australia, and *Sinocalycanthus* Cheng & S. Y. Chang and *Chimonanthus* L., in China [1]. Wintersweet (*Chimonanthus praecox*), also known as the 'wax shrub', is a hardy, fast-growing perennial shrub, native to China; it is dichogamous and diploid (2n=22) [2]. It is an important deciduous aromatic plant, and it is also one of the most precious epibiotic species dating back to the Tertiary period, being classified as a Class II protected wild plant in China [3]. Wintersweet has over 1000 years' history of cultivation and, as its name indicates, it blooms particularly in winter, from late November to March in central southern and south-western China. Its unique flowering time and strong fragrance make it one of the most popular ornamental plants in China; it is appreciated as a pot plant and for cut flowers and it has a high ornamental and economic value. It has also been introduced into Korea, Japan, Europe, America, and Australia [2]. Wintersweet flowers are used in traditional Chinese medicinal preparations to treat heatstroke, vomiting, coughs and measles [4]. They have also been recognized as the source of a natural essential oil, which can be used in perfumery, cosmetics and aromatherapy [5], [6]. Recently, these applications of wintersweet have received much attention in New Zealand [7].

There are several traits and properties of wintersweet flowers that are important in a commercial context. These include flower development, senescence, scent biosynthesis and emission, and resistance to biotic and abiotic stresses. Wintersweet blooms especially in winter and has a strong fragrance, and its molecular mechanisms of flower development may therefore be different from those of model species or of plants that bloom in the spring. The molecular and genetic processes that determine some of these flowering characteristics cannot be studied using model species such as *Arabidopsis thaliana*, or at least only to a limited extent. During the past decade, the genes and gene networks associated with important floral traits, including flower scent, color, morphology and senescence, have been identified and functionally characterized in several plants grown for their flowers, such as rose [8], carnation [9], [10], [11], and *Antirrhinum majus* [12], [13]. However, the genomic resources that are available for wintersweet or for other members of the Calycanthaceae family are scant and the transcriptional changes and molecular mechanisms that control these important traits and developmental processes in wintersweet flowers are still far from being elucidated. This hampers gene discovery and seriously hinders the improvement of wintersweet as a commercially important species.

This paucity of genomic information is indicated by the fact that although a normalized cDNA library has been constructed from flowers of wintersweet and 479 unigenes have been assembled [14], nevertheless as of late October 2013 only 867 expressed sequence tags (ESTs) for *Chimonanthus praecox* were available on the NCBI website (http://www.ncbi.nlm.nih.gov/). Because gene identification and characterization in *Chimonanthus praecox* remain so limited, the large-scale discovery and characterization of functional genes via genome sequencing or exploration of the transcriptome is essential. The high-throughput capacity of the latest generation of RNA sequencing (RNA-Seq) technology provides a unique opportunity for genomic exploration and gene discovery in non-model plant species for which there are no reference genome sequence data [15], [16]. The results of RNA-Seq also show high levels of reproducibility, for both technical and biological replicates [17], [18]. RNA-Seq generates absolute rather than relative gene expression measurements, providing greater insight and accuracy than microarrays [19], [20].

In this study, we used the Illumina RNA-Seq method to analyze the transcriptome of wintersweet flowers based on three cDNA libraries from different stages of flower development. A total of 106,995 transcripts were identified, and the sequences were annotated against the NR database using BLASTX. By using Illumina's digital gene expression platform, we investigated differential gene expression in open and in senescing flowers, and analyzed the selective expression of Kyoto Encyclopedia of Genes and

Genomes (KEGG) pathways. In particular, we identified genes of plant hormone signal transduction pathways that were differentially expressed during flower development, and also key genes of floral scent biosynthesis. To our knowledge, this is the first comprehensive transcriptomic study to identify genes and pathways that are differentially expressed during flower opening and senescence in wintersweet. It provides an important new bioinformatics resource for the further identification of genes and gene function involved in flower development in this species.

RESULTS

Illumina Sequencing and the Assembly of Sequence Reads

In this study, we divided wintersweet flower development into three essentially distinct stages: i) bud stage, with the flower buds enlarged and turned yellow and with a displayed petal (DP); ii) open flower stage (OF), with the flowers fully opened and emitting a strong fragrance; and iii) senescing flower stage (SF), with withering petals and loss of fragrance (Figure 1). Three individual flowers were used to prepare one pooled RNA sample for each of the three developmental stages (DP, OF and SF). Three cDNA libraries were then constructed and subjected to Illumina deep sequencing. We obtained 21,412,928, 26,950,404, and 24,912,954 qualified Illumina reads, respectively, for the DP, OF and SF stages, giving rise to 2,113,119,501, 2,653,944,208 and 2,451,552,573 bp total residues, respectively (Table 1). After eliminating primer and adapter sequences and filtering out the low-quality reads, we pooled together all the high-quality Illumina reads from the three different developmental-stage libraries. Using Trinity software, we then combined all the reads to form a transcriptome database for wintersweet [21]. We identified 106,995 transcripts, with total residues of 127,396,344 bp. The average length of each transcript was 1,190 bp, the shortest sequence being 351 bp and the longest being 16,998 bp (Table 1). The sequence length distribution of transcripts is indicated in Figure 2. Most of the transcripts (40.6%) were 401~800 bp in length (Table S1).

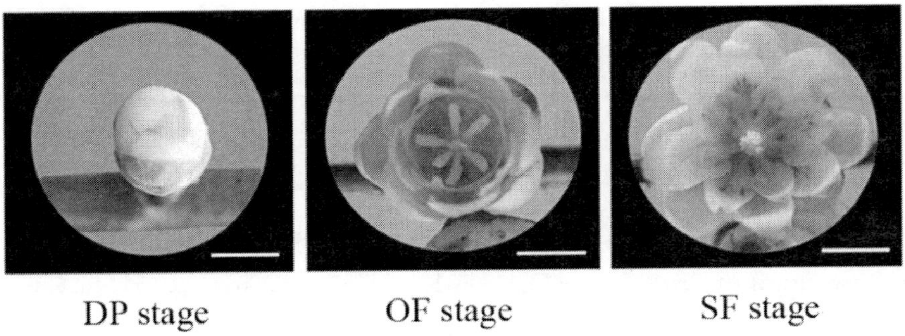

Figure 1. Flower development stages in wintersweet.

Pictures of wintersweet flowers: at flower bud stage with displayed petal (DP), at open flower stage (OF); and at senescing flower stage (SF). Scale bar=1 cm.

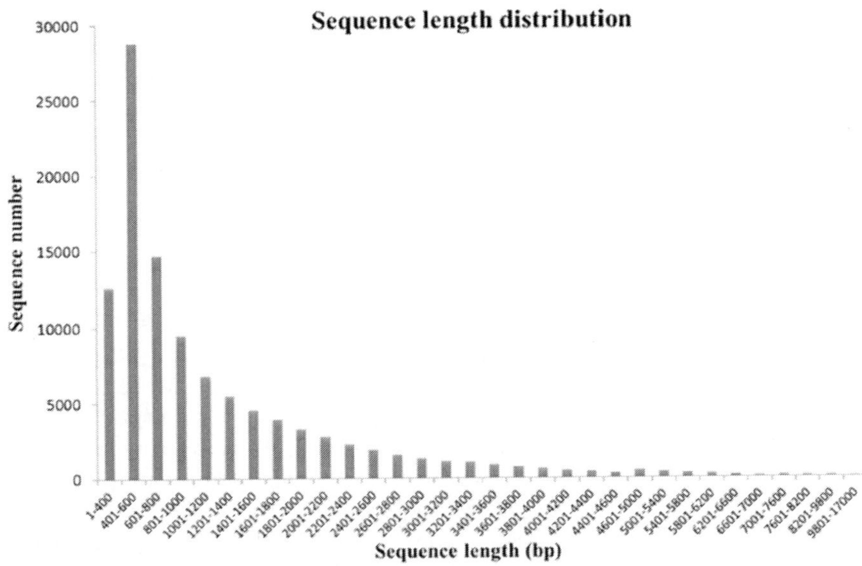

Figure 2. Sequence-length distribution of transcripts assembled from Illumina reads.

All the Illumina reads for each flower development stage (see Figure 1) were combined together (see text) and gave rise to 106,995 transcripts. The horizontal and vertical axes show the size and the number of transcripts, respectively.

Table 1. Summary of Illumina transcriptome sequencing for wintersweet flower.

	DP	OF	SF
Reads	21,412,928	26,950,404	24,912,954
Base number (bp)	2,113,119,501	2,653,944,208	2,451,552,573
Total residues (bp)		127,396,344	
Number of transcripts (n)		106,995	
Average length (bp)		1190.68	
Largest transcripts (bp)		16,998	
Smallest transcripts (bp)		351	

Sequence Annotation

For annotation, the sequences were searched against the NCBI non-redundant (NR) database using BlastX, setting a cut-off E-value of 10^{-5}. A total of 51,793 (48.4%) sequences showed significant similarity to known proteins in the NR database. The similarity distribution for all the search results showed that 54.7% of the matches were of high similarity, *i.e.* ranging from 80% to 100% similarity as reported in the BlastX results, whilst 39.2% of the matches were of similarity ranging from 60% to 80% (Figure 3A). In a further analysis of the matching sequences, we found that 37.7% of the sequences showed closest matches with sequences from *Vitis vinifera*. The next-closest matches were with sequences from *Theobroma cacao* and *Prunus persica*; 12.2% of the sequences showed closest matches with sequences from *Theobroma cacao*, whilst 7.5% of the sequences showed closest matches with sequences from *Prunus persica* (Figure 3B).

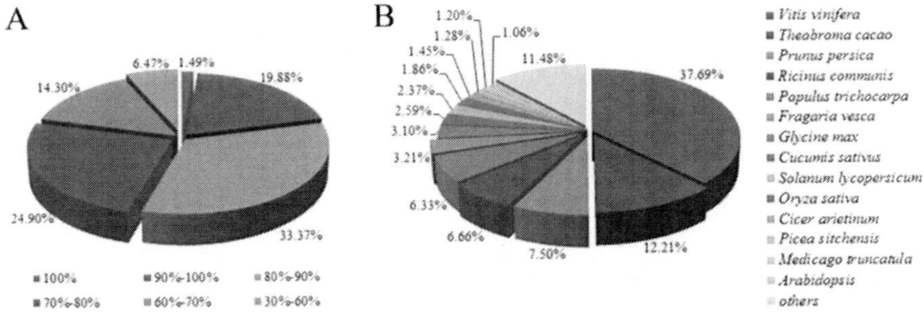

Figure 3. Results summary for sequence-homology search against NCBI NR database.

(A) Similarity distribution of the closest BLASTX matches for each sequence. (B) A species-based distribution of BLASTX matches for sequences. We used all the plant proteins in the NCBI NR database in performing the homology search and for each sequence we selected the closest match for analysis.

GO, COG and KEGG Classification

We added gene ontology (GO) terms using Blast2GO [22], which annotates high-score BLAST matches found to sequences in the NCBI NR proteins database. Genes of wintersweet were classified into three main GO categories and 57 sub-categories. Of the 106,995 assembled transcripts, 32,649 were successfully annotated by GO assignments, and some of them belonged to one or more of the three categories (Figure 4). Amongst the annotated sequences, biological process categories included cellular process (18,805; 57.6%), metabolic process (18,692; 57.3%), single-organism process (9,356; 28.7%), response to stimulus (6,010; 18.4%), biological regulation (4,880; 15.0%), regulation of biological process (4,417; 13.5%), and other categories (23,906; 73.2%). Furthermore, cellular component categories included cell part (16,837; 51.6%), cell (16,837; 51.6%), organelle (12,905; 39.5%), membrane (8,561; 26.2%), organelle part (5,671; 17.4%), macromolecular complex (4,709; 14.4%), membrane part (3,919; 12.0%), and other categories (2,551; 7.8%). In addition, molecular function categories included catalytic activity (18,042; 55.3%), binding (17,154; 52.5%), transporter activity (2,044; 6.3%), structural molecule activity (690; 2.1%), nucleic acid binding transcription factor activity (549, 1.7%), molecular transducer activity (530; 1.6%), and other categories (1,355; 4.2%) (Table S2).

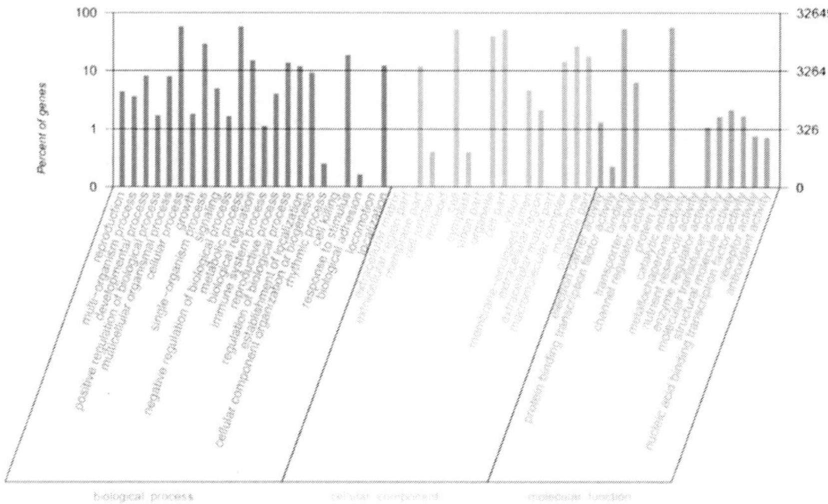

Figure 4. Gene ontology (GO) assignments for the flower transcriptome of wintersweet.

Results are summarized under three main GO categories: biological process, cellular component and molecular function. The left y-axis indicates the percentage of a specific category of genes in each main category. The right y-axis indicates the number of genes in the same category.

In a further analysis, the annotated sequences were subjected to a search against the Clusters of Orthologous Group (COG) database, for functional prediction and classification. Based on sequence homology, 21,893 unique sequences were assigned a COG functional classification. These sequences were classified into 25 COG categories, denoting involvement in cellular process, signal transduction, metabolism and other processes (Figure 5). The most common category was the non-specific category of 'general function prediction only' (4,341; 19.8%), followed by replication, recombination and repair (2,360; 10.8%), transcription (2,267; 10.4%), signal transduction mechanisms (2,022; 9.2%), posttranslational modification, protein turnover and chaperones (1,326; 6.1%), and carbohydrate transport and metabolism (1,138; 5.2%). Amongst other categories represented (Table S3) were translation, ribosomal structure and biogenesis (1,115; 5.1%) and amino acid transport and metabolism (851; 3.9%). The categories of cell motility (11; 0.05%) and nuclear structure (2; 0.01%) were the least represented.

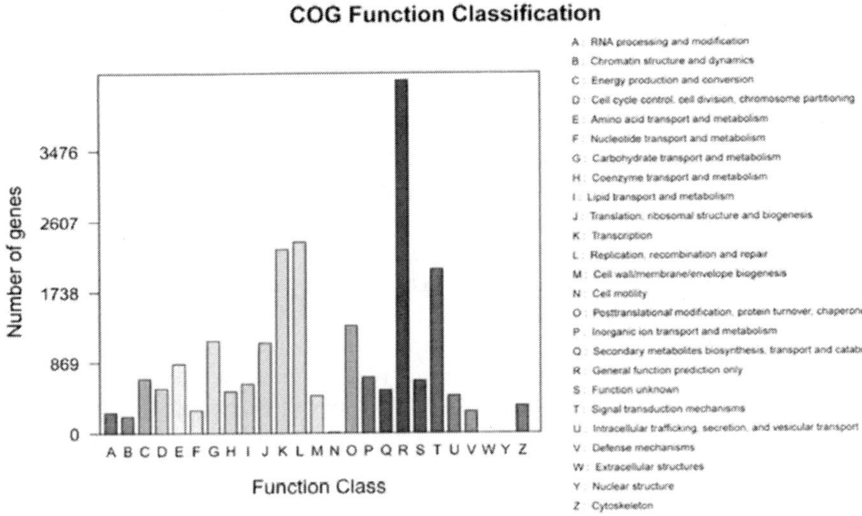

Figure 5. COG functional classification for the flower transcriptome of wintersweet.

From a total of 106,995 *de novo* assembled transcripts, 21,893 transcripts with significant homologies in the COG database (E-value ≤1.0 E-5) were classified into 25 COG categories.

In order to identify biochemical pathways, we mapped the annotated sequences onto the KEGG database, which is an alternative approach to the categorization of gene function and which places emphasis on biochemical pathways. In total, 15,587 transcripts were assigned to 312 KEGG pathways. A summary of the findings is presented in Table S4. The largest number of sequences were those associated with metabolic pathways (4,293; 27.5%), followed by sequences that were involved in the biosynthesis of secondary metabolites (2,049; 13.2%), in microbial metabolism in diverse environments (831; 5.3%), in the biosynthesis of amino acids (495; 3.2%), in plant hormone signal transduction (467; 3.0%), and in a number of other pathways. In particular, several important pathways of secondary metabolite biosynthesis, including the phenylpropanoid, flavonoid and carotenoid biosynthetic pathways, and several signaling pathways, including the p53, mitogen-activated protein kinase (MAPK) signaling pathway, were represented. These pathway assignments provide valuable information for the investigation of specific biochemical and development processes.

Differentially Expressed Genes (DEGs) during Flower Development

To study gene expression during flower development, a genome-wide expression analysis was carried out for each of the DP, OF and SF flower developmental stages (Figure 1). We mapped the Illumina reads from each stage onto our assembled transcriptome database. The expression of each gene was calculated using the numbers of reads mapping onto its transcripts. Comparisons of gene expression between DP and OF (DP *vs* OF) and between OF and SF (OF *vs* SF) showed that 2,706 and 1,466 genes were differentially expressed in DP *vs* OF and in OF *vs* SF, respectively (Log_2 fold changes: ≥2 or ≤−2; FDR ≤0.05). An overall view of the expression pattern for DP *vs* OF is depicted in Figure 6A. This shows the expression changes for 2,706 differentially expressed genes (DEGs), ranging from a 14-fold to a -12-fold change in expression. A corresponding view for OF *vs* SF is depicted in Figure 6B, which shows the changes for 1,466 DEGs, ranging from a 12-fold to an -8-fold change in expression. Figure 6C illustrates that for DP *vs* OF there were 1,681 up-regulated genes and 1,025 down-regulated genes, whilst in the case of OF *vs* SF, there were 1,067 and 399, respectively.

To investigate further, the 20 most up-regulated and the 20 most down-regulated genes were each selected from both the DP *vs* OF and the OF *vs* SF comparisons. Some of these genes showed matches to sequences in the NCBI NR database, such as cytochrome P450, α-terpineol synthase, lipid-transfer protein, cysteine protease, and GDSL esterase/lipase (Tables S5; S6). For example, we found that a cytochrome P450 gene was the most up-regulated gene (13-fold) at the OF stage but was down-regulated substantially at the SF stage (Table S5). Two cysteine protease genes were down-regulated with −10.9 and −8.9 fold changes respectively at the OF stage, and showed no expression at all at the SF stage (Table S5). Finally, we found that a GDSL esterase/lipase gene was induced specifically and strongly (10-fold change in expression) at the SF stage (Table S6).

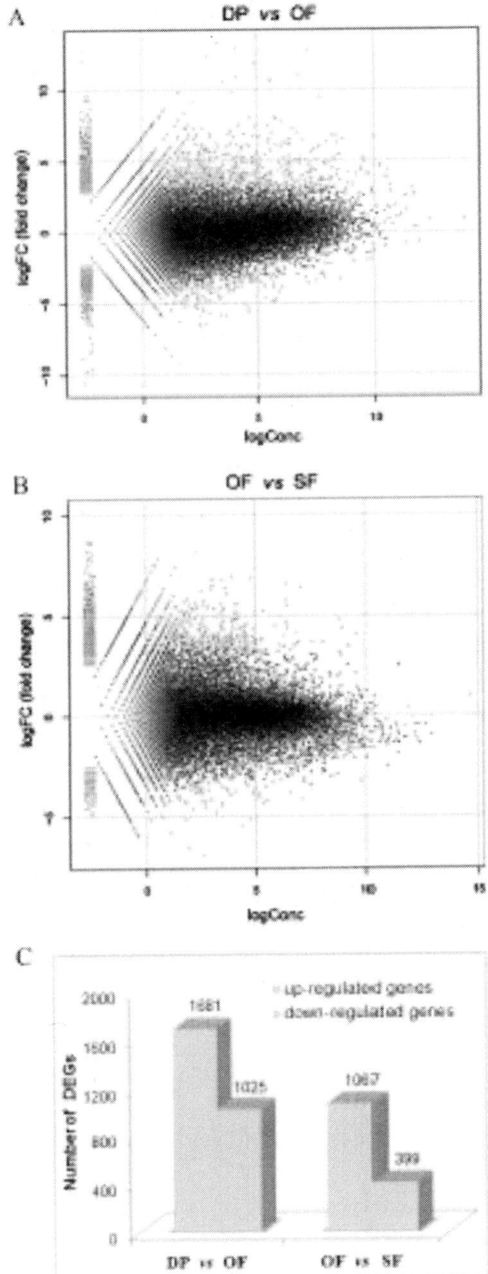

Figure 6. Differentially expressed genes (DEGs) in the flower development-stage comparisons DP vs OF and OF vs SF.

(A) Whole-study overview of log-fold changes in gene expression in DP *vs* OF. (B) Whole-study overview of log-fold changes in gene expression in OF *vs* SF. The x-axis indicates the absolute expression levels (LogConc). The y-axis indicates the log-fold changes between the two samples. Genes for which differential expression is significant are shown as red dots ($Log_2FC \geq 2$ or ≤ -2; FDR≤0.05). (C) The number of up- or down-regulated genes in DP *vs* OF and OF *vs* SF. DP *vs* OF refers to the comparison between the bud stage showing a displayed petal (DP) and the open flower stage (OF). OF *vs* SF refers to the comparison between the open flower stage (OF) and the senescing flower stage (SF).

Pathway Enrichment Analysis of DEGs

The KEGG Orthology Based Annotation System (KOBAS) [23] was used for the further identification of biosynthetic and other pathways and to explore in greater depth the functions of DEGs. Pathways that were found to be significantly up-regulated (corrected P-value ≤0.05) in both DP *vs* OF and OF *vs* SF were involved in cutin, suberin and wax biosynthesis; cyanoamino acid metabolism; pentose and glucuronate interconversions; phenylalanine metabolism; phenylpropanoid biosynthesis; plant hormone signal transduction; and starch and sucrose metabolism (Table 2). Since wintersweet, otherwise known as 'wax shrub', blooms in winter and displays a strong resistance to biotic and abiotic stress, we focused upon the pathway of cutin, suberin and wax biosynthesis, and obtained matches to key enzymes of cutin, suberin and wax biosynthesis, such as CYP86A1, CYP86A4, CYP86B1, ECERIFERUM (CER), and alcohol-forming fatty acyl-coenzyme A reductases (FARs). The expression of these genes was found to be differentially regulated during flower development in wintersweet. In addition, the *CER1* and *FAR* genes, which are related to wax biosynthesis, were found to be up-regulated at the OF stage and down-regulated at the SF stage (Figure 7).

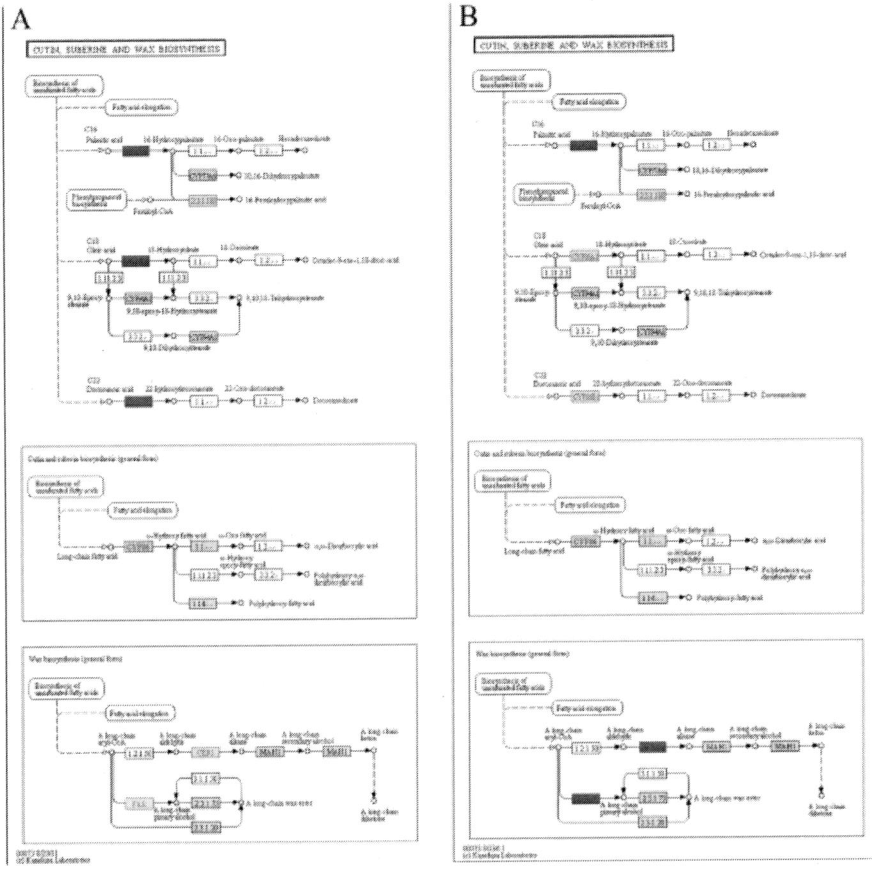

Figure 7. Regulatory changes in the pathway of cutin, suberin and wax biosynthesis during flower development.

(A) Gene expression changes between DP and OF stages. (B) Gene expression changes between OF and SF stages. Green boxes, genes identified in our data; light blue boxes, genes involved in the pathway present in the KEGG database but undetectable in our data; pink boxes, up-regulated genes; dark blue boxes, down-regulated genes.

Table 2. Statistically enriched pathways identified using KOBAS database in differentially expressed transcripts during flower development.

			DP *vs* OF			OF *vs* SF		
Pathway	**Pathway ID**	**Bn**	**Nt**	**P-Value**	**Corrected P-Value**	**Nt**	**P-Value**	**Corrected P-Value**
beta-Alanine metabolism	ko00410	114	27	0.000875	0.021922			
Biosynthesis of unsaturated fatty acids	ko01040	76				19	6.01E-06	0.000441
Cutin, suberine and wax biosynthesis	ko00073	61	28	3.07E-11	1.16E-08	19	6.96E-08	8.52E-06
Cyanoamino acid metabolism	ko00460	104	30	2.32E-06	0.000145	22	1.02E-05	0.000626
Flavone and flavonol biosynthesis	ko00944	11	6	0.001135	0.026674			
Flavonoid biosynthesis	ko00941	80	27	7.28E-09	9.12E-07			
Galactose metabolism	ko00052	175				26	0.001580	0.048309
Histidine metabolism	ko00340	57	17	0.000442	0.015111			
Limonene and pinene degradation	ko00903	36	13	0.000766	0.020561			
Methane metabolism	ko00680	213				39	1.53E-07	1.40E-05
Nicotinate and nicotinamide metabolism	ko00760	52	16	0.000333	0.012529			
One carbon pool by folate	ko00670	49	15	0.000672	0.019441			
Pentose and glucuronate interconversions	ko00040	124	43	9.51E-11	1.79E-08	20	0.001866	0.048919
Peroxisome	ko04146	238	48	0.000148	0.006188			
Phenylalanine metabolism	ko00360	111	25	0.001618	0.035787	22	4.44E-05	0.002327
Phenylpropanoid biosynthesis	ko00940	185	51	1.46E-08	1.37E-06	42	3.37E-10	1.24E-07
Plant hormone signal transduction	ko04075	467	92	3.65E-06	0.000196	72	4.62E-08	8.47E-06
Porphyrin and chlorophyll metabolism	ko00860	134				23	0.000525	0.024080
Starch and sucrose metabolism	ko00500	412	86	6.34E-08	4.77E-06	50	0.000638	0.025562
Stilbenoid, diarylheptanoid and gingerol biosynthesis	ko00945	21	10	0.000107	0.005008			

Bn (Back ground number), indicates the total number of transcripts present in each pathway. Nt (Number of transcripts), indicates the number of differentially expressed transcripts in each pathway.

We also focused upon plant hormone signal transduction pathways, apparently for the first time in wintersweet, almost all the key genes could be identified from our data (Figure S1). We identified 92 and 72 transcripts associated with plant hormone signal transduction pathways that were differentially expressed in DP *vs* OF and OF *vs* SF, respectively (Table 2). The DEGs were identified as phytohormone receptor or phytohormone-responsive genes known to be induced by various phytohormone signaling pathways, including those for auxins, cytokinins (CK), abscisic acid (ABA), ethylene (ET), brassinosteroids (BR), jasmonic acid (JA) and salicylic acid (SA), as indicated in Table 3. The early auxin-responsive genes (*i.e.* those associated with the early phase of a response to auxins) have been grouped broadly into three major classes: the auxin/indole-3-acetic acid (Aux/IAA), the GH3, and the small auxin-up RNA (SAUR) gene families. Our data showed that the *Aux/IAA*, *GH3* and *SAUR* genes were expressed differentially during wintersweet flower opening and senescence. Moreover, a number of JA-responsive genes, including *JAR* (JAR Jasmonate Resistant 1), *JAZ* (Jasmonates ZIM-domain protein) and *MYC2*, were all induced significantly at the OF stage, whereas the salicylic acid-responsive genes, *TGA* and *PR1*, were induced specifically and strongly at the SF stage (Table 3).

Table 3. Differentially expressed genes related to plant hormone signal transduction pathway during flower development.

Gene ID	Putative function	Nearest match	DP *vs* OF		OF *vs* SF	
			log_2 FC	FDR	log_2 FC	FDR
	Auxin signal transduction					
Cp13004_c0	auxin influx carrier protein (AUX1, LAX)	ABN81349.1	−2.51	0.013927	2.69	0.010811
Cp1471_c0	auxin influx carrier protein (AUX1, LAX)	CCF23026.1	4.25	2.69E-06		
Cp1637_c0	Aux/IAA	EMJ16790.1			2.28	0.035763
Cp6390_c0	Aux/IAA	XP_002304179.1	5.65	8.28E-09	−3.89	3.68E-05
Cp15215_c0	auxin response factor (ARF)	BAD19061.1	3.97	0.012847		
Cp17433_c0	auxin response factor (ARF)	XP_002326300.1	2.92	0.017690		
Cp39410_c0	auxin response factor (ARF)	XP_003634382.1			2.66	0.016236
Cp17028_c0	Auxin-responsive GH3 family protein (GH3)	AFC36444.1			3.06	0.006564
Cp19730_c0	Auxin-responsive GH3 family protein (GH3)	EOY12664.1	−2.59	0.020622		
Cp7344_c0	Auxin-responsive GH3 family protein (GH3)	AAD32141.1	5.14	1.01E-06	2.24	0.025821
Cp10520_c1	SAUR family protein	EOX96266.1			3.33	0.000578
Cp13060_c0	SAUR family protein	XP_002266248.1	3.78	0.000358		
Cp13833_c0	SAUR family protein	EOX96266.1			4.69	1.08E-05
Cp15265_c0	SAUR family protein	XP_002272614.1	3.10	0.005876		
Cp17085_c0	SAUR family protein	XP_002265932.1	6.72	0.001457		
Cp21781_c0	SAUR family protein	XP_002327252.1	−4.69	0.001700		
Cp24119_c0	SAUR family protein	XP_002968676.1			−4.21	0.017503
Cp3195_c0	SAUR family protein	EOX99720.1			3.18	0.001959
Cp53091_c0	SAUR family protein	CAN62213.1	6.14	0.010751		
	CK signal transduction					
Cp38923_c0	histidine phosphotransfer protein (AHP)	XP_002298243.1			4.16	0.020819
Cp21971_c0	type A response regulator (ARR-A)	XP_002302230.1	−8.09	4.17E-06		
Cp4567_c0	type A response regulator (ARR-A)	XP_002267404.1	2.33	0.012744		
	ABA signal transduction					
Cp13360_c0	abscisic acid receptor PYL-like protein (PYL)	XP_002516457.1	7.15	3.62E-08	−3.33	0.001215
Cp7252_c0	abscisic acid receptor PYL-like protein (PYL)	XP_002264158.1	9.81	5.39E-10		
Cp22532_c0	type 2C protein phosphatases (PP2C)	XP_002318387.1			5.65	2.51E-06
Cp9742_c0	type 2C protein phosphatases (PP2C)	XP_004291511.1			3.37	0.000739
Cp2918_c0	SNF1-related kinase 2 (SnRK2)	XP_002262726.1	2.67	0.003540		
Cp7416_c0	ABRE binding factors (ABF)	XP_002326942.1			2.59	0.014829
	Ethylene signal transduction					
Cp14969_c0	EIN3 binding F-box (EBF1/2)	AEK81539.1	6.49	1.98E-06		
	Brassinosteroid signal transduction					
Cp2974_c0	BRI1 KINASE INHIBITOR 1 (BKI1)	CAN78408.1	4.46	1.10E-06		
Cp11265_c0	CYCD3	BAA76478.1	−2.66	0.007958		
	JA signal transduction					
Cp930_c0	Jasmonate Resistant 1 (JAR1)	EOX95769.1	3.65	5.34E-05	−2.38	0.024910
Cp1074_c0	jasmonates ZIM-domain protein (JAZ)	XP_002529200.1	3.17	0.000610		
Cp12614_c0	jasmonates ZIM-domain protein (JAZ)	XP_002304118.1	3.95	0.000207		
Cp1550_c4	jasmonates ZIM-domain protein (JAZ)	XP_002534018.1	2.18	0.033780		
Cp2223_c0	jasmonates ZIM-domain protein (JAZ)	ADI39634.1	3.07	0.001190		
Cp2936_c0	jasmonates ZIM-domain protein (JAZ)	XP_002284855.1	3.75	4.30E-05	−2.59	0.007908
Cp12298_c0	MYC2	EOY23994.1	2.25	0.018227		
	SA signal transduction					
Cp22279_c0	BZIP transcription factor family protein (TGA)	XP_003554196.1	−4.21	0.000286		
Cp47428_c0	BZIP transcription factor family protein (TGA)	EOY09442.1			6.35	0.015379
Cp9043_c0	BZIP transcription factor family protein (TGA)	EOX92234.1			2.64	0.012478
Cp3374_c0	pathogenesis-related protein 1 (PR1)	BAF95881.1			8.03	6.50E-12

Differentially expressed genes with Log_2 fold changes ≥2 or ≤ −2, and FDR ≤0.05 were included.
The genes involved in the pathway are color-coded: green boxes, genes identified in our data; and light blue boxes, genes involved in the pathway present in KEGG database but undetectable in our data.

Genes of Floral Scent Biosynthesis

Wintersweet flowers develop a strong and specific fragrance during flower opening. In this study, we obtained from our transcriptomic data the key biosynthetic genes of floral scent production, including GPPS (geranyl diphosphate synthase), MRY (myrcene synthase), GES (geraniol synthase), LIS (S-linalool synthase), FPPS (farnesyl pyrophosphate synthase), TER (α-terpineol synthase), and SAMT (S-adenosyl-L-methionine: salicylic acid carboxyl methyltransferase). These included three homologous genes of *SAMT* and *MRY* and two homologous genes of *TER*. Based on the annotated sequences, we have therefore proposed a biosynthetic pathway for the formation of monoterpenes and sesquiterpenes in wintersweet (Figure 8A).

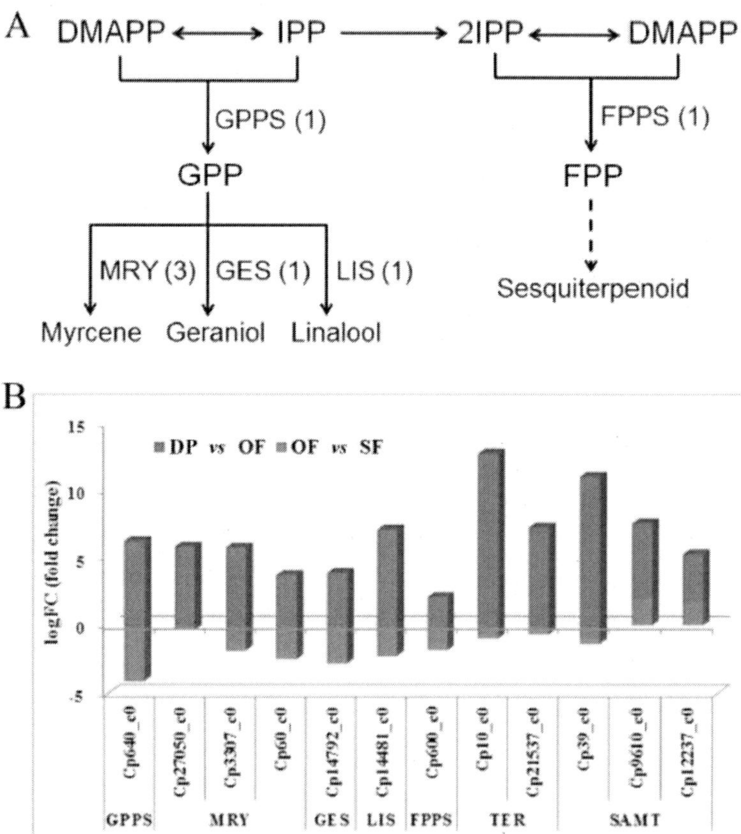

Figure 8. Proposed biosynthetic pathway for monoterpenes and sesquiterpenes in wintersweet.

(A) Proposed biosynthetic pathway for monoterpenes and sesquiterpenes. (B) Changes in expression (as –fold) of key genes of floral scent biosynthesis during flower development. DMAPP: dimethylallyl diphosphate; IPP: isopentenyl diphosphate; GPPS: geranyl diphosphate synthase; FPPS: farnesyl pyrophosphate synthase; MRY: myrcene synthase; GES: geraniol synthase; LIS: S-linalool synthase; TER: α-terpineol synthase; SAMT: S-adenosyl-L-methionine: salicylic acid carboxyl methyltransferase. Each enzyme name is followed in parentheses by the number of gene homologues encoding this enzyme. Solid lines indicate direct biochemical reactions, and broken lines indicate indirect reactions.

The expression of genes of floral scent biosynthesis was induced at the OF stage (Figure 8B). The expression of the *LIS* gene, which is responsible for α-linalool biosynthesis, was increased 7-fold at the OF stage; and it has been reported that α-linalool accounts for 36% of the total quantity of volatile compounds detected in wintersweet flowers [6]. Similarly, methyl salicylate has also been demonstrated to be one of the major volatile compounds in wintersweet flowers [6]; and the three *SAMT* gene homologues, responsible for the conversion of salicylic acid to methyl salicylate, were all induced significantly to varying extents at the OF stage.

Real-time Quantitative PCR Validation of RNA-Seq Results

To validate the findings from the sequencing data, an appropriate alternative methodology was chosen. Ten DEGs involved in plant hormone signal pathways were selected for validation using Real-time qPCR, with gene-specific primers designed using Primer Primer software (version 5.0), as shown in Table S7. The expression pattern of these genes at each of the three flower developmental stages is shown in Figure 9. Expression patterns determined by Real-time qPCR were consistent with those obtained by RNA-Seq, confirming the accuracy of the RNA-Seq results reported in this study.

Data were normalized against a reference of wintersweet actin and tubulin genes. All quantitative PCRs for each gene used three biological replicates, with three technical replicates per experiment; the error bars indicate SD.

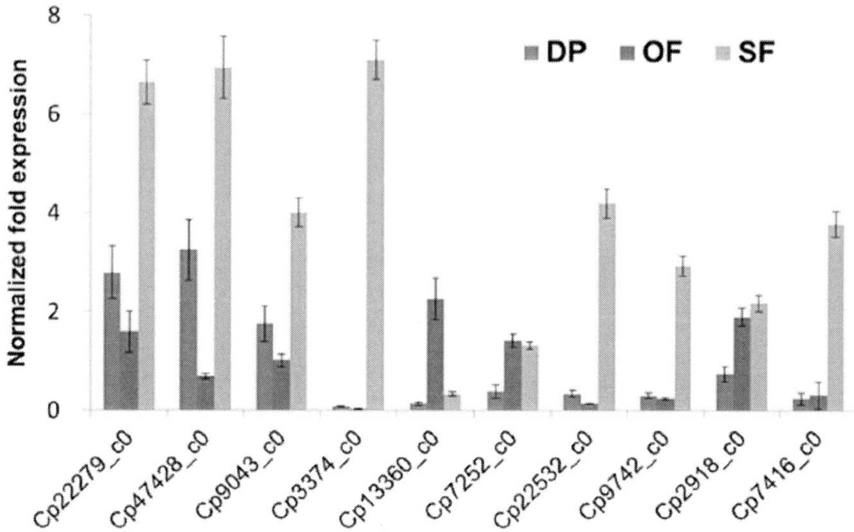

Figure 9. Real-time qPCR validation of ten genes involved in plant hormone signal transduction pathways.

DISCUSSION

Illumina Sequencing and Sequence Annotation

Wintersweet (*Chimonanthus praecox*) is mainly a garden plant and a traditional cut flower in China. However, little genomic information is available for this species. In this study, we have generated large amounts of cDNA sequence data, comprising long sequences of good quality, which will facilitate subsequent and more detailed studies. The availability of transcriptomic data for wintersweet will meet the basic requirements needed to undertake a thorough molecular-genetic and biochemical characterization of this species and its relatives.

For gene annotation, the sequences we had determined were searched against the NCBI NR database. Because of the limited genetic information available, no matches were obtained for as many as 55,202 (51.6%) of the sequences. Some of these sequences may represent novel genes in wintersweet, suggesting that flower development in wintersweet might involve some unique processes and pathways. Interestingly, the annotated sequences of wintersweet showed a higher homology to proteins of *Vitis vinifera* (Figure 3), which may reflect the evolutionary relationship between *Chimonanthus praecox* and *Vitis vinifera*. *Chimonanthus praecox* (Calycanthaceae) belongs to the order Laurales and to the clade magnoliids, which arose in evolution soon after the divergence of the eudicots. *Vitis vinifera* (Vitaceae) belongs to the clade rosids, which lies at the root of the eudicots, and which is closer phylogenetically to the clade magnoliids than to the order Malvales (e.g. *Theobroma cacao*), or to the order Rosales (e.g. *Prunus persica*).

In spite of the large proportion of sequences that showed no matches, a large number of transcripts were nevertheless assigned to a wide range of GO and COG classifications, which indicated that our transcriptome data represented a broad diversity of transcripts in wintersweet. KEGG predictions identified many genes associated with secondary metabolites, including genes of phenylpropanoid biosynthesis, flavonoid biosynthesis, phenylalanine metabolism and carotenoid biosynthesis; in addition, we identified genes related to three alkaloid biosynthesis pathways (Table S4). Alkaloids as a class often exhibit antibacterial and immunostimulatory properties, and this would appear to be consistent with the medicinal and health-promoting properties that are associated with wintersweet [24].

DEGs during Flower Development

Although the analysis of DEGs during flower opening and senescence has been performed using microarrays in *Iris* [25], *Alstroemeria* [26], and *Rosa* [27], nevertheless the availability of data relating to developmental expression profiles and comparing the transcriptomes of flowers or flower parts at different stages remains generally limited. The approach we used here identified >4000 DEGs, at a cut-off of 5% FDR, that showed differential expression patterns in the pairings, DP *vs* OF and OF *vs* SF. Further insights were obtained by examining the changes of gene expression for sets of related genes. Thus, amongst the 20 most up-regulated DEGs, we found one cytochrome P450 gene that was up-regulated significantly at the OF stage and down-regulated at the SF stage (Table S5); this is in apparent contrast to a report of increased expression of cytochrome P450 genes in senescing petals of *Alstroemeria*[26]. Cytochrome P450 genes are found in all kingdoms and show extraordinary diversity in their chemical reactions [28]; they are one of the largest gene families in plants, although the functions of cytochrome P450 genes in flowers are as yet unknown.

In our study, two cysteine protease genes were down-regulated significantly at the OF stage, and showed no expression at all at the SF stage (Table S5). This appears to contrast with a range of observations showing an up-regulation of cysteine proteases during flower senescence, in both ethylene-sensitive and ethylene-insensitive flowers, such as *Dianthus caryophyllus* (carnation) [29], *Hemerocallis spp.* (daylily) [30], *Alstroemeria peruviana* [31], *Sandersonia aurantiaca* [32], *Narcissus pseudonarcissus* (daffodil) [33], and *Gladiolus grandiflora* [34]. This suggests that there may be a different function, or regulation mechanism, for cysteine protease genes in wintersweet, compared to other species.

Genes Related to Plant Hormone Signaling Pathways

Plant hormones are involved in many different processes throughout the life of a plant, including growth, development and senescence [35]. Several plant hormones such as ET, ABA, GA have been reported to play important roles in flower opening and in senescence processes. The specific regulation of flower development is a complex process and is governed both by the endogenous levels of the hormones and by the sensitivity of the hormone response mechanisms. The expression of numerous genes has been found to be altered through the action of plant hormone signaling pathways, as well as by the diversity of interactions amongst plant hormones themselves [36].

Ethylene is one of the most important plant hormones, and it plays a major role in many flowers, such as rose, carnation and orchids, in regulating flower opening, senescence and abscission. A burst of endogenously produced ethylene in such flowers initiates senescence and coordinates the expression of genes required for the process [37]. In wintersweet, however, previous research indicated that no obvious production of ethylene could be detected from the flowers during flower development [38]. In the work we report here, we obtained families of genes encoding ethylene receptors (ETR1, ETR2, EIN4, ERS1 and ERS2), EIN3/EILs (EIN3-like proteins), and the ethylene response factor (ERF); in no cases, however, did we find that the expression of these genes was induced significantly during flower opening and senescence. These findings therefore indicated that flower opening and senescence in wintersweet may not be dependent upon an ethylene-mediated signal transduction process triggered by the endogenous production of small amounts of ethylene. On the other hand, it has been reported that ethephon treatment of wintersweet cut flowers can accelerate the process of flower opening, indicating that wintersweet flowers are sensitive to exogenous ethylene treatment [39].

Salicylic acid (SA) or its derivatives function in a diversity of plant processes, such as seed germination, respiration, stomatal responses, senescence and responses to abiotic stress [40]. Many of the components of the SA signaling pathway, including those of signal perception, have not yet been revealed [41]. However, it is known that SA or its signaling pathway is associated with the activation of diverse groups of defense-related genes, including those encoding pathogenesis-related (PR) proteins [42].

Moreover, the non-expresser of PR genes 1 (NPR1) and transcription factors such as TGACG-motif binding factors (TGAs) have been identified as key components of the SA response [43]. In our present study, the *TGA* and *PR1* genes were found to be up-regulated specifically and significantly at the SF stage (Table 3), suggesting that SA may be involved in the regulation of wintersweet flower senescence. The reported involvement of SA in defense responses and in reactions to abiotic stresses in plants may also be relevant, since wintersweet blooms in winter when temperatures are low.

In addition, we found that a range of auxin-related, ABA-related and JA signaling pathway-related genes isolated in our study, such as *Aux/IAA, ARF, GH3, SAUR,JAR1, JAZ, MYC2, SnRK2* and *ABF,* were differentially regulated during flower opening and senescence in wintersweet.

Taken together, these findings suggest that flower development in wintersweet is subject to complex regulation by multiple hormones. The regulatory mechanisms by which phytohormones are involved in wintersweet flower opening and senescence are still far from elucidated. The identification in this study of genes related to plant hormone signaling pathways is therefore an essential step in understanding fully the hormonal regulation of flowering and flower development in wintersweet and other winter-flowering species.

Floral Scent Biosynthesis and its Genetic Characterization

Floral scents are unique and distinct for each plant. A distinctive floral scent plays an important role in the reproductive processes of many plants and also enhances the aesthetic properties and economic value of ornamental plants and cut flowers. Flower scents vary between plant species on account of their characteristic profiles of volatile compounds of low molecular mass, such as monoterpenes, seisquiterpenes, benzenoids, phenylpropanoids, and fatty acid derivatives [44]. Many of these volatile compounds have already been identified in a number of flower species and in addition many of the biosynthetic enzymes involved have been identified and functionally characterized [45]. Some pathways have been characterized at the molecular-genetic level, primarily in model systems such as *Clarkia breweri* [46]–[50], *Petunia* [51], rose [52], [53], and *Antirrhinum* [50].

In a previous study, more than 30 volatile compounds had been detected in wintersweet flowers; these consisted almost exclusively of volatile terpenoids (monoterpenes and sesquiterpenes) and benzenoids [5], [6]. There was little molecular-genetic characterization of the biosynthesis of these compounds in wintersweet, however; and to the best of our knowledge, only the *FPPS* (farnesyl pyrophosphate synthase) gene and a homologue of the *SAMT*(S-adenosyl-L-methionine: salicylic acid carboxyl methyltransferase) gene had been cloned [54], [55].

In this study, we isolated additional key floral scent-synthesizing genes from wintersweet, based on sequence annotations and analyses of changes in gene expression during flower development. The expression of the *FPPS* gene was found to be induced significantly during flower opening, consistent with previous findings, and was correlated with the emission of sesquiterpenoids [54]. Terpineol synthases (TERs) have been isolated from plant species representing six genera: *Magnolia grandiflora* [56], *Pinus taeda* [57], *Santalum album* [58], *Vitis vinifera* [59], *Zea mays* [60] and *Nicotiana* species [61]. The TER enzyme of *M. grandiflora* produces the single product, α-terpineol, whilst the TER enzymes of the other species produce additional compounds, such as limonene, β-myrcene, α-pinene, and β-pinene [61]. In wintersweet flowers, both α-terpineol and α-pinene have been detected [4], [5].

The identification of genes of floral scent biosynthesis will facilitate future molecular-genetic and physiological studies of the regulation of floral scent production during flower development in wintersweet. It may also open new opportunities for the metabolic engineering of floral volatiles in plants [62].

CONCLUSIONS

This work presents the first *de novo* transcriptome sequencing analysis of wintersweet flower development using the Illumina RNA-Seq method. A total of 10,699 transcripts were assembled and 51,793 sequences were annotated. Our work provides an excellent platform for future genetic and functional genomics research in wintersweet and also provides an invaluable resource for genomics studies in other members of the Calycanthaceae family. Using Illumina sequencing-based gene analysis for the identification of DEGs, we identified a large number of candidate genes that were regulated either at the OF stage or at the SF stage, or at both, and we were able to identify a number of regulated pathways. In particular, we identified DEGs that were involved in plant hormone signal transduction pathways during flower development. The results indicated that flower opening and senescence may not be dependent upon the ethylene signal transduction pathway in wintersweet; in contrast, however, SA may be involved in the regulation of flower senescence. Finally, we obtained a number of key genes of floral scent biosynthesis and on this basis we have proposed a biosynthetic pathway for monoterpenes and sesquiterpenes in wintersweet flower based on annotated sequences. In summary, this comprehensive transcriptomic analysis has provided essential information on the genes and pathways which are involved in flower development in wintersweet. It will be fundamental to the development of wintersweet cultivars of improved horticultural or ornamental quality.

MATERIALS AND METHODS

Plant Materials

Flowers of wintersweet (*Chimonanthus praecox* (L.) Link) were collected from the nursery at Southwest University, Chongqing, China. Flowers at flower bud stage with displayed petal (DP), open flower stage (OF) and senescing flower stage (SF) were dissected from the plants and immediately frozen and stored in liquid nitrogen.

RNA Extraction, cDNA Library Construction and Illumina Deep Sequencing

For each flower developmental stage, three individual flowers from individual plants were randomly chosen and total RNA samples were extracted from each flower by using an RNAprep pure Plant RNA Purification Kit (Tiangen Biotech, Beijing, China). The quality and quantity of total RNA were analyzed using an UltrasecTM 2100 pro UV/Visible Spectrophotometer (Amersham Biosciences, Uppsala, Sweden) and by gel electrophoresis. For each developmental stage, the RNA samples from the three individuals were then pooled together in equal amounts to generate one mixed sample. These three mixed RNA samples (*i.e.* one for each flower stage) were subsequently used in cDNA library construction and Illumina deep sequencing.

Three cDNA libraries were prepared using a TruSeq™ RNA sample preparation Kit from Illumina (San Diego, CA, USA), and using in each case 5 μg of total RNA. Messenger RNA was isolated by polyA selection with oligo (dT) beads and fragmented using fragmentation buffer; cDNA synthesis, end repair, A-base addition and ligation of the Illumina-indexed adaptors were then all performed according to the Illumina protocol. Libraries were then size-selected on 2% Low Range Ultra Agarose for cDNA target fragments of 300–500 bp; this was followed by PCR amplification using Phusion DNA polymerase (NEB) for 15 PCR cycles. Following quantification by TBS380, paired-end libraries were sequenced using Illumina HiSeq™ 2000 system (2×100 bp read length) at Shanghai Majorbio Bio-pharm

Biotechnology Co., Ltd. (Shanghai, China). All raw-sequence reads data were deposited in NCBI Sequence Read Archive (SRA, http://www.ncbi.nlm.nih.gov/Traces/sra) with accession number SRA106143.

Transcriptome *De novo* Assembly

The raw reads were cleaned by removing low-quality reads (Q value <25), adapter sequences, reads with ambiguous bases 'N', and fragments of less than 20 bp in length, using SeqPrep (https://github.com/jstjohn/SeqPrep) and ConDeTri_v2.0.pl software(http://code.google.com/p/condetri/downloads/detail?name=condetri_v2.0.pl). *De novo* assembly of the *Chimonanthus praecox* transcriptome in the absence of a reference genome, using the three libraries obtained from the different flower stages, was accomplished using Trinity *de novo* transcriptome assembly software [21].

Sequence Annotation and Classification

For annotation, the sequences were searched against the NCBI non-redundant (NR) protein database [63] using BlastX, with a cut-off E-value of 10^{-5}. Gene ontology (GO) terms were extracted from the annotation of high-score BLAST matches in the NCBI NR proteins database (E value $\leq 1.0 \times 10^{-5}$) using blast2go (http://www.blast2go.com/b2ghome), and then sorted for the GO categories using in-house perl scripts [64]. Functional annotation of the proteome was carried out by a BlastP homology search against the NCBI Clusters of Orthologous Groups (COG) database (http://www.ncbi.nlm.nih.gov/COG/). KEGG pathway annotations were performed using Blastall software against the KEGG database.

Expression Analysis

After assembling the *Chimonanthus praecox* transcriptome, every RNA-seq library was separately aligned to the generated transcriptome assembly, using Bowtie [65]. The counting of alignments was performed using the RSEM package [66]. The DEGs were analyzed using the R Bioconductor package, edgeR [67]. The P-value set the threshold for the differential gene expression test. The threshold of the P-value in multiple tests was determined by the value for the false discovery rate (FDR) [68]. We used "FDR≤0.05 and the absolute value of Log_2 fold change (Log_2 FC)≥2" as the threshold to judge the significance of gene expression differences. For pathway enrichment analysis, DEGs were mapped to the terms in the KEGG database by using the KOBAS program (http://kobas.cbi.pku.edu.cn/home.do). Pathways were defined as significantly enriched pathways with respect to DEGs by taking a corrected P-value ≤0.05 as the threshold.

Real-time Quantitative PCR Validation of RNA-Seq Data

Ten DEGs involved in plant hormone transduction pathways were chosen for validation using Real-time qPCR. The primers, designed with the software Primer Primer 5.0, are listed in Table S7. Total RNA was extracted from flowers at each stage using an RNAprep pure Plant RNA Purification Kit (Tiangen Biotech, Beijing, China) and reverse transcribed into cDNA using a PrimeScript RT Reagent Kit with gDNA Eraser (TaKaRa, Otsu, Japan). The real-time polymerase chain reaction (PCR) was performed on 10 μL reactions using 5 μL of Ssofast EvaGreen Supermix (Bio-Rad, Hercules, CA), 0.5 μL of each primer (final concentration of 500 nM), 3.5 μL of water, and 0.5 μL of cDNA template. PCR was carried out using the Bio-Rad CFX96 RealTime PCR system (Bio-Rad, US), using a denaturation temperature of 95°C for 30 s, followed by 40 cycles of 95°C for 5 s and 59°C for 5 s. To control the specificity of the annealing of the oligonucleotides, dissociation kinetics were performed using the real-time PCR system at the end of the experiment (60 to 95°C, continuous fluorescence measurement). A comparative Ct method ($2^{-\Delta\Delta Ct}$) of relative quantification [69] was used to analyze the real-time quantitative PCR data, using Bio-Rad CFX Manager Software, Version 1.6. The genes for actin and tubulin were used for the calculation of relative transcript abundance [14]. The sizes of the amplified products were confirmed through gel electrophoresis. Negative controls without templates were carried out concurrently. All quantitative PCRs for each gene used three biological replicates, with three technical replicates per experiment.

ACKNOWLEDGMENTS

We wish to thank Professor Bo Hong (China Agricultural University) and Nan Ma (China Agricultural University) for their excellent advice on this paper.

AUTHOR CONTRIBUTIONS

Conceived and designed the experiments: MYL. Performed the experiments: DFL JM JFY. Analyzed the data: DFL ZNL DPL. Wrote the paper: DFL SZS YLG.

REFERENCES

1. Li Y, Li PT (2000) Origin, evolution and distribution of the Calycanthaceae. Guihaia 20: 295–300.
2. Zhang RH, Liu HE (1998) Wax Shrubs in World (Calycanthaceae). Beijing: China Science and Technology Press.
3. Zhao KG, Zhou MQ, Chen LQ, Zhang DL, Gituru WR (2007) Genetic diversity and discrimination of Chimonanthus praecox (L.) Link germplasm using ISSR and RAPD markers. Hortscience 42 (5): 1144–1148.
4. Zhao Y, Zhang Y, Wang ZZ (2010) Chemical Composition and Biological Activities of Essential Oil from Flower of Chimonanthus praecox (L.) Link. Lishizhen medicine and material medical research 21(3): 622–625.
5. Deng CH, Song GX, Hu YM (2004) Rapid determination of volatile compounds emitted from Chimonanthus praecox flowers by HS–SPME–GC–MS. Z. Naturforsch 59: 636–640.
6. Azuma H, Toyota M, Asakawa Y (2005) Floral scent chemistry and stamen movement of Chimonanthu preacox (L.) Link (Calycanthaceae). Acta Phytotax Geobot 56 (2): 197–201.
7. Feng JQ (2007) New Zealand flower industry–with special reference to wintersweet introduction and commercialization. Journal of Beijing Forestry University (suppl 1): 4–8.
8. Bendahmane M, Dubois A, Raymond O, Bris ML (2013) Genetics and genomics of flower initiation and development in roses. J Exp Bot 64(4): 847–857.
9. Iordachescu M, Verlinden S (2005) Transcriptional regulation of three EIN3-like genes of carnation (Dianthus caryophyllus L. cv. Improved White Sim) during flower development and upon wounding, pollination, and ethylene exposure. J Exp Bot 56: 2011–2018.
10. Harada T, Torii Y, Morita S, Masumura T, Satoh S (2010) Differential expression of genes identified by suppression subtractive hybridization in petals of opening carnation flowers. J Exp Bot 61(9): 2345–2354.
11. Tanase K, Nishitani C, Hirakawa H, Isobe S, Tabata S, et al. (2012) Transcriptome analysis of carnation (Dianthus caryophyllus L.) based on next-generation sequencing technology. BMC genomics 13(1): 292.
12. Davies B, Cartolano M, Schwarz-Sommer Z (2006) Flower Development: The Antirrhinum Perspective. Adv Bot Res 44: 279–321.
13. Shang Y, Venail J, Mackay S, Bailey PC, Schwinn KE, et al. (2011) The molecular basis for venation patterning of pigmentation and its effect on pollinator attraction in flowers of Antirrhinum. New Phytol 189(2): 602–615.
14. Sui SZ, Luo JH, Ma J, Zhu Q, Lei XH, et al.. (2012) Generation and analysis of expressed sequence tags from Chimonanthus praecox (Wintersweet) flowers for discovering stress-responsive and floral development-related genes. Comp Funct Genomics.
15. Wang Z, Gerstein M, Snyder M (2009) RNA-Seq: a revolutionary tool for transcriptomics. Nat Rev Genet 10: 57–63.
16. Ward JA, Ponnala L, Weber CA (2012) Strategies for transcriptome analysis in nonmodel plants. Am J B 99: 267–276.
17. Cloonan N, Grimmond SM (2008) Transcriptome content and dynamics at single-nucleotide resolution. Genome Biol 9: 234.
18. Nagalakshmi U, Wang Z, Waern K, Shou C, Raha D, et al. (2008) The transcriptional landscape of the yeast genome defined by RNA sequencing. Science 320: 1344–1349.
19. Hoen PA, Ariyurek Y, Thygesen HH, Vreugdenhil E, Vossen RH, et al. (2008) Deep sequencing-based expression analysis shows major advances in robustness,

resolution and inter-lab portability over five microarray platforms. Nucleic Acids Res 36: 141.
20. Marioni JC, Mason CE, Mane SM, Stephens M, Gilad Y (2008) RNA-seq: an assessment of technical reproducibility and comparison with gene expression arrays. Genome Res 18: 1509–1517.
21. Grabherr MG, Haas BJ, Yassour M, Levin JZ, Thompson DA, et al. (2011) Full-length transcriptome assembly from RNA-Seq data without a reference genome. Nat Biotechnol 29: 644–652.
22. Conesa A, Götz S (2008) Blast2GO: A comprehensive suite for functional analysis in plant genomics. Int J Plant Genomics 619–832.
23. Mao X, Cai T, Olyarchuk JG, Wei L (2005) Automated genome annotation and pathway identification using the KEGG orthology (KO) as a controlled vocabulary. Bioinformatics 21: 3787–3793.
24. Kitajima M, Mori I, Arai K, Kogure N, Takayama H (2006) Two new tryptamine-derived alkaloids from Chimonanthus praecox f. concolor. Tetrahedron Letters 47(19): 3199–3202.
25. Van Doorn WG, Balk PA, Van Houwelingen AM, Hoeberichts FA, Hall RD, et al. (2003) Gene expression during anthesis and senescence in Iris flowers. Plant Mol Biol 53(6): 845–863.
26. Breeze E, Wagstaff C, Harrison E, Bramke I, Rogers H, et al. (2004) Gene expression patterns to define stages of post-harvest senescence in Alstroemeria petals. Plant Biotechnol J 2(2): 155–168.
27. Dubois A, Remay A, Raymond O, Balzergue S, Chauvet A, et al. (2011) Genomic approach to study floral development genes in Rosa sp. PloS one 6(12): e28455.
28. Mizutani M, Ohta D (2010) Diversification of P450 genes during land plant evolution. Annu Rev Plant Biol 61: 291–315.
29. Jones ML, Larsen PB, Woodson WR (1995) Ethylene-regulated expression of a carnation cysteine protease during flower petal senescence. Plant Mol Biol 28: 505–512.
30. Valpuesta V, Lange NE, Guerrero C, Reid MS (1995) Up regulation of a cysteine protease accompanies the ethylene insensitive senescence of daylily (Hemerocallis) flowers. Plant Mol Biol 28: 575–582.
31. Wagstaff C, Leverentz MK, Griffiths G, Thomas B, Chanasut U, et al. (2002) Cysteine protease gene expression and proteolytic activity during senescence of Alstroemeria petals. J Exp Bot 53: 1–8.
32. Eason JR, Ryan DJ, Pinkney TT, O'Donoghue EM (2002) Programmed cell death during flower senescence: isolation and characterization of cysteine proteases from Sandersonia aurantiaca. Funct Plant Biol 29: 1055–1064.
33. Hunter DA, Steele BC, Reid MS (2002) Identification of genes associated with perianth senescence in daffodil (Narcissus pseudonarcissus L. 'Dutch Master'). Plant Science 163: 13–21.
34. Arora A, Singh VP (2004) Cysteine protease gene expression and proteolytic activity during floral development and senescence in ethylene-insensitive Gladiolus grandiflora. J Plant Biochem Biotech 13: 123–126.
35. Davies PJ (2010) The plant hormones: their nature, occurrence, and functions. In: Plant hormones, Springer Netherlands, pp1–15.
36. Depuydt S, Hardtke CS (2011) Hormone signalling crosstalk in plant growth regulation. Current Biology 21(9): R365–R373.
37. Shahri W, Tahir I (2011) Flower senescence-strategies and some associated events. The Botanical Review 77(2): 152–184.
38. Sheng AW, Guo WM, Sun ZH (1999) Study on dynamics of endogenous hormones and parameters concerned senescence in cut wintersweet flowers. Journal of Beijing Forestry University 2: 48–53.
39. Ma J, Li Z, Wang B, Sui SZ, Li MY (2012) Cloning of an expansin gene from chimonanthus praecox flowers and its expression in flowers treated with ethephon or 1-methylcyclopropene. Hortscience 47(10): 1472–1477.

40. Alonso-Ramírez A, Rodríguez D, Reyes D, Jiménez JA, Nicolás G, et al. (2009) Evidence for a role of gibberellins in salicylic acid-modulated early plant responses to abiotic stress in Arabidopsis seeds. Plant Physiol 150(3): 1335–1344.
41. Santner A, Estelle M (2009) Recent advances and emerging trends in plant hormone signalling. Nature 459(7250): 1071–1078.
42. Vlot AC, Dempsey DMA, Klessig DF (2009) Salicylic acid, a multifaceted hormone to combat disease. Annu Rev Phytopathol 47: 177–206.
43. Boyle P, Le Su E, Rochon A, Shearer HL, Murmu J, et al. (2009) The BTB/POZ domain of the Arabidopsis disease resistance protein NPR1 interacts with the repression domain of TGA2 to negate its function. The Plant Cell Online 21(11): 3700–3713.
44. Dudareva N, Pichersky E (2000) Biochemical and molecular genetic aspects of floral scents. Plant Physiol 122: 627–633.
45. Pichersky E, Noel JP, Dudareva N (2006) Biosynthesis of plant volatiles: nature's diversity and ingenuity. Science 311: 808–811.
46. Dudareva N, Cseke L, Blanc VM, Pichersky E (1996) Molecular characterization and cell type-specific expression of linalool synthase gene from Clarkia. Plant Physiol 111: 815.
47. Dudareva N, Cseke L, Blanc VM, Pichersky E (1996) Evolution of floral scent in Clarkia: novel patterns of S-linalool synthase gene expression in the C. breweri flower. The Plant Cell 8: 1137–1148.
48. Dudareva N, D'Auria JC, Nam KH, Raguso RA, Pichersky E (1998) Acetyl-CoA: benzylalcohol acetyltransferase–an enzyme involved in floral scent production in Clarkia breweri. Plant J 14: 297–304.
49. Dudareva N, Raguso RA, Wang J, Ross JR, Pichersky E (1998) Floral scent production in Clarkia breweri. III. Enzymatic synthesis and emission of benzenoid esters. Plant Physiol 116: 599–604.
50. Tholl D, Kish CM, Orlova I, Sherman D, Gershenzon J, et al. (2004) Formation of monoterpenes in Antirrhinum majus and Clarkia breweri flowers involves heterodimeric geranyl diphosphate synthases. The Plant Cell 16: 977–992.
51. Boatright J, Negre F, Chen F, Kish CM, Wood B, et al. (2004) Understanding in vivo benzenoid metabolism in petunia petal tissue. Plant Physiol 135: 1993–2011.
52. Guterman I, Shalit M, Menda N, Piestun D, Dafny-Yelin M, et al. (2002) Rose scent: genomics approach to discovering novel floral fragrance-related genes. The Plant Cell 14: 2325–2338.
53. Spiller M, Berger RG, Debener T (2010) Genetic dissection of scent metabolic profiles in diploid rose populations. Theor Appl Genet 120(7): 1461–1471.
54. Xiang L, Zhao K, Chen L (2010) Molecular cloning and expression of Chimonanthus praecox farnesyl pyrophosphate synthase gene and its possible involvement in the biosynthesis of floral volatile sesquiterpenoids. Plant Physiol and Biochem 48(10): 845–850.
55. Ma L, Li HF, Peng CC, Chen ZF, Long ZF (2012) Cloning of SAMT gene cDNA from Chimonanthus praecox and its expression in Escherichia coli. Agri Sci Techn 13(1): 88–93.
56. Lee S, Chappell J (2008) Biochemical and genomic characterization of terpene synthese in Magnolia grandiflora. Plant Physiol 147: 1017–1033.
57. Phillips MA, Wildung MR, Williams DC, Hyatt DC, Croteau RB (2003) cDNA isolation, functional expression, and characterization of (+)-α-pinene synthase and (–)-α-pinene synthase from loblolly pine (Pinus taeda): stereocontrol in pinene biosynthesis. Arch Biochem Biophys 411: 267–276.
58. Jones CG, Keeling CI, Ghisalberti EL, Barbour EL, Plummer JA, et al. (2008) Isolation of cDNAs and functional characterization of two multi-product terpene synthase enzymes from sandalwood, Santalum album L. Arch Biochem Biophys. 477: 121–130.
59. Martin DM, Bohlmann J (2004) Identification of Vitis vinifera (–) α-terpineol synthase by in silico screening of full-length cDNA ESTs and functional characterization of recombinant terpene synthase. Phytochem 65: 1223–1229.

60. Lin C, Shen B, Xu Z, Köllner TG, Degenhardt J, et al. (2008) Characterization of the monoterpene synthase gene tps26, the ortholog of a gene induced by insect herbivory in maize. Plant Physiol 146: 940–951.
61. Fähnrich A, Krause K, Piechulla B (2011) Product Variability of the 'Cineole Cassette' Monoterpene Synthases of Related Nicotiana Species. Molecular Plant 4 (6): 965–984.
62. Pichersky E, Dudareva N (2007) Scent engineering: toward the goal of controlling how flowers smell. Trends Biotechnol 25: 105–110.
63. Pruitt KD, Tatusova T, Maglott DR (2007) NCBI reference sequences (RefSeq): a curated non-redundant sequence database of genomes, transcripts and proteins. Nucleic Acids Res 35: 61–65.
64. Ashburner M, Ball CA, Blake JA, Botstein D, Butler H, et al. (2000) Gene ontology: tool for the unification of biology. Nat Genet 25(1): 25–29.
65. Langmead B, Trapnell C, Pop M, Salzberg SL (2009) Ultrafast and memory-efficient alignment of short DNA sequences to the human genome. Genome Biol 10(3): R25.
66. Li B, Dewey CN (2011) RSEM: accurate transcript quantification from RNA-Seq data with or without a reference genome. BMC Bioinformatics 12: 323.
67. Robinson MD, McCarthy DG, Smyth GK (2010) edgeR: a Bioconductor package for differential expression analysis of digital gene expression data. Bioinformatics 26: 139–140.
68. Benjamini Y, Yekutieli D (2001) The control of the false discovery rate in multiple testing under dependency. Ann Stat 29: 1165–1188.
69. Livak KJ, Schmittgen TD (2001) Analysis of relative gene expression data using real-time quantitative PCR and the 2–ΔΔCt method. Methods 25: 402–408.

CHAPTER 6

Provenancing Flower Bulbs by Analytical Fingerprinting: *Convallaria Majalis*

Saskia M. van Ruth [1,2,†,*] **and Ries de Visser** [3,†]

[1]RIKILT Wageningen UR, P.O. Box 230, 6700 EV Wageningen, The Netherlands

[2]Food Quality and Design Group, Wageningen University, P.O. Box 17, 6700 AA Wageningen, The Netherlands

[3]IsoLife B.V., P.O. Box 349, 6700 AH Wageningen, The Netherlands

ABSTRACT

The origin of agricultural products is gaining in appreciation while often hard to determine for various reasons. Geographical origin may be resolved using a combination of chemical and physical analytical technologies. In the present case of Lily of the Valley (*Convallaria majalis*) rhizomes, we investigated an exploratory set of material from The Netherlands, three other European (EU) countries and China. We show that the geographical origin is correlated to patterns of stable isotope ratios (isotope fingerprints) and volatile organic carbon (VOC) compounds (chemical fingerprints). These fingerprints allowed clear distinction using exploratory and supervised statistics. Isotope ratio mass spectrometry of $^{12}C/^{13}C$, $^{14}N/^{15}N$ and $^{16}O/^{18}O$ isotopes separated materials from Europe and China successfully. The VOC patterns measured by Proton Transfer Reaction Mass Spectrometry (PTR-MS) allowed distinction of three groups: material from The Netherlands, the other EU countries and China. This knowledge is expected to help developing a systematic and efficient analytical tool for authenticating the origin of flower bulbs.

Keywords: authenticity; fingerprint; isotope ratio mass spectrometry (IRMS); Lily of the Valley; origin; PTR-MS; stable isotopes

1. INTRODUCTION

Globalization of markets gives rise to a growing need for analytical tools capable of identifying origins of local products in a reliable and efficient way. Various bio-molecular, chemical and physical technologies have been tried in attempts to link agricultural products to their site of production or origin, also called authentication [1]. Analytical techniques for authenticating local products have been described for a wide range of agricultural and other products, like beef [2], Trappist beers [3], and pharmaceuticals [4]. However, much research is still needed towards the development of a comprehensive system of authentication based on scientific analytical methods. Little knowledge is currently available on the major issue of how to select the appropriate (combination of) analytical techniques that will be the most likely approach towards successful authentication of a particular product. Lilies are commonly kept ornamental flowering plants that are used in holiday celebrations, weddings, and funerals, and in various floral arrangements. Two thirds of the worldwide flower bulb production area is located in The Netherlands. The land area for flower bulb production in The Netherlands is *ca.* twenty-four thousand hectares in The Netherlands, including *ca.* twelve thousand hectares of tulips, and five thousand hectares of lilies. The Dutch have been known for their flower bulb production and export over the last 500 years [5]. By 1636, the tulip bulb became the fourth leading export product of The Netherlands—after gin, herring and cheese. The price of tulips skyrocketed because of speculation in tulip futures among people who never saw the bulbs. Many men made and lost fortunes overnight.

The *Liliaceae*, or lily family, is composed of 280 to 300 genera made up of 4000 to 4600 different species. The numbers vary because botanists differ in how to classify this diversity based on flowering type, ovary position, and distribution. There are ornamental plants within the group (lilies, tulips, hyacinths, daffodils, and amaryllis); food plants (onions, garlic, asparagus, leeks, shallots, and chives); and a variety of toxic species in the family, some of which are quite deadly [6].

Another floral plant associated with lilies is *Convallaria majalis*, commonly known as Lily of the Valley. Although part of the family *Asparagaceae*, in earlier classification systems the species were often treated as belonging to the family *Liliaceae*. It is a poisonous woodland flowering plant native throughout the cool temperate Northern Hemisphere in Asia, Europe and in the southern Appalachian Mountains in the United States. *C. majalis* is a herbaceous perennial plant that forms extensive colonies by spreading underground stems. These are called rhizomes. In botany, the term 'bulb' designates underground plant stems surrounded by modified leaves called scales which store nutrients, while in horticulture and gardening the term refers to any bulbous plant organ or underground stem, be it a corm, tuber, rhizome or true bulb. The *C. majalis* stems grow to 15–30 cm tall. Flowering stems have two leaves and 5–15 flowers on the stem apex. The flowers are usually white tepals, shaped like small bells, and sweetly scented (Figure 1). Flowering is in late spring (April/May) in the Northern Hemisphere.

C. majalis contains potent cardenolide glycosides, which are often toxic, and specifically heart-arresting [7]. In fact most of the literature on *Convallaria* deals with its toxic components and their biogenetic synthesis [8].

By nature, agricultural products have a land-based, and therefore geographical origin. Historically, application habits were shaped by socio-cultural factors and available local natural resources [9]. Such links between agricultural produce and territory have disappeared over time by various means. However, the last ten years consumers have a renewed interest in agricultural products strongly identified with a place of origin. The EU has recognized and supported the potential of differentiating quality products on a regional basis [10]. The EU regulation allows the application of the following geographical indications to a food product: Protected Designation of Origin (PDO), Protected Geographical Indication (PGI) and Traditional Specialties Guaranteed (TSG). PDO is the term used to describe foodstuffs that are produced, processed and prepared in a given geographical area using recognized methods. Examples are Roquefort (France), Traditional Balsamic Vinegar of Modena (Italy), and Farmers cheese from Leiden (The Netherlands). The use of geographical indications allows producers to obtain market recognition and often a premium price, not only for food products but also for flowers and bulbs. False use of geographical indications by unauthorized parties is detrimental to consumers and legitimate producers. From this point of view, the development of new and increasingly sophisticated techniques for determining the geographical origin of agricultural products is highly desirable for consumers, agricultural farmers, retailers and authorities. It is an analytically challenging problem that receives much attention in Europe. Reports on analytical methods for determining the geographical origin of agricultural products have been increasing since the 1980s [11]. Food and feed received considerable attention, but to the authors knowledge no authentication approaches have been reported for the provenance of flower bulbs in the scientific literature. Fingerprinting techniques combined with chemometrics are state-of-the-art analytical techniques in product authentication. These fingerprinting techniques aim to find a specific pattern for the authentic product (*i.e.*, for each geographical origin), which might allow discrimination between different products (*i.e.*, those from different geographical origins).

Stable isotope composition (or 'fingerprint') has been investigated using isotope ratio mass spectrometry (IRMS) for its utility in authenticating agricultural products, including wine, beer and cheeses [12], but also for pharmaceuticals [13] as well as in forensic studies on drugs [14]. These isotopic 'fingerprints' are intrinsic characteristics and built-in in all organic compounds and therefore largely insensitive to adulteration. Other successful applications have been described where stable isotopes serve as internal standards for quantitative Liquid Chromatography-MS [15], as tracers in source-sink studies [16,17,18] and in identifying functional micro-organisms [18,19].

It is well recognized that direct rapid headspace techniques measured by mass spectrometry without chromatographic separation can effectively represent a 'fingerprint' of the sample being analyzed and can provide distinct chemical information in relation to product odor, flavor, shelf-life, geographic or genetic origin, processing, and presence of micro-organisms. In the last decade, several non-chromatographic instrumental approaches, such as electronic noses with different types of chemical sensors, headspace mass spectrometry, or real time monitoring using techniques, such as Atmospheric pressure chemical ionization mass spectrometry (APCI-MS), proton transfer reaction mass spectrometry (PTR-MS) or selected ion flow tube mass spectrometry (SIFT-MS) have been applied for the characterization of the volatile compounds of food [20,21].

The aim of this study is a first exploration of two types of analytical fingerprints for their capabilities of differentiating rhizome samples of 'Lily of the Valley', *Convallaria majalis*, of different provenance for which isotope ratio analysis by IRMS and volatile organic compound (VOC) analysis by PTR-MS were selected. Isotope ratio analysis was selected because of the known impact of geology and climatology and thus geography on isotope ratio fingerprints of plant material. PTR-MS was selected because of the known impact of environmental conditions on volatile metabolites in plants and the technique's sensitivity, rapidity and non-destructive nature. In order to distinguish *C. majalis* cultivated in The Netherlands, other European countries and in China stable isotope ratios $^{13}C/^{12}C$, $^{15}N/^{14}N$ and $^{18}O/^{16}O$ determined by IRMS were determined as well as the integral VOC profiles, which were examined in conjunction with advanced statistical methods.

2. MATERIALS AND METHODS

2.1. Sample Material and Study Design

A total of 34 *C. majalis* rhizomes samples were collected. The set of samples consisted of samples from The Netherlands (18), three neighboring EU countries (2 from Belgium, 6 from France, 4 from Germany) and from China (4). Representative samples of dry rhizome materials were ground to a powder. These rhizome samples were subjected to stable isotope analysis by IRMS and VOC analysis by PTR-MS. For comparison flowers of a *C. majalis* plant grown in The Netherlands was analyzed for their volatile composition by PTR-MS as well.

2.2. Stable Isotope Analysis by IRMS

The principles of natural abundance stable IRMS have been described in detail [22], as well as the application of IRMS for geographical origin authentication of foods [23]. This method employs the natural variation in isotope ratios of the chemical elements and is suited for accurate (<0.0002 atom%) analysis of the ratios of isotopes (IR) occurring -in this case- in organic matter, like carbon (C, $^{13}C/^{12}C$; 1.1% ^{13}C), nitrogen (N, $^{15}N/^{14}N$; 0.3% ^{15}N) and oxygen (O, $^{18}O/^{16}O$; 0.2% ^{18}O). In other studies, hydrogen $^{2}H/^{1}H$ and sulfur $^{34}S/^{33}S/^{32}S$ have been used [12,23].

Oven-dry samples (1–2 mg) were packed in tin (^{13}C, ^{15}N) or silver foil (^{18}O) for oxidation and analysis using a PDZ Europa ANCA-GSL elemental analyzer interfaced to a PDZ Europa 20-20 isotope ratio mass spectrometer (Sercon Ltd., Cheshire, UK; C and N). Freeze-drying increased the $^{18}O/^{16}O$ ratio by ~5 per mil. Oxygen isotope analysis was performed using an elementar PyroCube (Elementar Analysensysteme GmbH, Hanau, Germany) interfaced to a PDZ Europa 20-20 isotope ratio mass spectrometer (Sercon Ltd., Cheshire, UK). The long-term standard deviation is 0.2 per mil for ^{13}C (0.0002 atom%) and 0.3 per mil for ^{15}N (0.0003 atom%).

In accordance with international agreements delta-values (δ, in per mil; ‰) provide the deviation relative to the ratio of isotopes of international standards, V-PDB (Vienna Pee Dee Belemnite), V-SMOW (Vienna Standard Mean Ocean Water) and Air for carbon, oxygen and nitrogen, respectively. For information on delta notation and the international references, see e.g., Sharp 2006 [22]. Single analyses were carried out on the 34 samples.

2.3. Volatile Organic Compound Analysis by PTR-MS

The VOC fingerprints were measured by PTR-MS. For this type of analysis the sample headspace is continuously introduced into a drift tube, where it is mixed with H_3O^+ ions formed in a hollow cathode ion source. VOCs that have proton affinities higher than water (>166.58 kcal/mol) are ionized by proton transfer from H_3O^+, mass analyzed in a quadrupole mass spectrometer and eventually detected as ion counts/s (cps) by a secondary electron multiplier. By using H_3O^+ as the proton source, the ionization of most of the common inorganic constituents of air (N_2, O_2 or CO_2) is avoided since they have proton affinities lower than H_2O. Furthermore, this soft ionization avoids excessive fragmentation of ions, which makes multicomponent analyte mass spectra simpler and easier to interpret [24].

A total of 2 g of sample (dry powder of rhizomes, or flowering stems) were placed in a 250 mL screw cap glass vial (Figure 1). Samples were equilibrated at 35 °C for at least 30 min in a water bath, in order to assure equilibrium of the VOC between sample and headspace. No further sample preparation was required. Measurements were performed using a commercial High Sensitivity PTR-MS system (Ionicon GmbH, Innsbruck, Austria). The headspace of the samples was delivered directly to the inlet of the PTR-MS system with a flow rate of 50 mL/min. The temperature of the inlet and drift chamber were both maintained at 60 °C to prevent loss of volatiles along the sampling inlet line for on-line analysis. A blank was measured before each sample. Measurements were carried out in the mass full-scan mode and the mass spectra were collected in the range of 20–160 atomic mass units (amu). A dwell time of 0.2 s/mass unit was used, resulting in a cycle time just under 30 s. Sample analyses were carried out in triplicate. For each replicate, a full mass scan was recorded. The data were background and transmission corrected, yielding one corrected mass spectrum per replicate. Then, the three mass spectra of the three replicates of each sample were averaged to obtain a mean mass spectrum per sample. In this manner, a dataset containing mean mass spectra per sample analyzed was compiled for the 34 samples.

Figure 1. Photograph of the sample flask and example mass spectrum of the flowers of *C. majalis*.

2.4. Statistical Analysis

Multi-factor analysis of variance (MANOVA, factors provenance and samples) and Fisher's least significant difference (LSD) tests were carried out to determine significant differences among groups for isotope ratio measurements using XLSTAT 2 March 2014 (Addinsoft, Paris, France). MANOVA can assess two or more independent variables for significance of effects on two or more metric dependents. It allows a joint analysis of each dependent rather than performing several univariate tests, thus avoiding multiple testing risks.

For further multivariate modeling and classification, Pirouette 4.0 (Infometrix, Seattle, WA, USA) was used. Principal Component Analysis (PCA) was performed on the PTR-MS data of the 34 samples to screen the multivariate data for outliers and to explore the presence of any natural clustering in the data. PCA performs a reduction in the data dimensionality in order to facilitate the visualization of the multivariate data retaining as much as possible the information present in the original data. Then, Partial Least Squares-Discriminant Analysis (PLS-DA) was used to develop a classification model for samples from The Netherlands *versus* other origins. PLS-DA is a supervised classification technique that is often used for high dimensional data, especially when the amount of variables greatly exceeds the number of samples. It performs a variable reduction on the data set by calculating new variables (called latent components or factors) combining the variables in the data set, in order to find the maximum correlation between them and the class variable, and thus, the maximum separation among two classes (The Netherlands *vs.* other provenance). Then, linear discriminant analysis is applied on the reduced variable set (the latent components) to provide the final classification model. Since data pre-processing can have a profound effect on the model results, several ways of data pre-processing were evaluated: none (raw data), auto-scaling (scaling to unit variance), mean centering, and log transformation. The optimal PLS-DA model was then selected and its performance examined by leave-one-out cross validation because of the limited size of the exploratory sample set.

3. RESULTS AND DISCUSSION

3.1. Stable Isotope Analysis by IRMS

The nitrogen, carbon, and oxygen isotope ratios (delta values) of the 34 samples were determined by IRMS. The isotope ratios were compared for the various provenances (Table 1). Overall, δ-^{15}N varied by 11 per mil, δ-^{13}C varied by 7 per mil, and δ-^{18}O by 12 per mil (sample extremes). MANOVA indicated significant differences in δ-^{15}N values between the *C. majalis* from China and their European counterparts (France, Germany, The Netherlands) whereas the Belgian products revealed intermediate values. For the δ-^{13}C values mostly overlapping ranges were observed. The δ-^{18}O values showed relatively high values for the rhizomes originating from The Netherlands and low values for those from China, with the other European countries showing intermediate values.

Naturally occurring stable isotope ratios of organic elements like N, C and O measured in this study have characteristic values in relation to their geographical origin. This is due to systematic effects of climate factors like temperature, humidity, precipitation, and geographical factors such as distance to the sea, plant physiological processes, plant genotype, and soil or substrate factors such as fertilizer type. The three isotope ratios that are dealt with here differ in their relation to the main environmental factors related to plant growth. $^{13}C/^{12}C$ is related plant water use efficiency, via water conditions, including water supply and relative humidity interacting with plant physiological characteristics [25]. $^{15}N/^{14}N$ varies with the chemical type and origin of the mineral N fertilizer [26], and increases with trophic level in N of organisms [27]. The $^{18}O/^{16}O$ isotope ratio of atmospheric water is directly related to temperature and inversely related to the dominant down-wind distance from the sea through the mechanism of preferential precipitation of the heavier water molecules (*i.e.*, $H_2{}^{18}O$; [28]), while this oxygen isotope ratio in leaf water (and thus, plant biomass) is affected by local climatic conditions affecting plant transpiration, *i.e.*, mainly temperature and humidity [29].

Table 1. Nitrogen, carbon and oxygen isotope ratio's (δ values, per mil) for *C. majalis* of various provenance [a]; means ± SE.

Provenance	δ-^{15}N	δ-^{13}C	δ-^{18}O
The Netherlands	1.5 ± 2.4 ^{x}	−26.6 ± 2.6 ^{y}	36.0 ± 3.2 ^{x}
Belgium	−0.9 ± 0.1 ^{xy}	−25.5 ± 3.4 ^{xy}	35.5 ± 2.7 ^{xy}
France	0.4 ± 1.2 ^{x}	−24.2 ± 0.7 ^{x}	34.7 ± 3.1 ^{xy}
Germany	1.8 ± 0.2 ^{x}	−24.5 ± 1.3 ^{xy}	33.0 ± 1.4 ^{xy}
China	−2.8 ± 2.2 ^{y}	−26.4 ± 2.0 ^{xy}	30.4 ± 4.0 ^{y}

[a] Different superscripts (x, y, xy) in a column indicate significant differences (Multi-factor analysis of variance (MANOVA) and Fisher's least significant difference (LSD) test, $p < 0.05$).

Since the individual isotopes would not allow full discrimination between *C. majalis* from The Netherlands, other European countries, and those of Chinese origin, multiple isotopes were compared for their discriminatory properties. The data of the two most discriminating isotopes, ^{15}N and ^{18}O were combined in a 2D plot (Figure 2). The two isotopes display a clear distinction between the rhizomes from Europe and China (red-colored symbols *versus* others). In the study sample numbers are low and do not represent all variation in real-life, but the results are promising though. Different nitrogen sources for fertilization and water sources are likely to have contributed to the consistencies observed.

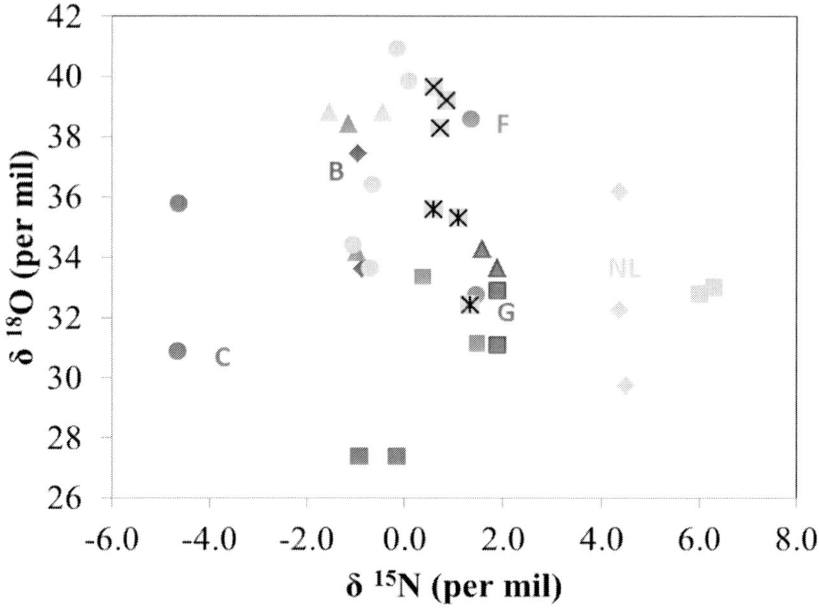

Figure 2. 2D-plot of N and O isotopic data of *C. majalis* rhizomes originating from five countries (Belgium (B, blue), China (C, red), France (F, green), Germany (G, purple) and The Netherlands (NL, yellow)), *i.e.*, stable isotope ratios $^{18}O/^{16}O$ ($\delta\ ^{18}O$) and $^{15}N/^{14}N$ ($\delta\ ^{15}N$). Different colors indicate different countries, symbol shapes indicate different origins within countries; identical symbols indicate replicate provenance samples; least significant difference (LSD) ($p < 0.05$) values: 0.7 ($\delta\ ^{15}N$), 4.6 ($\delta\ ^{18}O$).

3.2. Volatile Organic Compound Analysis by PTR-MS

VOCs in the headspace of the *C. majalis* rhizomes and flower samples were analyzed by PTR-MS without prior chromatographic separation and resulted in a spectrum of the VOCs (their mass-to-charge ratios and their intensities). The mean mass spectra of the samples from The Netherlands are presented in Figure 3 for both rhizomes and flowers. Some similarities in groupings of more predominant ions for both types of samples are observed in the lower molecular weight range. On the other hand, the flowers present clearly more higher molecular weight volatile compounds than the rhizomes. In decreasing order, the most abundant mass-to-charge ratios found in the flowers were m/z 37 (water cluster), m/z 45, m/z 33, m/z 81, m/z 137, m/z 91 and m/z 83.

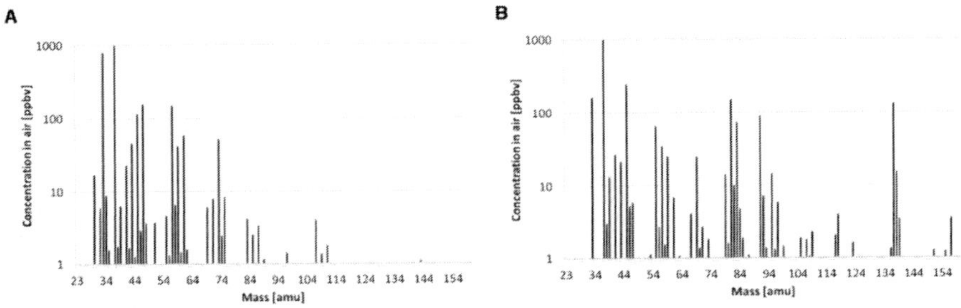

Figure 3. Mean PTR-MS (proton transfer reaction mass spectrometry) spectrum of *C. majalis* cultivated in The Netherlands: rhizomes (**A**) and flowers (**B**).

Flowers contain a large variety of volatile organic compounds, just as there are a multitude of colors and forms of flowers. Those compounds with a sufficiently high vapor pressure may have an odor, which is approximately for compounds with a low polarity and up to a molecular weight of around 300 [30]. The odor of Lily of the Valley plays an important role in perfumery. However, very few studies have identified the responsible volatile compounds of the plant. Brunke and co-workers [31] reported the following predominant volatile compounds in the extract of *C. majalis* analyzed: benzyl alcohol (35%), *cis*-3-hexen-1-ol (11%), citronellol (10%), geraniol (8%) and *cis*-3-hexenyl acetate (8%). Benzyl alcohol, citronellol, geraniol would carry floral-rosy-citrus notes, whereas the *cis*-3-hexen-1-ol and *cis*-3-hexenyl acetate would be responsible for green-grassy odor notes. They are all compounds with molecular weights of 100 and over, and therefore expected to appear (if present) in the VOC mass spectrum in the mass range of 100–160 amu. The response for mass peak 157 in the flower analysis is likely to be citronellol (mw = 156, ionized response expected at 157 amu). Takahiro *et al.* [32] identified 148 compounds. They investigated also the enantiomeric ratio of the chiral compounds in *C. majalis* using multidimensional GC-MS and determined predominantly the (S)-form of citronel, citronellyl acetate, citronellal, and dihydrofarnesol. Due to its pleasant odour various attempts have been made to synthesize Lily of the Valley-like fragrances. Lilial® (3-(4-*t*-butylphenyl)-2-methylpropanal), Lyrial®, and hydroxycitronellal are the commercially most important compounds of this class of odorants and they show apparent molecular similarities. Dupical® and Mugetanol®are two more recently developed Lily of the Valley odorants. Lillial® is a powerful, fresh, floral note reminiscent of lily of the valley, linden blossom, and cyclamen. Besides its use as a fragrance and fragrance intermediate it is used as an intermediate for the production of fenpropimorph, a biodegradable fungicide. It is produced on a kiloton scale by a multistep synthesis [28,33].

PTR-MS is not primarily used to identify the volatile compounds as isobaric compounds might yield the same m/z signal. Fragmentation in PTR-MS is mostly reduced due to the soft chemical ionization of the compounds; thus, the molecular structure of most volatile compounds is preserved. This allows the spectra to be used as fingerprints. Geographical provenancing using a univariate approach is not always a robust strategy. For instance, even if we found significant differences in the content of a particular volatile among the origins of the rhizomes, basing the verification of them on only this volatile could on the one hand easily lead to incorrect assignment to a particular origin and on the other hand may be the recipe to fraud. Therefore, a multivariate approach was selected for the VOC data.

The VOC fingerprints were examined using chemometric analysis, which is particularly suitable for handling large data sets. PCA was conducted on the VOC data of the 34 samples. The data matrix consisted of 34 rows (samples) and 138 variables (ions). The first three dimensions of the PCA on the normalized and auto-scaled data are presented in Figure 4.

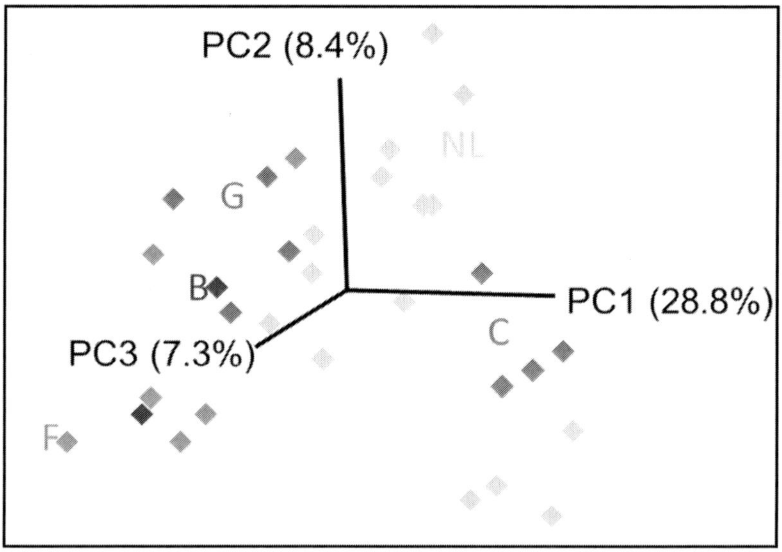

Figure 4. PCA scores plot of the PTR-MS (proton transfer reaction mass spectrometry) data (normalized and auto-scaled) of *C. majalis* from the five countries (Belgium (B, blue), China (C, red), France (F, green), Germany (G, purple) and The Netherlands (NL, yellow)).

Some natural clustering is observed: with groupings of the samples from The Netherlands, the other European samples, and the samples of Chinese origin. Subsequently the data for these three provenance groups were subjected to PLS-DA in order to optimize the classification of the samples by their VOC profiles. The data matrix was the same as for the PCA. A model was built to classify the samples from The Netherlands, from other European countries and from China using the normalized, auto-scaled data. The model's performance was evaluated by a leave-one-out cross validation and indicated 100% accuracy for the prediction of the origin for the samples from The Netherlands, other European countries and China (Table 2). In order to evaluate the masses contributing to the classification, a correlation spectrum for the masses and the model classifying samples from The Netherlands and the other origins was generated (Figure 5). It shows that many masses are positively or negatively correlated with the origin of the *C. majalis* samples origin.

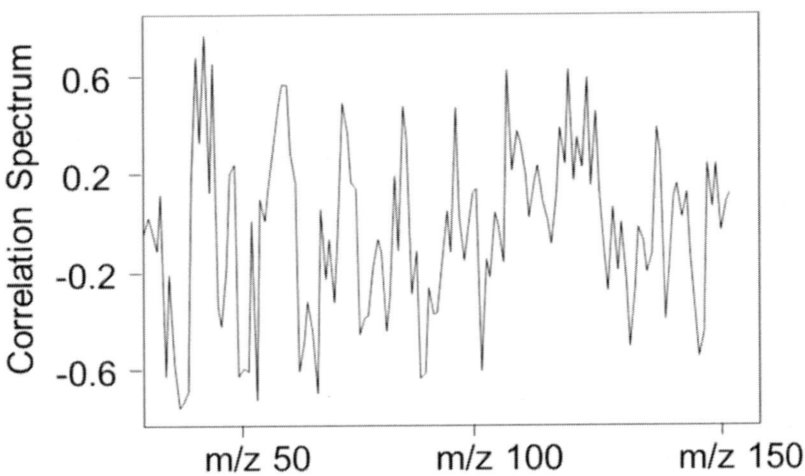

Figure 5. Correlation spectrum associated with the PLS-DA (Partial Least Squares-Discriminant Analysis) classification of *C. majalis* from The Netherlands and other countries presenting the relevance of particular masses analyzed by PTR-MS for the classification.

Table 2. PLS-DA (Partial Least Squares-Discriminant Analysis) leave-one-out cross validation results: classification of *C. majalis* from The Netherlands, other EU countries and China by their normalized and auto-scaled PTR-MS (proton transfer reaction mass spectrometry) spectral data.

Provenance	Prediction		
	The Netherlands	**Other EU Countries** [a]	**China**
The Netherlands	100%	0%	0%
Other EU countries	0%	100%	0%
China	0%	0%	100%

[a] *C. majalis* from Belgium, France, and Germany.

4. CONCLUSIONS

Consistent differences in composition between Lily of the Valley rhizomes of different provenances by both their nitrogen and oxygen isotope ratios and their VOC fingerprints were observed in the current set of samples. The isotope ratios allowed discrimination between samples cultivated in Europe and China. The VOC fingerprints differentiated between the samples from The Netherlands, from the other European countries and the samples from China. PTR-MS is a rapid, non-destructive technique that may be suitable as a first on site, screening technique, while the more time-consuming isotope ratio analysis could be used for further confirmation. Since origin may add value to agricultural produce, produce from particular origins may be susceptible to mix-up and substitution. In order to assure genuine trade, purchaser confidence, and therefore future production this type of analytical techniques may have a positive effect in addition to administrative controls. This study is a first step in that direction but the sample set would need to be further extended to build a reliable database for practical use.

ACKNOWLEDGMENTS

The authors are grateful to Naktuinbouw for providing the Lily-of-the-Valley sample material, to Alex Koot for technical help with the PTR-MS analyses, and to Sara N. de Visser for graphics support. Furthermore, the authors would like to acknowledge strategic Wageningen University and

Research Centre funding from the Dutch Ministry of Economic Affairs in the KB-project Authenticity-traceability, and the EU Operationeel Programma Oost-Nederland EFRO 2007-2013 (GO).

AUTHOR CONTRIBUTIONS

Ries de Visser contributed primarily to the generation and interpretation of the IRMS data and Saskia van Ruth was mostly involved in the PTR-MS part of the study as well as the statistical calculations. The current manuscript was drafted together.

CONFLICTS OF INTEREST

The authors declare no conflict of interest.

REFERENCES

1. Olsen, P.; Borit, M. How to define traceability. Trends Food Sci. Technol. 2012, 29, 142–150.
2. Monahana, F.J.; Moloneyb, A.P.; Osorioa, M.T.; Röhrlea, F.T.; Schmidta, O.; Brennanc, L. Authentication of grass-fed beef using bovine muscle, hair or urine. Trends Food Sci. Technol. 2012, 28, 69–76.
3. Mattarucchi, E.; Stocchero, M.; Moreno-Rojas, J.M.; Giordano, G.; Reniero, F.; Guillou, C. Authentication of Trappist beers by LC-MS fingerprints and multivariate data analysis. J. Agric. Food Chem. 2010, 58, 12089–12095.
4. Berkowitz, S.A.; Engen, J.R.; Mazzeo, J.R.; Jones, G.B. Analytical tools for characterizing biopharmaceuticals and the implications for biosimilars. Nat. Rev. Drug Discov. 2012, 11, 527–540.
5. Schama, S. The Embarrassment of Riches: An Interpretation of Dutch Culture in the Golden Age; Alfred A. Knopf: New York, NY, USA, 1987; p. 698.
6. Burrows, G.E.; Tyrl, R.I. The Liliaceae. In Toxic Plants of North America; Burrows, G.E., Tyrl, R.I., Eds.; Iowa State Press: Ames, IA, USA, 2001; pp. 271–295.
7. Fitzgerald, K.T. Lily toxicity in the cat. Top. Companion Anim. Med. 2010, 25, 213–217.
8. Ogorodnikova, A.V.; Latypova, L.R.; Mukhitova, F.K.; Mukhtarova, L.S.; Grechkin, A.N. Detection of divinyl ether synthase in Lily-of-the-Valley (Convallaria majalis) roots. Phytochemistry 2008, 69, 2793–2798.
9. Delamont, S. Appetites and Identities: An Introduction to the Social Anthropology of Western Europe; Routledge: London, UK, 1994; p. 272.
10. Dimara, E.; Skuras, D. Consumer evaluations of product certification, geographic association and traceability in Greece. Eur. J. Market. 2003, 37, 690–705.
11. Luykx, D.M.A.M.; van Ruth, S.M. A review of analytical methods for determining the geographical origin of food products. Food Chem. 2008, 107, 897–911.

12. Kelly, S.; Heaton, K.; Hoogewerff, J. Review: Tracing the geographical origin of food: The application of multi-element and multi-isotope analysis. Trends Food Sci. Technol. 2005, 16, 555–567.
13. Jasper, J.P.; Westenberger, B.J.; Spencer, J.A.; Buhse, L.F.; Nasr, M. Stable isotopic characterization of active pharmaceutical ingredients. J. Pharm. Biomed. Anal. 2004, 35, 21–30.
14. Aguilera, R.; Hatton, C.K.; Catlin, D.H. Detection of epitestosterone doping by isotope ratio mass spectrometry. Clin. Chem. 2002, 48, 629–636.
15. Nakabayashi, R.; Sawada, Y.; Yamada, Y.; Suzuki, M.; Hirai, M.Y.; Sakurai, T.; Saito, K. Combination of liquid chromatography-Fourier transform ion cyclotron resonance-mass spectrometry with 13C-labeling for chemical assignment of sulfur-containing metabolites in onion bulbs. Anal. Chem. 2013, 85, 1310–1315.
16. De Visser, R.; Vianden, H.; Schnyder, H. Kinetics and relative significance of remobilized and current C and N incorporation in leaf and root growth zones of Lolium perenne after defoliation: Assessment by 13C and 15N steady-state labelling. Plant Cell Environ. 1997, 20, 37–46.
17. Frost, G.; Sleeth, M.L.; Sahuri-Arisoylu, M.; Lizarbe, B.; Cerdan, S.; Brody, L.; Anastasovska, J.; Ghourab, S.; Hankir, M.; Zhang, S.; et al. The short-chain fatty acid acetate reduces appetite via a central homeostatic mechanism. Nat. Commun. 2014, 29.
18. Tannock, G.W.; Lawley, B.; Munro, K.; Sims, I.M.; Lee, J.; Butts, C.A.; Roy, N. RNA-stable isotope probing (RNA-SIP) shows carbon utilization from inulin by specific bacterial populations in the large bowel of rats. Appl. Environ. Microbiol. 2014, 80.
19. Schmidt, O.; Horn, M.A.; Kolb, S.; Drake, H.L. Temperature impacts differentially on the methanogenic food web of cellulose-supplemented peatland soil. Environ. Microbiol. 2014, 16.
20. Reineccius, G.; Peterson, D. Principles of food flavor analysis. In Instrumental Assessment of Food Sensory Quality; Kilkast, D., Ed.; Woodhead Publishing Ltd.: Cambridge, UK, 2013; pp. 53–102.
21. Heenan, S.; van Ruth, S.M. Emerging flavour analysis methods in food authentication. In Instrumental Assessment of Food Sensory Quality; Kilkast, D., Ed.; Woodhead Publishing Ltd.: Cambridge, UK, 2013; pp. 284–312.
22. Sharp, Z.D. Principles of Stable Isotope Geochemistry; Prentice Hall: Upper Saddle River, NJ, USA, 2006; p. 344.
23. Drivelos, S.A.; Georgiou, C.A. Multi-element and multi-isotope-ratio analysis to determine the geographical origin of foods in the European Union. Trends Anal. Chem. 2012, 40, 38–51.
24. Hansel, A.; Jordan, A.; Holzinger, R.; Prazeller, P.; Vogel, W.; Lindinger, W. Proton transfer reaction mass spectrometry: On-line trace gas analysis at the PPB level. Int. J. Mass Spectrom. Ion Process. 1995, 149, 609–619.
25. Seibt, U.; Rajabi, A.; Griffiths, H.; Berry, J.A. Carbon isotopes and water use efficiency: Sense and sensitivity. Oecologia2008, 155, 441–454.
26. Flores, P.; Fenoll, J.; Hellín, P. The feasibility of using δ 15N and δ 13C values for discriminating between conventionally and organically fertilized pepper (Capsicum annuum L.). J. Agric. Food Chem. 2007, 55, 5740–5745.
27. De Visser, S.N.; Freymann, B.P.; Schnyder, H. Trophic interactions among invertebrates in Termitaria in the African savanna: A stable isotope approach. Ecol. Entomol. 2008, 33, 758–764.
28. Miller, D.L.; Mora, C.I.; Grissino-Mayer, H.D.; Mock, C.J.; Uhle, M.E.; Sharp, Z. Tree-ring isotope records of tropical cyclone activity. Proc. Natl. Acad. Sci. USA 2006, 103, 14294–14297.
29. Helliker, B.R. On the controls of leaf-water oxygen isotope ratios in the atmospheric crassulacean acid metabolism epiphyte Tillandsia usneoides. Plant Physiol. 2011, 155, 2096–2107.
30. Fráter, G.; Bajgrowicz, J.A.; Kraft, P. Fragrance chemistry. Tetrahedron 1998, 54, 7633–7703.

31. Takahiro, I.; Tsuyoshi, K.; Kazuhiko, T.; Fumio, M. Volatile components of Lily of the Valley (Convallaria majalis)-CHIRAROMA analysis. Koryo, Terupen oyobi Seiyu Kagaku ni kansuru Toronkai Koen Yoshishu 2006, 50, 79–81.
32. Brunke, E.J.; Ritter, F.; Schmaus, G. New data on trace components with sensory relevance in flower scents. Dragoco Rep. 1996, 1, 5–21.
33. Forsyth, S.A.; Gunaratne, H.Q.N.; Hardacre, C.; McKeown, A.; Rooney, D.W.; Seddon, K.R. Utilisation of ionic liquid solvents for the synthesis of Lily-of-the-Valley fragrance {β-Lilial®; 3-(4-t-butylphenyl)-2-methylpropanal}. J. Mol. Cat. A Chem. 2005, 231, 61–66.

CHAPTER 7

Efficient And Heritable Transformation of *Phalaenopsis* Orchids

Hong-Xian Hsing[1,2]†, Yi-Jyun Lin[1,2†,] Chii-Gong Tong[1,2], Min-Jeng Li[1,2], Yun-Jin Chen[1,2] and Swee-Suak Ko1[,2*]

[1] *Academia Sinica Biotechnology Center in Southern Taiwan, Tainan 741, Taiwan.*

[2] *Agricultural Biotechnology Research Center, Academia Sinica, Taipei 115, Taiwan.*

ABSTRACT

Background

Phalaenopsis orchid (*Phal.* orchid) is visually attractive and it is important economic floriculture species. *Phal.* orchids have many unique biological features. However, investigation of these features and validation on their biological functions are limited due to the lack of an efficient transformation method.

Results

We developed a heritable and efficient *Agrobacterium*- mediated transformation using protocorms derived from tetraploid or diploid *Phal.* orchids. A T-DNA vector construct containing eGFP driven by ubiquitin promoter was subjected to transformation. An approximate 1.2–5.2 % transformation rate was achieved. Genomic PCR confirmed that hygromycin selection marker, *HptII* gene and target gene *eGFP* were integrated into the orchid genome. Southern blotting indicated a low T-DNA insertion number in the orchid genome of the transformants. Western blot confirmed the expression of eGFP protein in the transgenic orchids. Furthermore, the GFP signal was detected in the transgenic orchids under microscopy. After backcrossing the pollinia of the transgenic plants to four different *Phal.* orchid varieties, the BC1 progenies showed hygromycin resistance and all surviving BC1 seedlings were *HptII* positive in PCR and expressed GFP protein as shown by western blot.

Conclusions

This study demonstrated a stable transformation system was generated for *Phal.* orchids. This useful transformation protocol enables functional genomics studies and molecular breeding.

Keywords: *Phalaenopsis aphrodite Agrobacterium tumefaciens* Transformation Protocorm eGFP Hygromycin selection

BACKGROUND

Phalaenopsis orchids, commonly known as moth orchid, is one of the most popular and profitable ornamental plants in the global floral market. *Phal.* orchids have very unique biological features such as crassulacean acid metabolism (CAM) photosynthesis that uptake CO_2 at night (Guo and Lee 2006), epiphytic habitat with high water and nutrient usage efficiency, unique flower pattern formation (Su et al. 2013b), symbiosis with mycorrhizae (Yumiko et al. 2007), immature embryos (seeds without endosperm) in mature capsules (Yu and Goh 2001). To elucidate these interesting and novel biological features, validation of gene function using stable transformation approach must be helpful. Moreover, the recent technological advances in next generation sequencing (NGS) combining bioinformatics have generated highly abundant sequence information. Several orchid databases based on the transcriptomic information are constructed and available in the web, such as Orchidstra (Su et al. 2013a) and OrchidBase (Fu et al. 2011). In the era of big data and functional genomics, efficient transformation of *Phal.* orchid can contribute to the new discoveries with strong impact.

Over the years consumers have sought ever more attractive and versatile morphological traits in flowers, such as colors, shapes, fragrance. Although conventional breeding has been fruitful in last decades, to meet the demands of the market for novel floral colors or fragrances, control of flowering time, resistant to diseases and environmental stresses and so on, transgenic technology is needed. Introgression of foreign genes from distant species may create more dramatic outcomes in transgenic lines than conventional breeding.

According to previous studies, the most reliable genetic transformation platforms successfully applied to *Phal.* orchids are the *Agrobacterium*-mediated approach and particle bombardment (Chai et al. 2002; Mii and Chin 2010; Teixeira da Silva 2011). Target explants for transformation using protocorms (Chin et al. 2007; Mishiba et al. 2005; Semiarti et al. 2007, 2010), protocorm-like bodies (PLB) (Anzai et al. 1996; Anzai and Tanaka 2001; Chai et al. 2002; Chan and Lin 2005; Chen et al. 2011; Hsieh et al. 1997; Julkifle et al. 2010; Liao et al. 2004; Subramaniam et al. 2009; Subramaniam and Xavier 2010), callus (Belarmino and Mii 2000; Sjahril and Mii 2006); and pollen tube pathways (Hsieh and Huang 1995; Tsay et al. 2012) have been reported. So far, the two most-often used host orchid tissues, protocorms and PLBs have been genetically transformed equally well. PLBs are presumed to be genetically uniform and can be induced efficiently from various somatic tissues including flower stalks, young leaves and stem segments (Chen et al. 2010; Yee et al. 2008). However, protocorm transformations have been proposed to be simpler than subculture of PLBs (Mishiba et al. 2005; Semiarti et al. 2007).

In the present study, we developed an alternative transformation procedure using protocorms generated by germination of seeds derived from diploid *Phal. aphrodite* and tetraploid *Phal.* cultivars. A T-DNA vector construct containing eGFP driven by ubiquitin promoter was used for transformation. The transformed protocorms were selected by hygromycin and subsequently regenerated successfully. We backcrossed the transgenic line and the BC1 progenies showed hygromycin resistant and the transgene is heritable. Finally, molecular analyses using PCR and western blot showed that all surviving backcross F1 explants were positive transformants.

METHODS

Plant materials and growth condition

Three different *Phalaenopsis* orchid varieties were used for *Agrobacterium*-mediated transformation in this study. A popular big white flower tetraploid orchid of Dpts. Join Angel 'TH274-1' was bought from Join Orchids. It was self-pollinated and single seed descendant line (denoted L3) was used in our transformation. A commercial tetraploid orchid of *Phal. aphrodite* cv. M1663 (denoted Ama 4n) was bought from Chain Port Orchid Co. Ltd. A native Ama diploid of *Phal. aphrodite* subsp. *formosana* (denoted

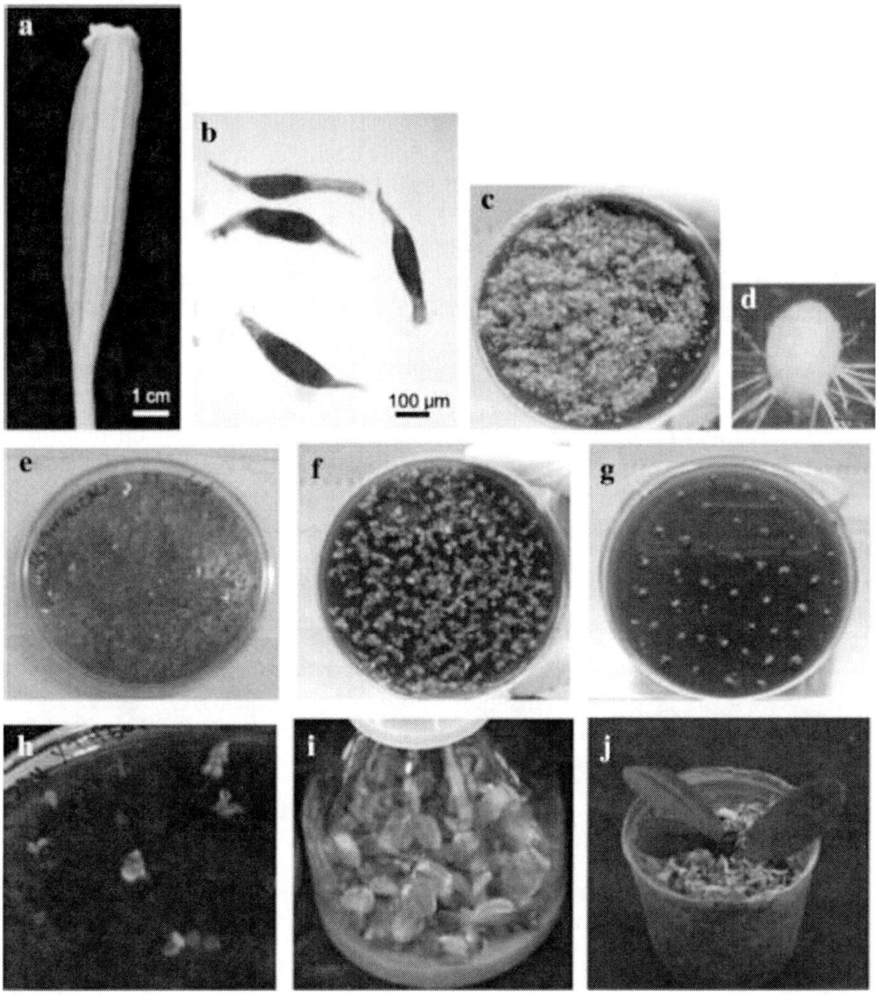

Figure 1. Procedures for *Agrobaterium*-mediated transformation of protocorms in *Phalaenopsis* orchid. **A** Mature orchid capsule at 4 months after pollination. **B** Stage of mature seeds ready for sowing. **C** Growth of protocorms sown on 1/2 MS agar plates for 1 month. **D** Healthy protocorm ready for infection. **E** *Agrobacteria* infection. **F** The 1st round selection for hygromycin resistance on T2 media containing 25 ppm hygromycin and 40 ppm meropenem (T2 + MH). **G** The 2nd round selection for hygromycin resistance on T2 + MH. **H** Growth of explants on 1/2 MS plate. **I** Growth of transgenic lines in jar. **J** Transplanted transgenic orchid to 1.7″ pot containing sphagnum moss

Ama 2n) was bought from the market of Shanhua. Plant-type and flower morphology of the three *Phalaenopsis* orchids is shown in the Additional file 1: Figure S1. *Phal.* orchid L3 with large white flowers more than 10 cm in diameter; whilst Ama (4n) and Ama (2n) have medium size white flowers around 7 cm in diameter. When blooming, flowers were self-pollinated to produce capsules. It took about 4–5 months for capsules to mature. Capsules were harvested when they turned slightly yellowish green at the tip of the capsule (Figure 1a). Mature capsules could be stored at room temperature or 4 °C for 2 weeks and seeds still maintained good viability. Orchid seeds were sown on a 1/2 MS (Duchefa Bi ochemia) agar plate containing 0.1 % activated charcoal and placed in a tissue culture room at constant temperature of 24 °C under a 12 h photoperiod.

Plasmid vectors and bacterial strain

A green fluorescence marker gene eGFP driven by ubiquitin promoter was constructed into pZP200 backbone (http://www.biotech.unl.edu/pzp200) containing the selection marker *HptII* gene that encodes *hygromycin phosphotransferase,* was driven by cauliflower mosaic virus (CaMV) 35S promoter (Figure 2a). The vector, pCAMBIA1305.1, which contained a β-glucuronidase (*GUS*) was driven by CaMV 35S promoter (http://www.cambia.org/daisy/cambia/585.html). The plasmid was separately subjected to transformation and subsequent hygromycin selections. *Agrobacterium tumefaciens* strain EHA105 was used for transfection in this study.

Inoculation, co-cultivation, and selection

Agrobacterium strain EHA105 containing eGFP target gene was spread on lysogeny broth (LB) plate containing 50 ppm kanamycin and 100 ppm rifampicin, and cultured overnight. Single colonies were selected and then cultured in 5 mL LB solution containing 50 ppm kanamycin and 100 ppm rifampicin, shaking at 250 rpm in a 28 °C incubator for 2 days. Agrobacteria were transferred to 20 mL LB containing 50 ppm kanamycin and 100 ppm rifampicin, and incubated at 28 °C to reach an optical density (OD) of a suspension of cells at 600 nm of around 1.0. Inoculum was centrifuged at 3500 rpm for 15 min, then the pellet was re-suspended with 1/2 MS containing 5 % sucrose, adjusted to pH 5.7, then 100 μM acetosyringone (AS) was added to the inoculum.

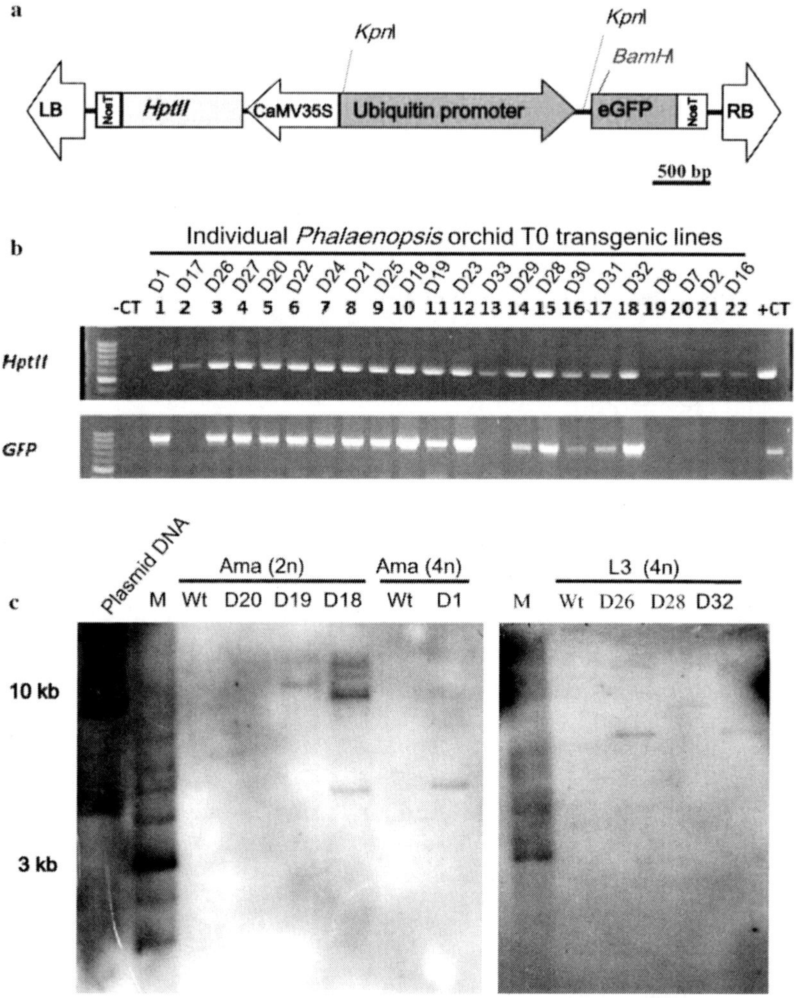

Figure 2. Confirmation of *Phalaenopsis* transgenic lines. **A** Construction map of Ubi:eGFP used in this study. Schematic structure of the T-DNA region in the vector of pZp200. CaMV 35S, promoter of CaMV 35S; *HptII*, coding region of *hygromycin phosphotransferase* gene; eGFP, coding region of *eGFP*; NosT, terminator of *nopaline synthase* gene; *LB* left border; *RB* right border. **B** Genomic PCR analysis of *eGFP* and *HptII* genes. −CT, negative control without addition of the genomic DNA template; +CT, with added plasmid DNA. **C** Southern blot analysis of *HptII* showing integration of T-DNA in *Phalaenopsis* orchid genome. Transgenic lines and their respective Wt plants were analyzed side by side

Healthy *Phal.* protocorm of size ~1.2 mm were selected, placed on a petri dish, and immersed in the *Agrobacterium* inoculum for 25 min. *Agrobacterium* inoculum was removed and the infected protocorms were spread on New Dogashima medium (NDM) (Belarmino and Mii 2000) containing 10 g/l maltose, 0.1 ppm naphthaleneacetic acid, 0.4 ppm benzyladenine (BA), 100 μM acetosyringone (AS), and 0.29 % phytagel. Infected protocorms were co-cultivated in the dark for 2–3 days until growth of *A. tumefaciens* was observed. Protocorms were washed with distilled water containing 40 ppm meropenem (China Chemical & Pharmaceutical) 3–4 times, and afterwards the protocorms were spread on T2 plates (Chen et al. 2009) containing 25 ppm hygromycin and 40 ppm meropenem (T2 + MH). Antibiotic selection of putative transgenic explants was conducted on T2 + MH three times, at 2-week intervals.

Transformation rate

To test the amenability of this protocol, a batch of experiments was carried out in triplicate using Ubi:eGFP construct. The numbers of the initially germinated protocorms of TH274 × TH601 hybrid and surviving protocorms that were resistant to hygromycin from the first to the third round of selection were recorded. Parental lines of TH274 and TH601 were purchased from Join Orchids. The percentage transformation rate was calculated based on the explants surviving after the third round of selection.

DNA isolation

Genomic DNA was isolated from a small piece of leaf tissue about 2 × 1 cm taken from putative transgenic lines using the cetyltrimethylammonium bromide (CTAB) method (Lee et al. 2015). The genomic DNA was used for PCR as well as for Southern blot analysis.

PCR determined transgenic lines

Polymerase chain reaction (PCR) was used to determine transgenic lines using *HptII* primers 5′-GATGTAGGAGGGCGTGGATA-3′ and 5′-CGTCTGCTGC TCCATACAAG-3′ and the target gene was confirmed by using primers eGFP 5′-ATGGTGAGCAAGGGCGAGGA-3′ and Nos-R 5′-ATCGCAAG ACCG GCAACAGGA-3′. The amplicon sizes are predicted to be 621 and 930 bp, respectively. The PCR amplification was carried out using following conditions: 95 °C, 3 min (1 cycle), 95 °C, 1 min, 58 °C, 1 min, 72 °C, 1 min (34 cycles), 72 °C, 10 min. Annealing temperature for *HptII* and *eGFP* was 58 and 60 °C, respectively.

For early detection of gene insertion, a small portion of leaf tissue (1 × 1 mm) was excised from small seedlings after 2 months recovery from the 3rd round hygromycin selection. DNA was directly extracted using Phire Plant Direct PCR Kit (Thermo Scientific) and PCR was performed using the *HptII* primers.

Southern blot analysis

Genomic DNA was extracted from leaf tissues (1 g fresh weight) using the CTAB method. After quantification by gel electrophoresis, 10 µg of each DNA was digested with *BamHI* and fractionated on a 0.8 % agarose gel. Southern hybridization and detection were carried out using a digoxigenin-labeled *HptII* probe following the manufacturer's instructions (Roche, http://www.rocheapplied-science.com).

Western blot analysis

Total protein of leaf samples was extracted with Culture Cell Lysis Reagent (CCLR) buffer (100 mM K_2HPO_4, 100 mM KH_2PO_4, pH 7.8, containing 1 % Triton X-100, 10 % glycerol, 1 mM EDTA, and 7 mM 2-mercaptoethanol). The protein amount was measured by using the Bio-Rad Protein Assay Kit with bovine serum albumin as a standard. For western blot analysis, 30 µg total protein from each sample was loaded and separated by SDS-PAGE with a 10 % acrylamide gel and transferred onto polyvinylidene fluoride (PVDF) membrane for antibody probing. The following primary antibodies were used: anti-eGFP, rabbit polyclonal antibody (Yao-Hong Biotechnology, Cat#YH-80005) at 1:10,000 dilution and anti-Actin mouse monoclonal antibody (Sigma, A0480) at 1:2500 dilution. Horseradish peroxidase (HRP)-conjugated secondary antibodies were obtained from GeneTex. Secondary anti-rabbit antibody was diluted at 1:10,000; anti-mouse antibody was diluted at 1:12,000.

GFP fluorescence microscopy

Leaf and root tissues of the D20 transgenic line and the non-transgenic wild-type were used for GFP florescence observation. Root tissue excised about 8 mm from the root tips and leaves was sampled from the mid vein of young developing leaves. Tissues were embedded in 4 % agarose II low gelling temperature (AMERESCo, Lot# 1624co42). The agarose sample block was trimmed to a desired size and adhered to a specimen stage. Vertical and transverse sections of the roots were made, and leaves were transverse sectioned. Tissues were sectioned to ~120 µm thickness using a blade installed in the Vibratome 1000 plus sectioning system (Fedelco, SL).

A GFP signal was detected and an image was taken using a Zeiss LSM710 confocal microscope equipped with a T-PMT under an FITC filter at excitation of 488 nm and emission wavelength of 500–560 nm.

GUS assay

T0 explants of 35S:GUS and non-transgenic wild type were stained with X-Gluc and incubated at 37 °C overnight according to a previously described protocol (Jefferson et al. 1987). Chlorophyll was removed by immersing in 95 % ethanol for 24 h. Pictures of whole explants were taken using a Nikon SMZ1500 dissecting microscope at 7.5× magnification. For detailed observation of the GUS expression pattern in transgenic cells, explants were sectioned using a vibratome according to the protocol mentioned in the previous section.

Inheritance of transgene

Transgenic orchid was crossed to non-transgenic wild types to generate BC1 segregating populations to determine if the transgene is heritable. One T0 plant (line D1) bloomed giving two flowers. A total of four D1 pollinia were backcrossed to four independent female parental non-transgenic *Phal.* orchids in a distinct genetic background, i.e., Ama (4n, the original Wt of D1), Ama (2n), C2-4 hybrid (TH274 × *Phalaenopsis equestris*), and V65 × V15 (hybrid orchid), respectively. Parental lines of *Phalaenopsis equestris*, V65 and V15 were purchased from Jumbo Orchids, Kaohsiung Guanyinshan and Chain Port Orchid Co. Ltd., respectively. Approximately four months after pollination, orchid capsules were harvested and the BC1 seeds were sown on a 1/2 MS plate containing 25 ppm hygromycin. Two months after sowing, the numbers of explants that were resistant or susceptible to hygromycin were calculated. The surviving green seedlings were considered to be transgenic lines that were resistant to hygromycin, whilst those wilting or undeveloped protocorms were hygromycin susceptible wild-type like progenies. In order to confirm introgression of the transgene to the next generation, 22 BC1 s of Ama (4n)/D1, Ama (4n) Wt as a negative control and D1 (the original T0 transgenic plant) as a positive control were applied to PCR detection using *HptII* primers and western blot using eGFP antibody.

RESULTS

Transformation processes and obtained transgenic seedlings

Mature orchid capsules were harvested (Figure 1a) and surface sterilized with 70 % EtOH. The capsule was cut longitudinally with a sterilized scalpel and tiny seeds were spread evenly on 1/2 Murashige and Skoog (MS) agar plates. Seed germination and growth of protocorms took place on agar plates at constant temperature of 24 °C (Figure 1b). Five weeks after sowing, protocorms had grown substantially and had turned light green in color (Figure 1c). Healthy globular shape protocorms of approximately 1.2 mm in length (Figure 1d) were selected and infected with *Agrobacterium* inoculum (Figure 1e). Selection of putative transgenic lines was performed by three successive hygromycin selections at 2-week intervals per selection (Figure 1f–h). The greenish explants after the 3rd round of antibiotic selection (Figure 1h) were recovered in a 1/2 MS agar plate for 2 months and then transferred to a jar for approximately 3 months before transplanting the transgenic orchid to a 1.7" pot containing sphagnum moss. Finally, a total of 74 independent transgenic orchid seedlings overexpressing Ubi:eGFP were obtained and grown in a GMO greenhouse and all had apparently normal morphology. The phenotypes of the 20 transgenic lines are shown in the Additional file 1: Figure S2.

Timeline for transformation

It took about 32 weeks to obtain transformation results, starting from sowing seeds to PCR analysis on transformed seedlings right before transplanting to 1.7″ pots. However, the whole process could be reduced to 15 weeks with modifications. Firstly, always maintaining viable protocorms can save about 5 weeks, the amount of time needed for seeds to develop. Secondly, if seedlings are transferred to 1/2 MS not containing hygromycin; after the 3rd hygromycin selection, seedlings can grow rapidly. After 2 months of growth, a small piece of leaf tissue (1 × 1 mm) was excised for early PCR analysis. These modifications save time and increase efficiency of transformation.

Frequency of transformation

Transformation rate is critical for the success of a transformation system. In order to know the transformation efficiency of this protocol, three batches of experiments were tested using Ubi:eGFP vector. As can be seen in

Additional file 1: Table S1, transformation rate varied in different batches ranging from 1.2 to 5.2 %. Normally, healthy *Phal.* orchid capsule contain tens of thousands tiny seeds that appear as dust. A single capsule of *Phal. aphrodite* can be sown on 8–10 agar plates of 9 cm diameter. Each plate can germinate more than 1000 protocorms for *Agrobacteria* infection. Thus, the selection is rather easy and less labor intensive. Therefore, increase initial *Agrobacteria* infected protocorms plate numbers can apparently obtain more transgenic *Phal.* orchid lines.

Molecular confirmation of transgenic lines

A total of 74 individual transgenic lines of *Phal.* orchids were obtained and planted in the greenhouse. Among them, 22 transgenic orchid pot seedlings (Additional file 1: Figure S2) were randomly selected, genomic DNA was isolated, and genomic PCR was performed. There was no *hygromycin phosphotransferase* gene (*HptII*) and enhanced green fluorescence protein (*eGFP*) amplification with the DNA from the Wt plant but obvious PCR banding was obtained in the positive control of the vector DNA (Figure 2a). All of the 22 tested T0 transgenic lines showed *HptII* banding after the high stringency hygromycin selection (Figure 2b). However, five T0 plants did not show *eGFP* bands (Figure 2b). This might due to a problem with DNA insertion of eGFP in those lines. Overall, the T-DNA integration was accomplished normally in most of the transgenic lines.

To determine the T-DNA insertion number, Southern blot analysis was performed by randomly selecting some transgenic lines developed from the genetic background of Ama (2n) i.e., D18, D19 and D20; from Ama (4n) such as D1, and from L3 (4n) background such as D26, D28, and D32. Only the transgenic lines showed a hybridization signal, with no signal being seen in the Wt plants of Ama (2n), Ama (4n), and L3 (4n). Southern blot data indicated that most of the transgenic lines contained one or two copies of T-DNA insertion (Figure 2c).

Total protein was isolated from leaves of Ubi:eGFP transgenic orchid lines and their respective three wild types, L3 (4n), Ama (4n), and Ama (2n). Thirty micrograms of total protein was loaded and western blot analysis was performed using anti-eGFP antibody. Results indicated that none of the three wild types expressed the GFP signal but all the transgenic lines expressed eGFP protein in the T0 transgenic orchids (Figure 3a). The lower panel showed anti-Actin antibody to ensure equal protein loading (Figure 3a).

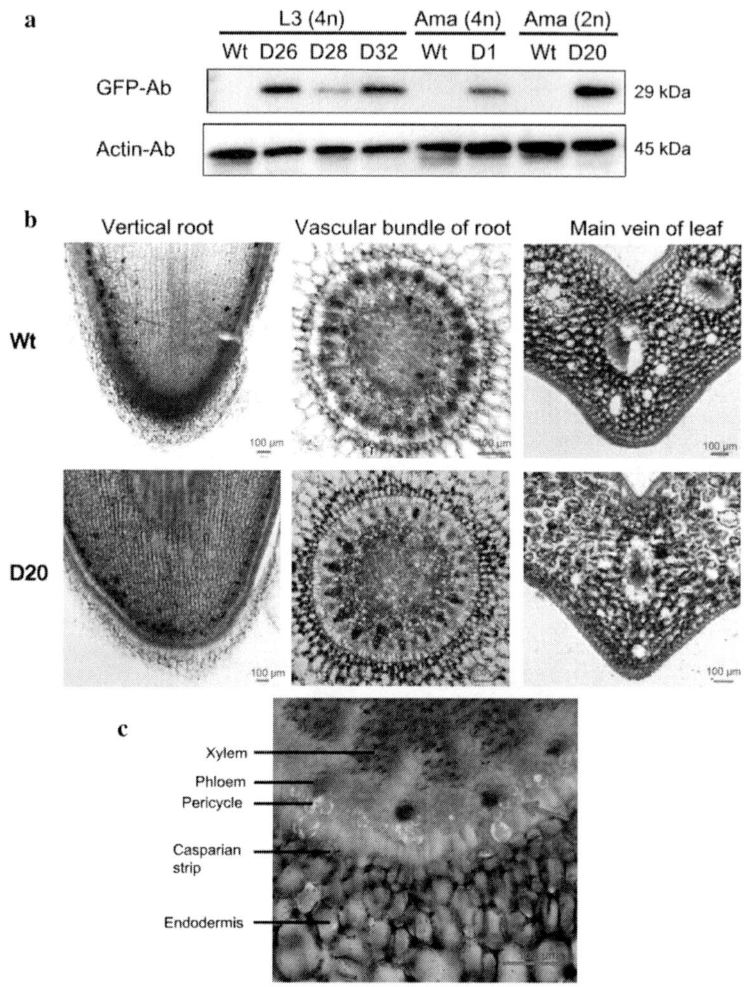

Figure 3. Western blot and detection of the GFP fluorescence signal in transgenic orchids. **A** Western blot analysis of the expression of GFP protein in transgenic orchids. Total protein was isolated from the leaf of orchid and 30 µg total protein was loaded into 10 % acrylamide gel, electrophoresed, and transferred onto polyvinylidene fluoride (PVDF) membrane for antibody probing. The membrane was probed using an anti-GFP antibody and anti-actin antibody served as the loading control. *Wt* wild-type; *4n* tetraploid; *2n* diploid. **B** GFP fluorescence signal was detected in D20 transgenic *Phalaenopsis aphrodite*. **C** A magnified view of the GFP signal in partial tissue of **B**. Root tissues of the D20 transgenic line and Wt were vertical and transversely sectioned, and leaves were transverse-sectioned using a vibratome and an image was obtained using a confocal microscope

Green florescent signal in orchid transgenic lines

Green florescent protein (GFP) is a marker for monitoring gene expression in transgenic plants (Hraska et al. 2006). Under confocal microscopy, as expected, there was an absence of GFP detection in the root and leaf of wild type but the GFP fluorescence signal was clearly observed in the D20 transgenic line, with more expression in the outer layers of the vertical root section (Figure 3b). A strong GFP signal was seen in the cortex tissues in the transverse root and a very strong GFP signal was accumulated in the pericycle cells of the transgenic line as shown in Figure 3c. The pericycle that was located between the endodermis and phloem is the tissue that enables plants to grow roots and facilitates plant development. Cells in the pericycle actively undergo cell division and proliferation (Dubrovsky et al. 2000). We observed more GFP signals around the pericycle under fluorescence microscope. Here we found the root pericycle close to the root tip has less endogenous GFP and it is good tissue to detect the expression of the GFP transgene (Figure 3c).

Uniform GUS expression in transgenic tissues

Non-chimeric transgenic plants are essential for stable transformation. In order to check that the transformation protocol established in this study did not produce chimerism, T0 transgenic explants of 35S:GUS were stained with X-Gluc solution. All tissues showed strong GUS positive signals, including the leaf, root, and hypocotyl (Figure 4a). Vibratome tissue section showed uniform GUS positive signals in all tissues and cells (Figure 4b). These results confirmed that the transformation system established in this study did not produce chimeric transformants.

Stable inheritance of transgene

It is very important to develop transgenic orchids with stable inheritance and consistent expression of target genes in the next generation. Therefore, pollinia of the D1 transgenic line were backcrossed to four distinct genetic backgrounds, Ama (4n, the original Wt of D1), Ama (2n), and the hybrids of C2-4 and V65 × V15, respectively, and produced BC1 seeds. Some BC1 progenies showed resistance in hygromycin selection (Figure 5a). The hygromycin resistant explants varied from 43 to 56 % in four BC1 progenies (Figure 5b). Southern blotting data indicated that D1 is a single T-DNA insertion transgenic line (Figure 2c). BC1 progenies were heritable (Figure 5b). Genomic PCR analysis showed that 22 hygromycin resistant BC1 of Ama (4n)/D1 seedlings all contained *HptII* amplicons, indicating that the introgression of transgenes is stable in the next generation

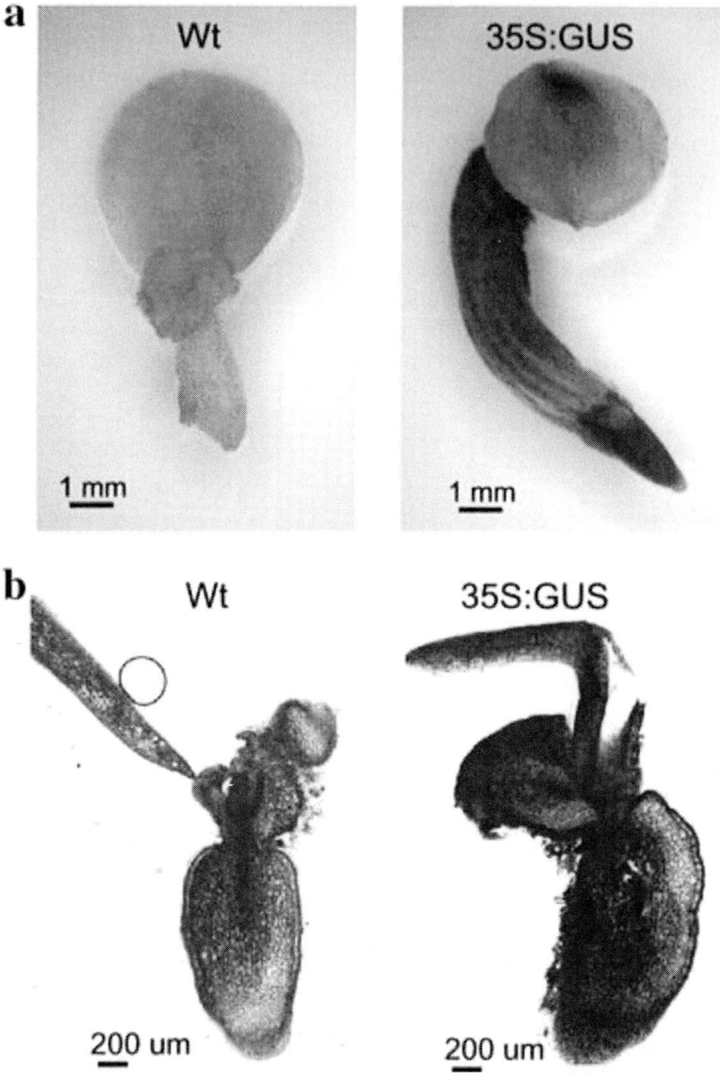

Figure 4. Histochemical Gus staining in transgenic *Phalaenopsis* orchid lines. Transgenic orchid explant harboring 35S:GUS and a Wt non-transgenic control were stained with X-Gluc and incubated at 37 °C overnight. **A** Image obtained using a Nikon SMZ1500 dissecting microscope (×7.5 magnification). **B** Explant tissue sections of approximately 120 μm in thickness were photographed on a Zeiss Axio Scope A1 microscope equipped with an Axio-Cam HRc camera (Zeiss, Germany) (×25 magnification)

a

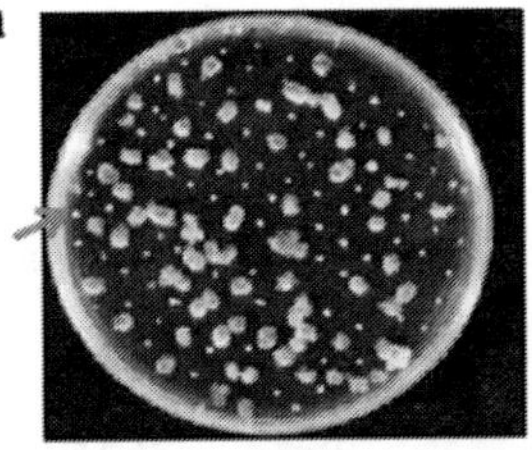

b

BC1	Number of seeds	Hygromycin resistant	Transgenic line (%)
Ama (2n)/ D1	110	52	47%
Ama (4n)/ D1	159	68	43%
C2-4/ D1	137	74	54%
V65xV15/ D1	194	109	56%

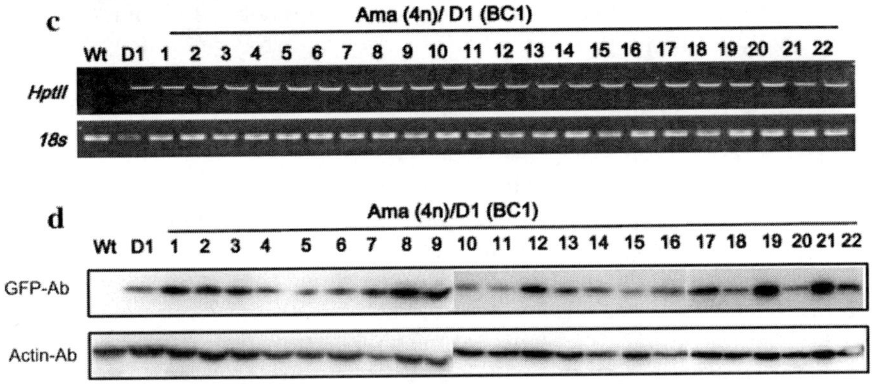

Figure 5. Stable inheritance of transgene in BC1 orchid progeny. **A** Pollinia of D1 transgenic line was crossed-pollinated to four distinct genetic backgrounds of Wt orchids and generated BC1 progenies. Ama (4n)/D1 BC1 seeds were sown on 1/2 MS agar plates containing 25 ppm hygromycin. Some BC1 progenies were resistant to hygromycin and some were susceptible showing white and wilt symptoms. *Red arrows* indicate hygromycin susceptible explants. **B** Record of survival rate in BC1 transgenic lines. **C** Genomic PCR showed *HptII* gene integration in the hygromycin-resistant BC1 progenies. **D** Western blot showed GFP protein expression in the hygromycin-resistant BC1 progenies. Ama (4n) was used as the non-transgenic wild-type negative control. D1, the original T0 transgenic parent was used as a positive control

of *Phal.* orchid (Figure 5c). Moreover, western blot also indicated that all hygromycin survival progenies positively expressed eGFP protein (Figure 5d). We concluded that the transgenic lines generated using the *Agrobacterium*-mediated young protocorm method presented in this study is stable and efficient. To be noted, the expression and introgression of transgenes to the next generation was demonstrated in this study.

DISCUSSION

Phalaenopsis orchids have been a mainstay of the global ornamental flower trade in recent years. Genetic engineering has emerged as a method with high potential to modify traits more precisely and enable functional genomics studies more efficiently. Although *Agrobacterium*-mediated transformation has been successfully applied to many agricultural or horticultural crops, it still remains difficult to apply to some plants that are not easily clonally propagated. Here, we reported an efficient method of *Agrobacterium*- mediated transformation using protocorms of *Phal. aphrodite.* Its advantages over previous methods can be summarized as follows:

We achieved a stable, efficient, yet less labor-intensive transformation method to genetically transform monocotyledonous *Phal.* orchids. Stable integration, expression, and inheritance of the transgenes were confirmed using molecular and genetic studies. To date, the transformation rates reported for *Phal.* orchids have been rather low and have never reached 2 % (Mishiba et al. 2005; Semiarti et al. 2007). However, this protocol could generate up to 5.2 % transformation efficiency (Additional file 1: Table S1).

Choose of appropriate size of target explants for transformation is very important for stable transformation. The orchid protocorms which are similar to immature embryos with extensive cell proliferation are good target for transformation. Chimerism could occur if large orchid embryos or PLBs are used as target tissues (Kuehnle and Sugii 1992). In the present study, we found small-size protocorms of 1.2 mm are suitable for transformation and we did not observe chimerism in transgenic explants after GUS staining (Figure 4). Moreover, the inheritance of transgenes is stable in all the BC1 transformation lines tested in our genetic study as shown in Figure 5. These data demonstrates that transformation protocol developed in this study is stable and heritable.

To speed up transformation process, we adopted three successive 25 ppm hygromycin selections at 2-week interval in this protocol. The total of 6 weeks in hygromycin selection media is much shorter than other transformation methods that usually take up to several months or longer. Selection under high concentration of hygromycin can reduce false positive transformants. In addition, using the Direct PCR Kit enables us to detect gene insertion in early stage to validate the positive transformants. These modifications significantly save time and increase efficiency of transformation. It is known that slow in growth is one of the bottlenecks for *Phal.* orchid production as well as for researches. Therefore, it is suggested once PCR confirms the transformants, immediately recover the transgenic seedlings in 1/2 MS agar without subjecting the pressure from antibiotics could speed up their growth dramatically.

It is known that GFP fluorescence has some weak points, such as the diminishing green fluorescence in older tissues, variation in fluorescence levels among different tissues, and interference with autofluorescence in plants (Hraska et al. 2006; Molinier et al. 2000). In the present study, we successfully detected the GFP fluorescent signal of transgenic *Phal.* orchid using a florescence microscope and enabled the determination of the presence of the transgene (Figure 3b, c). Moreover, the transgene can be monitored non-destructively using a dissecting fluorescence microscope. The root tip can be partially cut to identify the GFP signal and will not affect the survival of the seedling. Interestingly, we found that pericycle cells close to root tip express high levels of GFP in this study (Figure 3c). Higher cytoplasmic density in young tissues causing higher levels of expression of GFP than in older tissues has been previously described (Molinier et al. 2000). Therefore, it is suggested that the pericycle close to the root tip of orchid that contains dense proliferating cells could be a good tissue type to detect introgression of transgenes. We suggest that eGFP is a suitable reporter system that provides reliable results of introgression of transgenes and it is a good marker for the genetic transformation of *Phal.* orchids.

The *Agrobacterium*-mediated transformation system in this study may provide a practical tool for studying functional genomics, which can further assist the gene discovery and molecular breeding processes. GMO of ornamental crops that are not edible is more acceptable to consumers than GMO of food. Genome editing technology such as CRISPR (Liu et al. 2016; Zaidi et al. 2016; Zhu et al. 2016) provides the prospect of modification of orchids using highly efficient transformation techniques in *Phal.* orchids.

CONCLUSIONS

This study established an *Agrobacterium*-mediated transformation system using orchid protocorm as a target explant. This protocol works well in tetraploid as well as in diploid *Phal. aphrodite* varieties. Moreover, this method is simple, stable, efficient, heritable, time saving, and less labor-intensive. In the era of big data and functional genomics, this transformation method can significantly contribute to the gene discoveries, genetic engineering and general R&D of *Phal.* orchid.

NOTES

Hong-Xian Hsing and Yi-Jyun Lin are co-first authors

AUTHORS' CONTRIBUTIONS

SSK designed the experiments, HXH, YJL, MJL, SSK, YJL conducted the experiments, SSK and CGT wrote the manuscript. All authors read and approved the final manuscript.

ACKNOWLEDGEMENTS

We are grateful to the support of the Academia Sinica Innovative Translational Agricultural Research Grant to Swee-Suak Ko project (iTAR#2015CP02-3). We express our sincere appreciation for technical support from Ms. Chi Fang, Ms. Whe-re Shen, and Ms. Yi-Chen Lien. We thank Ms. Miranda Loney for English editing.

COMPETING INTERESTS

The authors declare that they have no competing interests.

REFERENCES

1. Anzai H, Tanaka T (2001) Transgenic Phalaenopsis (a moth orchid). In: Bajaj YPS (ed) Biotechnology in agriculture and forestry transgenic crops III 48. Springer, Berlin, pp 249–264.

2. Anzai H, Ishii Y, Shichinohe M, Katsumata K, Nojiri C, Morikawa H, Tanaka M (1996) Transformation of Phalaenopsis by particle bombardment. Plant Tissue Cult Lett 13:265–272.
3. Belarmino MM, Mii M (2000) Agrobacterium-mediated genetic transformation of a Phalaenopsis orchid. Plant Cell Rep 19:435–442.
4. Chai ML, Xu CJ, Senthil KK, Kim JY, Kim DH (2002) Stable transformation of protocorm-like bodies in Phalaenopsis orchid mediated by *Agrobacterium tumefaciens*. Sci Hort 96:213–224.
5. Chan YL, Lin KH, Liao LJ, Chen WH, Chan MT (2005) Gene stacking in Phalaenopsis orchid enhances dual tolerance to pathogen attack. Transgenic Res 14:279–288.
6. Chen WH, Kao YL, Tang CY (2009) Method for producing polyploid plants of orchids. US Patent 8,383,881 B2, 26 Feb 2013
7. Chen YC, Yang YC, Chen YH, Chao YP, Chang LZ, Chen YW (2010) Modification of Phalaenopsis metabolism by genetic engineering. Acta Hort 878:473–480.
8. Chen WH, Hsu CY, Cheng HY, Chang H, Chen HH, Ger MJ (2011) Downregulation of putative UDP-glucose: flavonoid 3-*O*-glucosyltransferase gene alters flower coloring in Phalaenopsis. Plant Cell Rep 30:1007–1017.
9. Chin DP, Mishiba K, Mii M (2007) Agrobacterium-mediated transformation of protocorm-like bodies in Cymbidium. Plant Cell Rep 26:735–743.
10. Dubrovsky JG, Doerner PW, Colon-Carmona A, Rost TL (2000) Pericycle cell proliferation and lateral root initiation in *Arabidopsis*. Plant Physiol 124:1648–1657.
11. Fu CH et al (2011) OrchidBase: a collection of sequences of the transcriptome derived from orchids. Plant Cell Physiol 52:238–243.
12. Guo WJ, Lee N (2006) Effect of leaf and plant age, and day/night temperature on net CO2 uptake in *Phalaenopsis amabilis* var. formosa. J Amer Soc Hort Sci 131:320–326
13. Hraska M, Rakousky S, Curn V (2006) Green fluorescent protein as a vital marker for non-destructive detection of transformation events in transgenic plants. Plant Cell Tissue Organ Cult 86:303–318.
14. Hsieh RM, Huang PL (1995) Studies on genetic transformation of Phalaenopsis via pollen tube pathway. J Chin Soc Hort Sci 41:309–324
15. Hsieh RM et al (1997) *Agrobacterium tumefaciens*-mediated transformation of Phalaenopsis orchid. Rept Taiwan Sugar Inst 155:41–54
16. Jefferson RA, Kavanagh TA, Bevan MW (1987) GUS fusions: beta-glucuronidase as a sensitive and versatile gene fusion marker in higher plants. EMBO J 6:3901–3907
17. Julkifle AL, Rathinam X, Sinniah UR, Subramaniam S (2010) Optimisation of transient green fluorescent protein (GFP) gene expression in *Phalaenopsis violacea* orchid mediated by *Agrobacterium tumefaciens*-mediated transformation system. Aust J Basic Appl Sci 4:3424–3432
18. Kuehnle AR, Sugii N (1992) Transformation of Dendrobium orchid using particle bombardment of protocorms. Plant Cell Rep 11:484–488.
19. Lee SH, Li CW, Liau CH, Chang PY, Lioa LJ, Lin CS, Chan MT (2015) Establishment of an Agrobacterium-mediated genetic transformation procedure for the experimental model orchid *Erycina pusilla*. Plant Cell Tiss Organ Cult 120:211–220.
20. Liao LJ, Pan IC, Chan YL, Hsu YH, Chen WH, Chan MT (2004) Transgene silencing in Phalaenopsis expressing the coat protein of Cymbidium mosaic virus is a manifestation of RNA-mediated resistance. Mol Breed 13:229–242.
21. Liu D, Hu R, Palla KJ, Tuskan GA, Yang X (2016) Advances and perspectives on the use of CRISPR/Cas9 systems in plant genomics research. Curr Opin Plant Biol 30:70–77.
22. Mii M, Chin DP (2010) Genetic transformation of orchids. Acta Hort 878:461–466.
23. Mishiba K, Chin DP, Mii M (2005) Agrobacterium-mediated transformation of Phalaenopsis by targeting protocorms at an early stage after germination. Plant Cell Rep 24:297–303.
24. Molinier J, Himber C, Hahne G (2000) Use of green fluorescent protein for detection of transformed shoots and homozygous offspring. Plant Cell Rep 19:219–223.

25. Semiarti E et al (2007) Agrobacterium-mediated transformation of the wild orchid species *Phalaenopsis amabilis*. Plant Biotech 24:265–272.
26. Semiarti E et al (2010) High-frequency genetic transformation of *Phalaenopsis amabilis* orchid using tomato extract-enriched medium for the pre-culture of protocorms. J Hort Sci Biotechnol 85:205–210.
27. Sjahril R, Mii M (2006) High-efficiency Agrobacterium-mediated trasnformation of Phalaenopsis using meropenem, a novel antibiotic to eliminate Agrobacterium. J Hort Sci Biotech 81:458–464.
28. Su CL, Chao YT, Yen SH, Chen CY, Chen WC, Chang YC, Shih MC (2013a) Orchidstra: an integrated orchid functional genomics database. Plant Cell Physiol 54:e11.
29. Su CL, Chen WC, Lee AY, Chen CY, Chang YC, Chao YT, Shih MC (2013b) A modified ABCDE model of flowering in orchids based on gene expression profiling studies of the moth orchid Phalaenopsis aphrodite. PLoS ONE 8:e80462.
30. Subramaniam S, Xavier R (2010) Emerging factors that influence efficiency of T-DNA gene transfer into *Phalaenopsis violacea* orchid via *Agrobacterium tumefaciens*—mediated transformation system Int. J Biol 2:64–73
31. Subramaniam S, Samian R, Midrarullah Xavier R (2009) Preliminary factors influencing transient expression of Gusa in Dendrobium Savin White protocorm-like bodies (PLBs) using Agrobacterium-mediated transformation system. World Appl Sci J 7:1295–1307
32. Teixeira da Silva JA, Tanaka M (2011) Optimization of particle bombardment conditions for hybrid Cymbidium: Part II. Transgenic Plant J 5:78–82
33. Tsay HS, Ho HM, Gupta SK, Wang CS, Chen PT, Chen ECF (2012) Development of pollen mediated activation tagging system for Phalaenopsis and Doritaenopsis Electron. J Biotechnol.
34. Yee NC, Abdullah JA, Mahmood M, Basiron N (2008) Co-transfer of gfp, CHS and hptII genes into Oncidium Sharry Baby PLB using the biolistic gun. Afr J Biotechnol 7:2605–2617.
35. Yu H, Goh CJ (2001) Molecular genetics of reproductive biology in orchids. Plant Physiol 127:1390–1393.
36. Yumiko B, Brundrettb M, Batty A, Dixon K, Kochd J, Sivasithamparam K (2007) Diversity of mycorrhizal fungi of terrestrial orchids: compatibility webs, brief encounters, lasting relationships and alien invasions. Mycol Res 111:51–61.
37. Zaidi SS, Mansoor S, Ali Z, Tashkandi M, Mahfouz MM (2016) Engineering plants for Geminivirus resistance with CRISPR/Cas9 system. Trends Plant Sci 21:279–281.
38. Zhu J, Song N, Sun S, Yang W, Zhao H, Song W, Lai J (2016) Efficiency and inheritance of targeted mutagenesis in maize using CRISPR-Cas9. J Genet Genomics 43:25–36.

CHAPTER 8

Flower Induction, Microscope-Aided Cross-Pollination, and Seed Production in the Duckweed *Lemna Gibba* with Discovery of a Male-Sterile Clone

Lili Fu[1], Meng Huang[1], Bingying Han[1], Xuepiao Sun[1], K. Sowjanya Sree[2], Klaus-J. Appenroth[3] & Jiaming Zhang[1]

[1]*Institute of Tropical Bioscience and Biotechnology, MOA Key Laboratory of Tropical Crops Biology and Genetic Resources; Hainan Bioenergy Center, CATAS, Haikou, Hainan Province, 571101, China.*

[2]*Department of Environmental Science, Central University of Kerala, RSTC, Padanakkad, 671314, Kerala, India. 3 Institute of Plant Physiology, University of Jena, Dornburger Str. 159, 07743, Jena, Germany.*

ABSTRACT

Duckweed species have a great potential to develop into fast-growing crops for water remediation and bioenergy production. Seed production and utilization of hybrid vigour are essential steps in this process. However, even in the extensively-studied duckweed species, *Lemna gibba*, flower primordia were often aborted prior to maturation. Salicylic acid (SA) and agar solidification of the medium promoted flower maturation and resulted in high flowering rates in *L. gibba* 7741 and 5504. Artificial cross-pollination between individuals of *L. gibba* 7741 yielded seeds at high frequencies unlike that in *L. gibba* 5504. In contrast to clone 7741, the anthers of 5504 did not dehisce upon maturation, its artificially released pollen grains had pineapple-like exine with tilted spines. These pollens were not stained by 2,5-diphenylmonotetrazoliumbromide (MTT) and failed to germinate. Therefore, clone 5504 is male sterile and has potential application with respect to hybrid vigour. Moreover, pollination of flowers of 5504 with 7741 pollen grains resulted in intraspecific hybrid seeds, which was confirmed by inter-simple sequence repeat (ISSR) markers. These hybrid seeds germinated at a high frequency, forming new clones.

INTRODUCTION

Lemnaceae, commonly known as duckweeds, is a family of small aquatic monocotyledonous plants that comprise five genera and 37 species[1,2]. They represent the smallest flowering plants with reduced morphology[3]. Duckweeds can rapidly assimilate nutrients such as nitrogen and phosphate from municipal, industrial, and agricultural wastewaters and have shown great value in wastewater remediation[4,5,6,7,8,9]. The biomass of duckweeds may contain high starch content and could be used to produce bioethanol and biobutanol[10,11,12,13,14]. Some duckweed clones contain high protein content, and have been used as high quality livestock feed[15,16,17]. Even the duckweed biomass accumulated on wastewaters may be used for feed production[18,19,20].

Lemna gibba is one of the duckweed species that has been widely used in research and applications. *L. gibba* is very efficient in removal of organic and inorganic nutrients from domestic and industrial wastewater[21,22,23,24], as well as heavy metals such as lead and uranium[25]. *L. gibba* was also found to be efficient in removal of harmful organisms such as *Giardia cysts*, *Cryptosporidium oocysts*, faecal coliforms and coliphage from wastewater[26]. It can also be used in bioassays of toxicity of industrial waste[21].

The potential of duckweeds in commercial applications has brought forward a need for domestication and the breeding of elite varieties serving different purposes. Seed production and application of hybrid vigour have an enormous importance in making duckweed a future crop plant. However, duckweeds reproduce mainly vegetatively. In this process, mother fronds form genetically identical daughter fronds that are therefore called a clone. Plants collected from different places are in most cases genetically different from one another and form different clones. Flowering is a rare event in nature and in the laboratory cultures in most duckweed species27, 28. Natural flowers are morphologically identical to those induced artificially (Figure 1). Artificial induction of flowering and seed production in duckweeds is therefore needed for future efforts to breed new duckweed varieties by cross-pollination to improve their efficiency.

Figure 1.*L. gibba* plant with natural flowers. Red arrows indicate flowers; M, mother frond; D, daughter frond; G, granddaughter frond.

Many efforts have been made to induce flowering in different duckweed species. Salicylic acid (SA) is one of the most important plant hormones used in duckweed flower induction. SA induced flower formation in *L. gibba* G3 and *L. paucicostata* LP6 (valid term *L. aequinoctialis*), and the continuous presence of SA was found to be essential to obtain maximum flowering29. SA induced flowering in *L. gibba* G3 in coordination with day length30, 31. Flower promotion was obtained with day lengths of more than 9 h, but never with less than 8 h. Salicylic acid treatment caused a shift in the critical day length of about 2 hours31. Benzoic acid is another chemical that is effective in flower induction. When *L. gibba* G3 was grown on 1/10 M medium or 1/2 H medium under long day conditions, addition of benzoic acid to either media led to a flowering response, which was inhibited by the plant hormones IAA, GA3, ABA, and zeatin[32,33].

Analysis of flower induction in previous reports was based on dissecting the budding pockets of *L. gibba*. Fronds with flower primordia, i.e. immature flowers, visible under a dissecting microscope were usually considered as flowering[31,32,34], whereas no data on induction of mature flowers are available up to date to the best of our knowledge. However, only mature flowers can be used for cross-pollination for the purpose of breeding. In this study, we present efficient methods to induce the formation of mature flowers and production of intraspecific hybrid seeds in *L. gibba*. For that purpose, two geographically distinct *L. gibba*, clones 7741 (origin: Italy, better known by the older term G3) and 5504 (origin: China), and one clone of *L. minor*, clone 7210 (origin: Japan) were investigated. The genetic background of these clones was investigated by barcoding according to the described methods[35,36,37](Supplementary material, Figure S1). From clone 7741, seeds were produced by artificial self-pollination. Moreover, intraspecific hybrid seed production by pollen transfer between different clones of *L. gibba* under a microscope was successfully carried out for the first time. This strategy will be used in breeding programmes in the future to make duckweed a crop plant. Moreover, the analysis of pollen viability and cross-pollination led to the identification of a male-sterile line of *L. gibba* which is the first report in Lemnaceae family.

RESULTS

Flower maturation in *L. gibba* is limited in conventional culture media

Three commonly used duckweed media including liquid and/or semi-solid media H, E, and MH were used to study their influence on the rate of flowering. Two clones of *L. gibba*, 7741 and 5504, and one clone of *L. minor*, 7210 were incubated on these media under 16 h light period according to the previous reports for *L. gibba* G3[31,38]. The fronds of all clones grew well on all media (Figure 2a, controls). Immature flowers were found in more than 50% of the colonies of *L. gibba* 5504 and 7741, but no flower primordia could be observed in *L. minor* 7210 (data not shown). Mature flowers were hardly induced in any of the clones (Table 1, Figure 2b,c), the highest rate of mature flowers obtained in solidified MH medium was approximately 2% for both of the *L. gibba* clones. First, we investigated photoperiod as a possible reason for the low percentage of mature flower formation. The clones were cultured under 8–16 h photoperiods. Immature flowers were induced at high frequencies in *L. gibba* 7741 and 5504 but not in *L. minor* 7210 under photoperiod ≥10 h (data not shown). However, most of these flower primordia in *L. gibba* were aborted and only a very low number of mature flowers were developed when grown in MH medium. Mature flowers were not observed at all in H and E media (Table 1). Flower primordia often co-existed with frond primordia in the same budding pocket (Figure 2c).

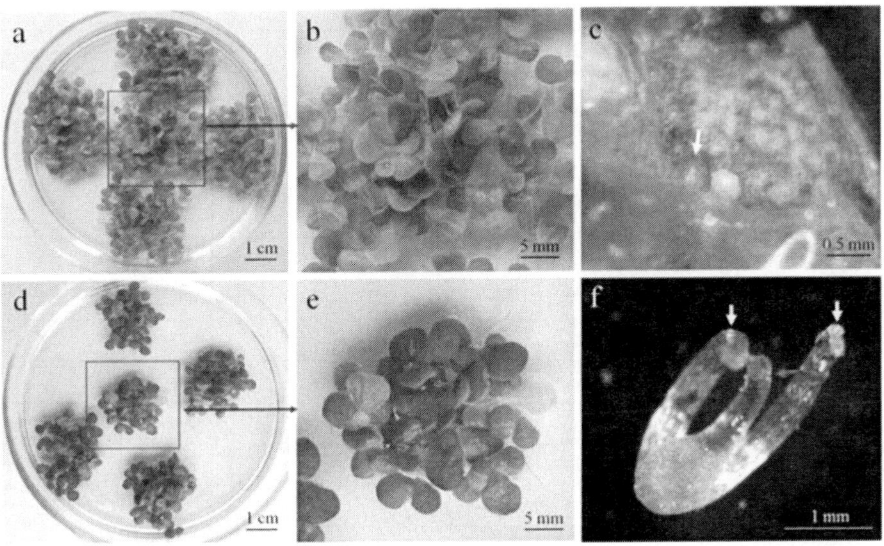

Figure 2. Comparison of *L. gibba* clone 7741 plants grown in media without ((**a–c**); upper row as a control) and with ((**d–f**); bottom row) salicylic acid (SA). (**a**) Control fronds grown in medium without SA showing high amount of biomass 15 days after inoculation. (**b**) Magnification of the *red-squared* region in (**a**) showing the absence of flowers. (**c**) Dissected budding pocket of a frond growing on SA-free medium; the white arrow indicates frond primordium, and the red arrow indicates flower primordium. (**d**) Fronds growing on medium supplemented with 20 μM SA showing lower amount of biomass after 15 days. (**e**) Magnification of the *red-squared* region in (**d**), showing flowers (red arrows); (**f**) Mature flower dissected from a budding pocket with pistil (red arrow), and two stamen (white arrows; anthers have shed).

Table 1: Influence of different basic media and salicylic acid on the rate of mature flower formation.

Clone	Nutrient medium	Time to first flower emergence (d)	Flowering rate (%)
Lemna gibba 7741	E	n/a	0
	H	n/a	0
	MH	20	1.7 ± 0.5
	E + 20 µM SA	10	51.0 ± 1.1
	MH + 20 µM SA	12	73.5 ± 3.5
Lemna gibba 5504	E	n/a	0
	H	n/a	0
	MH	20	2.3 ± 0.8
	E + 20 µM SA	14	6.5 ± 3.0
	MH + 20 µM SA	12	54.1 ± 4.2
Lemna minor 7210	E	n/a	0
	H	n/a	0
	MH	n/a	0
	E + 20 µM SA	n/a	0
	MH + 20 µM SA	n/a	0

SA promotes flower maturation

When SA (e.g. 20 µM) was supplemented to E and MH media, the vegetative growth of all three clones, regardless of flower formation, slowed down although at different concentrations (cf. Figure 2a,d; Figure 3A). Mature flowers with one pistil and/or two stamens bending upward were observed under stereomicroscope (for *L. gibba* 7741 see Figure 2e,f). Salicylic acid (SA) in MH medium resulted in higher flower induction rates than in E medium (Table 1). *Lemna gibba* 7741 had flowering rates of 51% and 74% in E + SA and MH + SA, respectively, while clone 5504 had flowering rates of 6.5% and 54%, respectively (Table 1). Even in the presence of SA, *L. minor* 7210 did not show any flower development.

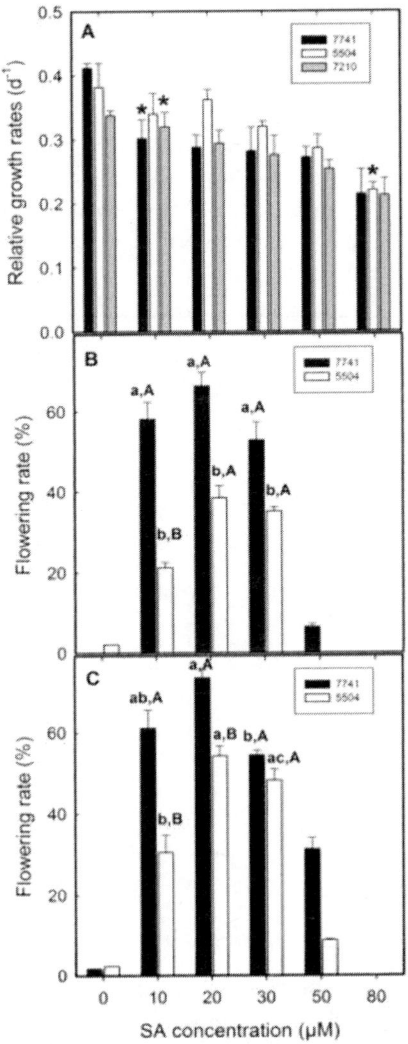

Figure 3 Dose-response effects of salicylic acid (SA) on growth and flowering rates. (**A**) Inhibitory effect of SA on the growth of *L. gibba* clones 7741 and 5504 and *L. minor* clone 7210 in semi-solid medium. Asterisks indicate the lowest concentration that significantly inhibited the growth of the three clones (students's t-test, $\alpha < 0.05$). (**B**) Effects of SA concentration in liquid medium on the flowering rates of *L. gibba* 7741 and 5504. (**C**) Effects of SA on solidified medium on the flowering rates of *L. gibba* 7741 and 5504. Note: Data are means of three independent experiments and are presented as mean ± standard error. Significance of differences in (**B**) and (**C**), were tested by one-way ANOVA ($\alpha < 0.05$); Lower case letters mark significance of the effect of different SA concentrations, capical letters evaluate the differences in the response of both *L. gibba* clones. The results on *L. minor* 7210 in (**B**) and (**C**) were not presented because there was no stimulating effect.

The first mature flowers were observed approximately 10–14 days after application of SA (Table 1, Figure 4). The number of newly developed mature flowers scored daily, increased steadily in the following days until day 20, and declined thereafter in the two *L. gibba* clones. Approximately 70% of total mature flowers were obtained between day 17 and 21, with a peak around day 20 (Figure 4). Approximately 17 and 31 flowers per plate on an average were obtained at day 20 in *L. gibba* 5504 and 7741, respectively. The flowering rates in the following experiments were presented always at day 20.

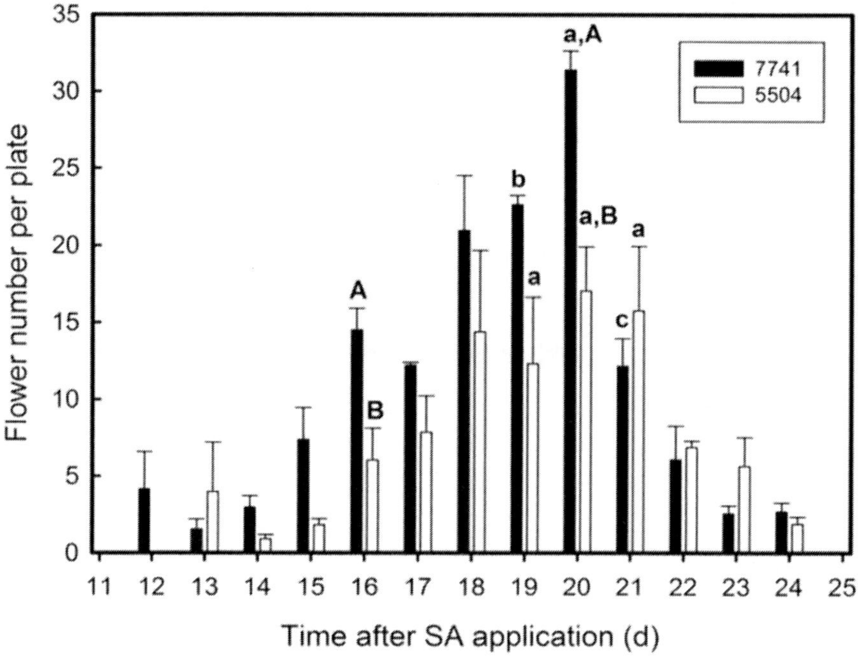

Figure 4. Temporal distribution of the number of flowering fronds. The fronds were cultured in 1/2 MH medium supplemented with 20 µM salicylic acid. Significance of differences from day 15 to day 21 after SA application were evaluated by ANOVA at a significance level of 0.05. Lower case letters indicate differences in the flower number over time, capital letters mark differences between *L. gibba* clones 7741 and 5504.

Effect of SA concentration and agar solidification on flowering rate

Flowering rates were very low without SA in the nutrient medium (Table 1). When the lowest tested concentration of SA (10 µM) was supplemented to the liquid medium, the flowering rates already abruptly increased (Figure 3B). The optimum SA concentration was 20 μM, although hardly any significant difference could be detected between 10 and 30 μM (in clone 5504 the rate was lower at 10 μM). At 20 μM SA flowering rates were 66% and 38% for *L. gibba* 7741 and 5504, respectively. When the SA concentration was increased further, flowering rates decreased (Figure 3B). Almost no flowers were induced when the SA concentration was increased to 80 μM and frond death rate increased to approximately 70% (data not shown).

Semi-solid medium has not been tested previously for flower induction in duckweeds. The general effect of SA on the flowering rates in this medium was similar to that in the liquid medium, and the optimum SA concentration was also 20 μM (Figure 3C). However, compared to the liquid media, the flowering rates in the semi-solid media were slightly higher in general and significantly higher at 20, 30 and 50 μM SA in clone 5504 (Figure 3B,C). The highest flowering rate in the semi-solid media reached 74% and 54% for clones 7741 and 5504 at 20 μM SA, respectively.

Difference in emergence of flowers and pollen characteristics in *Lemna gibba* 7741 and 5504

The stamens of *L. gibba* 7741 usually appeared ahead of the pistil (Figure 5a), and the anthers released pollen grains rapidly after the appearance of pistil (Figure 5b). Each anther of this clone contained approximately 210 pollen grains. In contrast, in *L. gibba* 5504 the pistil emerged first (Figure 5c), and the stamens could be observed only by dissecting the budding pocket at this moment (Figure 5d). A dew droplet was often observed on top of the pistil before stamens emerged (Figure 5e), which was suggested to help in pollination. Moreover, the anthers of *L. gibba* 5504 were never observed to release pollen grains until they withered (Figure 5f). Their pollen grains were released artificially with a dissecting needle, having a smaller number of approximately 180 grains per anther.

Scanning electron microscopy indicated that the slits of the mature anther of *L. gibba* 5504 remained closed (Figure 6a), while that in the anthers of *L. gibba* 7741 was widely open (Figure 6b). The artificially released pollen grains of *L. gibba* 5504 had pineapple-like exine with tilted spines of 0.6 to 1.3 μm in length (Figure 6c). The sizes of the pollen grains varied in diameter from 15 to 21 μm. In contrast, the pollen grains of *L. gibba* 7741 had smooth exine with uniformly and vertically distributed spines (Figure 6d). The sizes of 7741 pollen grains were relatively uniform with a diameter between 20 to 22 μm.

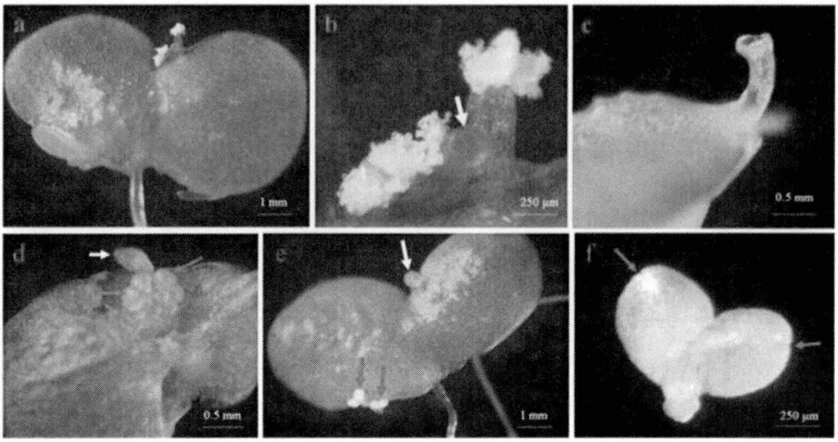

Figure 5. Flowering of *L. gibba* clones 7741 (**a**,**b**) and 5504 (**c**–**f**) under the influence of salicylic acid. (**a**) Flowering colony of clone 7741. (**b**) Magnification of the flower in (**a**), showing the stigma (yellow arrow) and pollens on the stamens. (**c**) Pistil emerged from the budding pocket of clone 5504 bending upward. (**d**) the same frond in (**c**) dissected from the ventral side to show the two stamens in the budding pocket (red arrows), the white arrow indicates the pistil. (**e**) Colony of 5504 with two stamens (red arrows) in one frond after the pistil withered, and a liquid droplet secreted by the pistil in another frond (white arrow). (**f**) Anther of clone 5504 that did not dihisce before withering (red arrows indicate the closed slits).

Pollen viability test by chemical staining and germination

Several chemical staining methods were tested to distinguish viable and non-viable pollen grains in duckweeds. The MTT method was most effective, staining viable pollen grains of *L. gibba* 7741 violet-purple whereas dead pollen grains (killed by heating) lacking almost any stain (Supplementary Figure S2, Supporting Information). Pollen grains of *L. gibba* 5504 and 7741 were then stained in parallel experiments (Figure 7a,b). Most of the pollens of clone 7741 were stained dark violet (approximately 67 ± 3%; Figure 7a), and were considered viable. Approximately 20% of the pollen grains of clone 7741 were not stained and were considered non-viable. The rest of the pollen grains (approximately 10%) were stained pale violet, and their viability was therefore uncertain. In contrast, none of the pollen grains of clone 5504 were stained dark violet (Figure 7b).

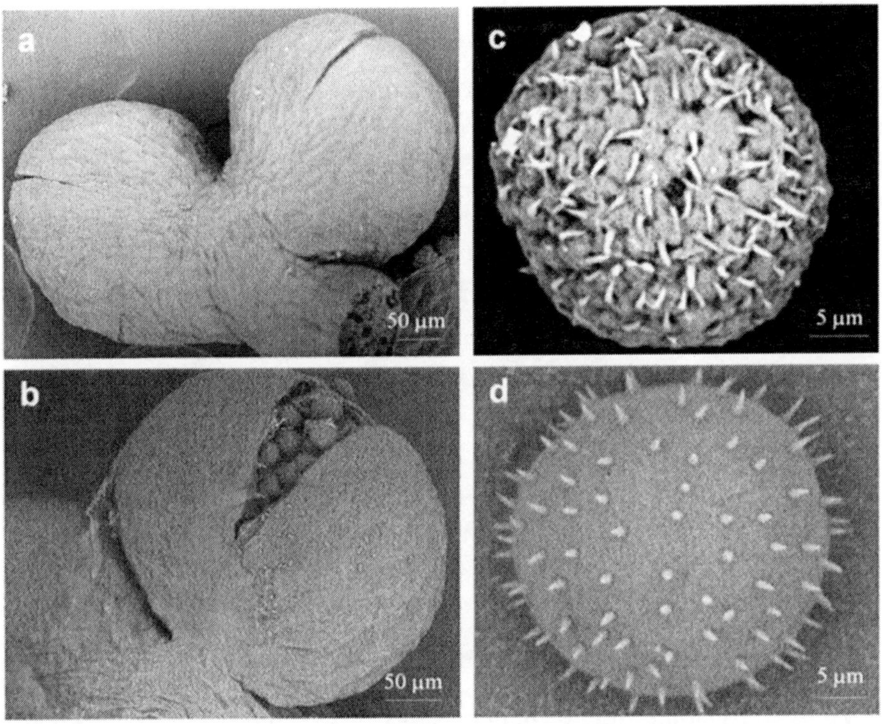

Figure 6. Scanning electron micrographs of anthers and pollen grains. (**a**) Anther of *L. gibba* 5504 showing the closed slit. (**b**) Anther of *L. gibba* 7741 with opened slit. (**c**) Pollen grain of clone 5504 showing pineapple-like exine and tilted spines. (**d**) Pollen grain of clone 7741 showing the smooth exine and uniformly distributed spines.

The viability of pollens was further tested by germination experiments. Some of the pollen grains of clone 7741 germinated within five minutes, and germination frequency was approximately 40% as counted 3 h later (Figure 7c), whereas none of the pollen grains of clone 5504 germinated even if the incubation time was prolonged to 24 h (Figure 7d). Therefore, we concluded from both experiments (Figure 7b,d) that pollen of clone 5504 was sterile.

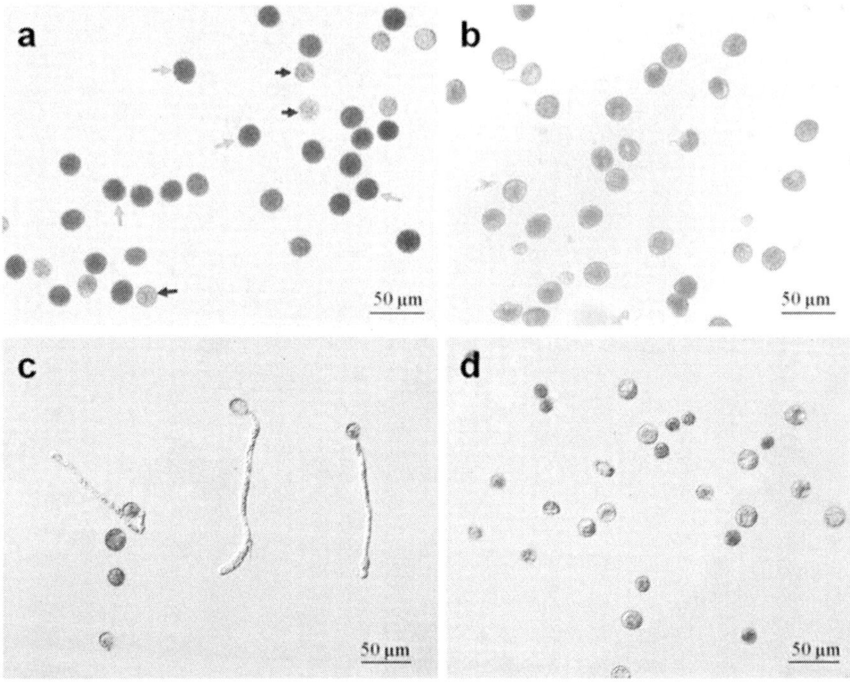

Figure 7. Viability test of pollen grains from *L. gibba* 7741 and 5504. (**a**) Pollen grains from clone 7741 stained with MTT solution. (**b**) Pollen grains from clone 5504 stained with MTT solution. (**c**) Germinating pollen grains of clone 7741. (**d**) Lack of germination of pollen grains of clone 5504 after 16 h of incubation under germination conditions. The green and the blue arrows indicate the viable and the non-viable pollen grains, respectively.

Seed formation following pollination

Natural pollination of flowers of *L. gibba* in the culture plates rarely produced seeds. Artificial pollination between individuals of *L. gibba* 7741, however, produced seeds (Figure 8a) with a setting rate of approximately 20–30%. Two or three seeds were most often observed enclosed in the thin-layer pericarp of the fruit (Figure 8b), one or four seeds were observed occasionally in one fruit. The mature seeds were white and oval with a purple operculum to the posterior end (Figure 8c). Scanning electron microscopy showed that the seed coat was covered with longitudinal-ribs (Figure 8d).

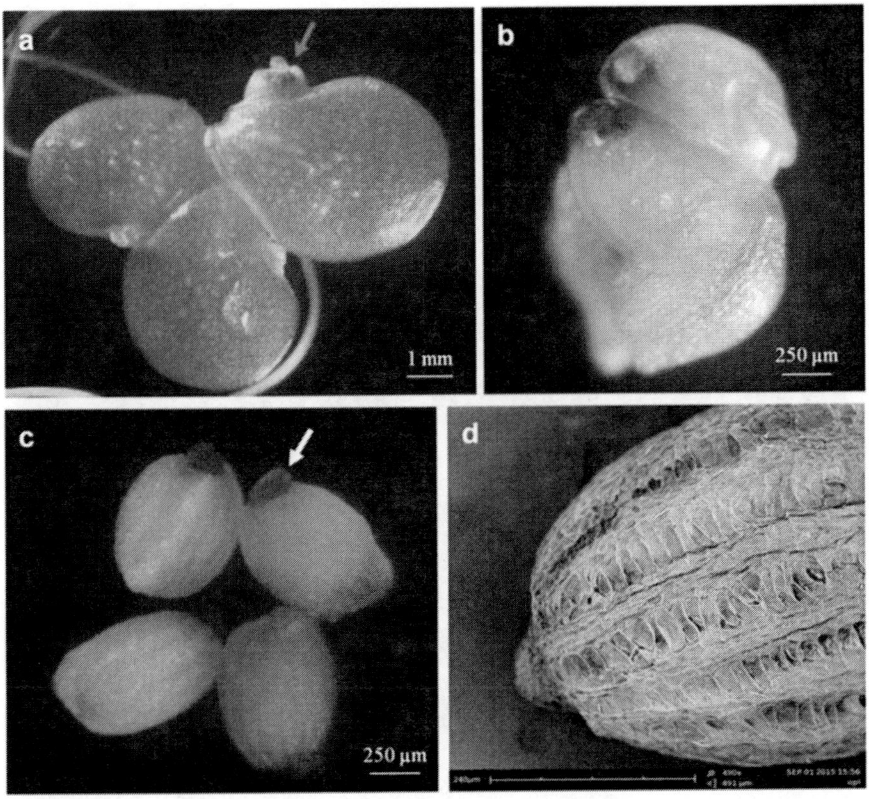

Figure 8. Seeds of *L. gibba* 7741. (**a**) Colony with an immature fruit (red arrow). (**b**) Immature fruit with two seeds. (**c**) Mature seeds with purple operculum (white arrow). (**d**) Scanning electron graph of a seed showing the longitudinal-ribs.

In contrast to clone 7741, no seeds were obtained from clone 5504 after natural and artificial pollination between individuals of this clone.

Artificial cross-pollination between *L. gibba* clones and validation of hybrids

Artificial cross-pollination of the stigmas of clone 7741 using pollen grains of clone 5504 did not result in any seed formation. However, artificial pollination of the stigmas of clone 5504 using the pollen grains of clone 7741 produced intraspecific hybrid seeds. These seeds were morphologically identical to those resulted from self-pollination of clone 7741 (Figure 9a). Also, the hybrid seeds germinated normally at the same rate as those formed after self-fertilization: seeds from clone 7741 alone germinated to 100% (100 seeds tested), and hybrid seeds germinated to 97% (35 seeds tested). Hybrid colonies were formed after germination as shown in Figure 9. The initial hybrid fronds were small compared to their parents (Figure 9b); however, the subsequent fronds grew larger gradually until they reached the size of their parents.

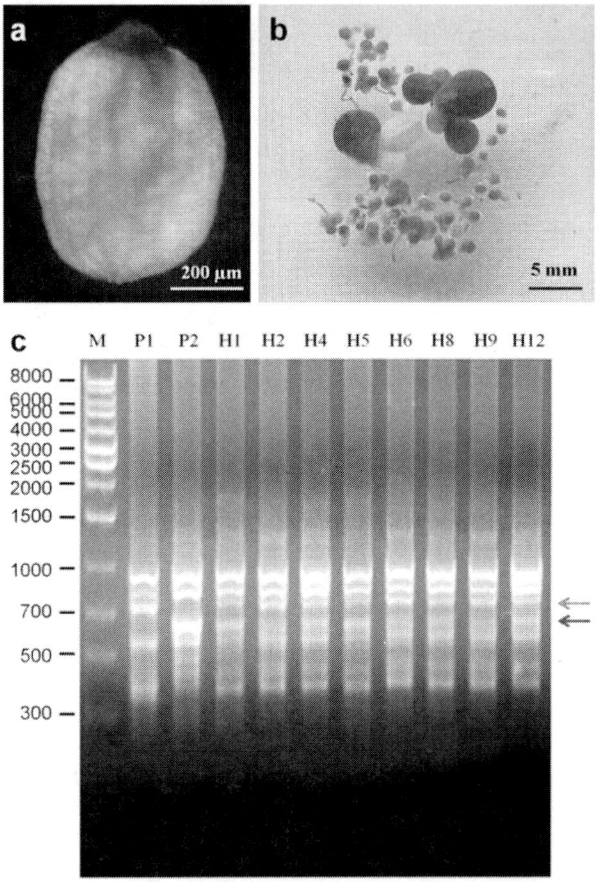

Figure 9. Germination of hybrid seeds and ISSR validation. (**a**) A typical hybrid seed from crossing between clones 5504 and 7741. (**b**) Hybrid clones five days after germination, their female (F) and male (M) parents are used as size controls. (**c**) ISSR validation of the hybrid clones; M, molecular weight marker (Axygen 1 kb ladder); P1, male parent 7741; P2, female parent 5504; H1 through to H12, hybrid clones; the green arrow indicates 7741-specific band, the red arrow indicates 5504-specific band.

In order to prove that indeed hybrids were formed between the paternal clone 7741 and the maternal clone 5504, the parental and hybrid clones were investigated by ISSR by testing 50 primers. Primer B825 showed parent-specific bands and was used to test the assumed hybrid clones (Figure 9c). Results indicated that both parental bands could be clearly detected in the hybrid lines, therefore, the hybrid lines inherited the genetic properties of both parents (Figure 9c).

DISCUSSION

Flower induction in *Lemna gibba* but not in *Lemna minor*

In order to develop a method for efficient flower induction, which serves the purpose of cross-pollination, we measured the flowering rate as the number of colonies with mature flowers divided by the number of total colonies. The previously used number of immature flowers[30, 38, 39] is not useful for subsequent pollination and seed formation experiments. We successfully induced mature flowers at high frequencies for *L. gibba* clones 7741 and 5504, respectively. No flower induction was observed, under any of the conditions tested, in *L. minor* 7210, which suggested that different duckweed species responded differently to SA induction and may need different methods to induce flowering, e.g., in the case of *L. minor*, ethylenediamine-N,N'-bis (2-hydroxyphenylacetic acid) (EDDHA) is a successful flower-inducer[40]. Although no flowering was induced in *L. minor* 7210 by SA, its growth rate was reduced significantly, which indicated that growth inhibition was not a result of flowering. Flowering and vegetative growth are not always competitive in duckweeds, for example, *Wolffia microscopica* was reported to grow very rapidly during continuous flowering season[41].

In the previous reports, authors had considered immature flowers to quantify flower induction[30, 38, 39]. However, this parameter is not useful for subsequent pollination and seed formation experiments as most of the immature flowers were aborted during subsequent development. In the present paper a method was developed to induce the formation of mature flowers at high rates, calculated as the number of colonies with mature flowers divided by the number of total colonies. Without SA in the MH or E media, flower primordia (i.e. immature flowers) were induced in *L. gibba* 7741 and 5504 at high frequencies. However, mature flowers were obtained only at very low percentages (ca. 2% in MH medium). Supplementation of media with SA increased the rates of formation of mature flowers dramatically, for both *L. gibba* 7741 and 5504. The results in semi-solid MH medium with SA were even slightly higher than in liquid medium. Formation of high percentage of mature flowers is a prerequisite for the production of seeds.

It was reported that ammonium in the Hutner's medium inhibited the flower primordia formation in *L. gibba* G3[30,38]. This inhibition effect could be reversed by addition of SA and hexacyanoferrate (III) to the medium[30]. However, we obtained high percentage of flower primordia in *L. gibba* 7741 and 5504 even in the ammonium containing MH medium, which was more than 50%, respectively. Therefore, ammonium may not be an inhibition factor for flower primordial differentiation at least in clones 7741 and 5504.

Moreover, the flowering rates presented here were based on the mature flower counts at the end of the experiment (20 days), when flower percentage was at the climax (Figure 3), ignoring the flowers matured before the end of the experiments. Therefore, the actual flowering rates were even higher than what are presented here. We concluded that the induction of mature flowers with SA is a useful technique for later cross-pollination and breeding.

Evaluation of pollen viability

In order to produce seeds, pollen must be viable. Chemical staining is a widely used method to examine pollen viability[42,43,44]. We used several staining reagents to test the viability of *L. gibba* pollen grains and found that the MTT staining was the most effective (Supplementary Figure S2). Heat-killed pollens were recommended previously as a negative control[44]. We adopted this method to prepare dead pollens. However, the spines or echinates were removed from the exine by the heat-killing process (Supplimentary material, Figure S2), which resulted in varied morphology, and therefore need to be improved in the future. Regardless of this, pollens of *L. gibba* 7741 clearly showed viability as tested by staining and pollen germination. This was not the case for pollens of *L. gibba* 5504 (Figure 7). Thus, vital pollens were produced from mature flowers in clone 7741 but not in clone 5504.

Hybrid vigor and male sterility

It has been shown in the present paper that the success of spontaneous self-fertilization can be greatly improved by artificial self-fertilization and that even intraspecific cross-fertilization between different clones of *L. gibba* is possible to succeed with high rate. Artificial pollen transfer from clone 7741 to stigmas of clone 5504 resulted in fertilizaiton and formation of seeds. These seeds germinated normally and formed new vegetative plants. To prove that these plants were intraspecific hybrids formed by clone 7741 as pollen-donor and 5504 as pollen acceptor, parent plants and seedlings were analyzed by ISSR. The presence of parent-specific bands in the seedlings demonstrated that these plants are indeed hybrids. Artificial transfer of pollens from clone 5504 to stigmas of clone 7741 did not result in fertilization. We conclude that *L. gibba* 5504 is male-sterile. To the best of our knowledge, this is the first report about male-sterility in the duckweed family. Our conclusion is based on the following results: First, the emergence of the stamens of clone 5504 lagged behind the pistil. Second, the anthers of clone 5504 were not ruptured and pollen grains were not released. Third, the pollen grains of clone 5504 had an underdeveloped and pineapple-like exine with tilted spines. Fourth, they were not stained by MTT, and finally, these pollens did not germinate, and did not result in seeds by pollinating on stigmas of both clones 5504 and 7741.

Male sterility is one of the bases in commercialization of hybrid vigour and has been widely used in hybrid seed production of maize45, 46, rice47, 48, wheat49, cotton50, oilseed rape51, and many other crops52. The discovery of a male-sterile line of duckweed may be useful in future application of hybrid vigour in different applications of duckweed.

METHODS

Plant materials and culture media

Three axenic clones, *Lemna gibba* L. 5504, 7741 and *L. minor* 7210, were used. Clone 5504 was isolated from the Innovation River in Pudong, Shanghai, China, with an original accession number SH020453 and maintained at the Biological Collection Centre of the Institute of Tropical Bioscience and Biotechnology, Chinese Academy of Tropical Agricultural Sciences, as well as at the Rutgers Duckweed Stock Cooperation at Rutgers University, New Jersey, USA under international accession number 5504. Clone 7741 was a standard *L. gibba* clone derived from clone G335, the later was originally isolated from Sicilia, Siracusa, Italy by R. Kandeler, Vienna, Austria. Clone 7210 was identified as *L. minor* (Supporting Information, Figure S1). It was originally isolated in Japan, maintained in Landolt's collection, Zurich, Switzerland and obtained from the Rutgers Duckweed Stock Cooperative. The identities of these clones were verified by phylogenetic analysis based on their *atpF-atpH* intergenic sequences and their *rps16* intron sequences as suggested by Borisjuk *et al.*36 (Supporting Information, Figure S1).

The components in the three basic nutrient media used in this study are provided in Table S1 (Supporting Information). All nutrient media were supplemented with 29 mM sucrose with or without 6 g L^{-1} agar (Biotechnology grade, Beijing Solarbiao Science & Technology co., Ltd, Beijing, China) for solidification. Media were autoclaved and the SA stock solution was filter sterilized (0.22 μm filter units; Millipore, USA).

Observation of flower primordium and determination of critical day length

To observe flower induction and to determine critical day length, fronds were cultured in MH medium under 8 h light/16 h darkness for 20 days, then transferred to fresh MH or E-medium and continued to culture under 8, 9, 10, 12, 14, and 16 h photoperiod for 20–30 days. The budding pockets were dissected and observed under a dissecting microscope.

Effects of SA on flower induction

Three-frond colonies of the above mentioned clones were cultured in three replicates on semi-solid MH medium supplemented with 29 mM sucrose and 20 μM SA, and cultivated under 16 h photoperiod at 40 $\mu mol \cdot m^{-2} \cdot s^{-1}$ photosynthetic active radiation (PAR) at 25 ± 1 °C. Each replicate contained 10 plates, and each plate contained five colonies. In contrast to previous reports, only fronds with mature flowers were considered as flowering, which is in accordance with the purpose of the investigation. Mature flowers were counted every day from their first emergence until day 24.

To study the effect of the SA concentration on the flowering rate, three-frond colonies were inoculated with threefold replicates in the MH medium supplemented with different concentrations of SA (0, 10, 20, 30, 50, and 80 μM) and cultivated under 16 h daily photoperiod at 40 $\mu mol \cdot m^{-2} \cdot s^{-1}$ PAR at 25 ± 1 °C. Flowering rates were calculated 20 days after inoculation. The flowering rate was defined as the total number of plants (frond colonies) with mature flowers at the time of counting divided by the total number of plants inspected.

Effects of SA on frond growth rate

Three-frond colonies were inoculated in three replicates on the basal medium supplemented with different concentrations of SA (0, 10, 20, 30, 50, and 80 μM) and incubated under 16 h day photoperiod at 40 $\mu mol \cdot m^{-2} \cdot s^{-1}$ PAR at 25 ± 1 °C. Frond number of each sample was counted seven days after inoculation. The relative growth rate (RGR) was calculated as described previously54: $RGR = (\ln N_2 - \ln N_1)/t$, where N_1represents the initial frond number, N_2 represents the final frond number, and t represents the days of incubation.

Estimation of pollen grains in the anthers

To estimate the quantity of pollen grains in the anthers, 30 mature anthers from randomly chosen colonies were put in three 1.5 mL centrifuge tubes with 10 anthers in each tube and broken with a vaccination needle. The broken anthers were then treated with 0.5 mL 2.5% pectinase (Y23, Yakult Pharmaceutical Industry Co., Ltd., Tokyo, Japan) solution for 4 h at 28 °C, followed by dilution with 0.5 mL 2.5% sucrose solution. The centrifuge tubes were then vortexed for 10 sec. The resulting pollen suspension was spread on glass slides in 5 μL droplets, and the pollen grains in each droplet were counted under a light microscope (Axioskop 40, Zeiss, Jena, Germany) with triplicates for each sample. Pollen grain quantity in each anther was calculated as the average number of pollen grains in each 5 μL droplet × the total volume of pollen suspension in one tube (1000 μL)/the volume of each droplet (5 μL)/the number of anthers in each tube (10).

Pollen viability tests by chemical staining

Four different staining methods were tested for their effectiveness to distinguish between viable and non-viable pollen (Trognitz 1991; Khatun & Flowers 1995). As control, pollens incubated at 80 °C for 2 h were considered dead55. The staining reagents included: (1) 2,3,5-triphenyl tetrazolium chloride (TTC), (2) 2,5-diphenyl monotetrazolium bromide (MTT), (3) 5-bromo-4-chloro-3-indoyle-β-galactoside (X-gal) (Rodriguez-Riano & Dafni 2000), and (4) fluorescein diacetate (FDA) (Kearns & Inouye 1993).

The MTT reagent was finally selected. Mature anthers were placed on a glass slide and incubated with 5 μL MTT solution. The working solutions of

MTT contained 1% staining reagent with 5% sucrose that was added to the anthers. Pollen grains were released into the solution by breaking the anthers with a vaccination needle. After incubating for 10 min, the pollen grains were observed under a light microscope (Axioskop 40, Zeiss, Jena, Germany), and the viability was evaluated by the colour of the pollen. Ten anthers were examined per clone. The pollen viability is given as percentage of the number of viable pollen grains to the total number of pollen grains observed.

Pollen viability test by germination

Germination medium (MH medium supplemented with 1% sucrose) was spread on glass slides to a thin layer. Pollen grains were then inoculated by scratching the anthers on the surface of the medium. The germination of the pollen grains was observed under microscope at intervals of five minutes.

Scanning electron microscopy

Mature anthers were fixed in formaldehyde-acetic acid-ethanol solution (FAA, 100 ml solution contains 5 ml 40% formaldehyde, 5 ml glacial acetic acid, and 90 ml 70% ethanol) at 4 °C overnight, and then dehydrated in a series of tert-butyl alcohol (30%, 50%, 70%, 80%, and 90%) for 15 min at each concentration. The anthers were dehydrated with 100% tert-butyl alcohol three times for 1 hour. The dehydrated anthers were then freeze-dried and inspected with a scanning electron microscope (FEI Phenom, Phenom Wold, Eindhoven, Netherlands). Pollen grains were released from the dehydrated anthers with a vaccination needle and suspended in tert-butyl alcohol. Aliquots of the pollen suspension were then spread over glass slides, adhered to electron microscopy sample tables (FEI Phenom), and observed with a scanning electron microscope (FEI Phenom). Twenty pollen grains from each clone were randomly chosen to measure their size and to study their morphology.

Artificial cross-pollination under stereomicroscope

Flowering fronds of *L. gibba* 7741 and 5504 were transferred to fresh semi-solid MH medium supplemented with 29 mM sucrose and were incubated at 40 µmol . m^{-2} . s^{-1} PAR and 25 ± 1 °C for 24 h. Artificial pollination was then performed in the tissue culture hood under a stereomicroscope. Pollen grains of clone 7741 were pollinated on the stigma of clone 5504 under a stereomicroscope by moving the whole colony with tweezers towards the stigma. The pollinated fronds were then transferred to fresh plates containing MH medium supplemented with 29 mM sucrose and 20 μM SA and incubated under 16 h daily photoperiod of 40 μmol · m^{-2} · s^{-1} PAR at 25 ± 1 °C. Seeds were harvested after 20 days. Similar experiments were carried out by transferring pollens of clone 7741 on to the stigma of clone 5504 in order to test the possibility for intraspecific cross-pollination between different clones.

Seed germination and validation of hybrid formation

Fresh seeds were punctured on the seed coat with a scalpel, and were incubated in MH medium under 16 h daily photoperiod of 40 μmol · m^{-2} · s^{-1} PAR at 25 ± 1 °C. Each line that was vegetatively propagated from one seed was considered as a new clone.

DNA was isolated using a genomic DNA extraction kit (Tiangen Biotech Co. LTD, Beijing, China) according to the reccommended protocol. Fifty primers that were selected from the universal inter-simple sequence repeat (ISSR) primers56 were synthesized and used to distinguish genetic differences between parental clones 7741 and 5504. One primer (B825, 5′ ACACACACACACACACT 3′) showed clone-specific bands and was used to validate the genetic background of the hybrid clones. The PCR reactions were performed at 94 °C for 4 min, followed by 30 cycles of 94 °C for 30 sec, 48 °C for 30 sec, and 72 °C for 1 min. The PCR products were separated on 1.5% agarose gel, stained with Goldview Nucleic Acid Gel Stain (Solarbio, Beijing, China), and visualized using the gel imager Tanon 410 (Tanon, Shanghai, China).

ADDITIONAL INFORMATION

Publisher's note: Springer Nature remains neutral with regard to jurisdictional claims in published maps and institutional affiliations.

REFERENCES

1. Sree, K. S., Bog, M. & Appenroth, K.-J. Taxonomy of duckweeds (Lemnaceae), potential new crop plants. Emir. J. Food Agric. 28, 291–302,

2. Appenroth, K.-J., Borisjuk, N. & Lam, E. Telling duckweed apart: Genotyping technologies for the Lemnaceae. Chin. J. Appl. Environ. Biol. 19, 1–10,
3. Landolt, E. Biosystematic investigations in the family of duckweeds (Lemnaceae) Vol. 2: The family of Lemnaceae: a monographic study. – Morphology, karyology, ecology, geographic distribution, nomenclature, descriptions., (Eidgenössische Technische Hochschule Zürich, 1986).
4. Ziegler, P., Sree, K. S. & Appenroth, K. J. The use of duckweed in relation to water remediation. Desalination and Water Tretament 63, 327–342,
5. Ziegler, P., Sree, K. S. & Appenroth, K. J. Duckweeds for water remediation and toxicity testing. Toxicol. Environ. Chem. 98, 1127–1154,
6. El-Shafai, S. A., El-Gohary, F. A., Nasr, F. A., van der Steen, N. P. & Gijzen, H. J. Nutrient recovery from domestic wastewater using a UASB-duckweed ponds system. Biores. Technol. 98, 798–807,
7. Ge, X., Zhang, N., Phillips, G. C. & Xu, J. Growing Lemna minorin agricultural wastewater and converting the duckweed biomass to ethanol. Biores. Technol. 124, 485–488,
8. Xu, J. & Shen, G. Growing duckweed in swine wastewater for nutrient recovery and biomass production. Biores. Technol. 102, 848–853,
9. Zhao, Y. et al. Pilot-scale comparison of four duckweed strains from different genera for potential application in nutrient recovery from wastewater and valuable biomass production. Plant Biol. 17, 82–90,
10. Li, X., Jin, Y., Gao, X., Zhang, G. & Zhao, H. Fermentation method of high ratios of biobutanol with Landoltia punctata. China Brewing 31, 85–88 (2011).
11. Qian, C. et al. Improving production of bioethanol from duckweed (Landoltia punctata) by pectinase pretreatment. Energies 5, 3019–3032,
12. Tao, X. et al. Comparative transcriptome analysis to investigate the high starch accumulation of duckweed (Landoltia punctata) under nutrient starvation. Biotechnol. Biofuels 6, 72,
13. Liu, Y. et al. Uniconazole-induced starch accumulation in the bioenergy crop duckweed (Landoltia punctata) I: transcriptome analysis of the effects of uniconazole on chlorophyll and endogenous hormone biosynthesis. Biotechnol. Biofuels 8, 57,
14. Liu, Y. et al. Uniconazole-induced starch accumulation in the bioenergy crop duckweed (Landoltia punctata) II: transcriptome alterations of pathways involved in carbohydrate metabolism and endogenous hormone crosstalk. Biotechnol. Biofuels 8, 64,
15. Anderson, K. E., Lowman, Z., Stomp, A. M. & Chang, J. Duckweed as a feed ingredient in laying hen diets and its effect on egg production and composition. Intern. J. Poultry Sci. 10, 4–7,
16. Leng, R. A., Stambolie, J. H. & Bell, R. Duckweed - a potential high-protein feed resource for domestic animals and fish. Livestock Res. Rural. Dev. 7, Article 5 (1995).
17. Bairagi, A., Sarkar Ghosh, K., Sen, S. K. & Ray, A. K. Duckweed (Lemna polyrhiza) leaf meal as a source of feedstuff in formulated diets for rohu (Labeo rohitaHam.) fingerlings after fermentation with a fish intestinal bacterium. Biores. Technol. 85, 17–24,
18. Iqbal, S. Duckweed aquaculture: potentials, possibilities and limitations for combined wastewater treatment and animal feed production in developing countries. SANDEC Report No. 6/99 (1999).
19. Islam, M. S. et al. Wastewater-grown duckweed may be safely used as fish feed. Can. J. Microbiol. 50, 51–56,
20. Cheng, J. J. & Stomp, A.-M. Growing duckweed to recover nutrients from wastewaters and for production of fuel ethanol and animal feed. CLEAN – Soil, Air, Water 37, 17–26,

21. Cayuela, M. L., Millner, P., Slovin, J. & Roig, A. Duckweed (Lemna gibba) growth inhibition bioassay for evaluating the toxicity of olive mill wastes before and during composting. Chemosphere 68, 1985–1991,
22. Ran, N., Agami, M. & Oron, G. A pilot study of constructed wetlands using duckweed (Lemna gibbaL.) for treatment of domestic primary effluent in Israel. Water Res. 38, 2240–2247 (2004).
23. Zimmo, O. R., Al-Sa'ed, R. M., van der Steen, N. P. & Gijzen, H. J. Process performance assessment of algae-based and duckweed-based wastewater treatment systems. Water Sci. Technol. 45, 91–101 (2002).
24. Zimmo, O. R., van der Steen, N. P. & Gijzen, H. J. Nitrogen mass balance across pilot-scale algae and duckweed-based wastewater stabilisation ponds. Water Res. 38, 913–920,
25. Mkandawire, M., Taubert, B. & Dudel, E. G. Capacity of Lemna gibbaL. (duckweed) for uranium and arsenic phytoremediation in mine tailing waters. Int. J. Phytoremediation 6, 347–362,
26. Falabi, J. A., Gerba, C. P. & Karpiscak, M. M. Giardia and Cryptosporidium removal from waste-water by a duckweed (Lemna gibbaL.) covered pond. Lett. Appl. Microbiol. 34, 384–387,
27. Landolt, E. & Kandeler, R. The Family of Lemnaceae - A Monographic Study. Vol. 4, Biosystematic Investigations in the Family of Duckweeds(Lemnaceae). Vol. 4 (Veröffentlichungen des Geobotanischen Institutes der ETH, 1987).
28. Pieterse, A. H. Is flowering in Lemnaceae stress-induced? A review. Aquat. Bot. 104, 1–4,
29. Khurana, J. P. & Cleland, C. F. Role of salicylic acid and benzoic acid in flowering of a photoperiod-insensitive strain, Lemna paucicostataLP6. Plant Physiol. 100, 1541–1546,
30. Tanaka, O. & Cleland, C. F. Comparison of the ability of salicylic acid and ferricyanide to induce flowering in the long-day plant, Lemna gibbaG3. Plant Physiol. 65, 1058–1061,
31. Cleland, C. F. & Tanaka, O. Effect of daylength on the ability of salicylic acid to induce flowering in the long-day plant Lemna gibbaG3 and the short-day plant Lemna paucicostata6746. Plant Physiol. 64, 421–424,
32. Cleland, C. F. & Briggs, W. S. Gibberellin and CCC effects on flowering and growth in the long-day plant Lemna gibbaG3. Plant Physiol. 44, 503–507,
33. Fujioka, S. et al. The role of benzoic acid and plant hormones in flowering of Lemna gibbaG3. Plant Cell Physiol. 26, 655–659, d
34. Tanaka, O. et al. Flower-Inducing Activity of Lysine in Lemna paucicostata6746. Plant Cell Physiol. 38, 124–128,
35. Bog, M. et al. Genetic structure of the genus LemnaL. (Lemnaceae) as revealed by amplified fragment length polymorphism. Planta 232, 609–619,
36. Borisjuk, N. et al. Assessment, validation and deployment strategy of a two-barcode protocol for facile genotyping of duckweed species. Plant Biol. 17, 42–49,
37. Xu, Y. et al. Species distribution, genetic diversity and barcoding in the duckweed family (Lemnaceae). Hydrobiologia 743, 75–87,
38. Cleland, C. F. & Briggs, W. R. Flowering responses of the long-day plant Lemna gibbaG3. Plant Physiol. 42, 1553–1561,
39. Cleland, C. F. & Ajami, A. Identification of the flower-inducing factor isolated from aphid honeydew as being salicylic acid. Plant Physiol. 54, 904–906,
40. Krajncic, B. & Nemec, J. Mechanisms of EDDHA effects on the promotion of floral induction in the long-day plant Lemna minor(L.). J. Plant Physiol. 160, 143–151,
41. Sree, K. S. et al. The duckweed Wolffia microscopica: A unique aquatic monocot. Flora 210, 31–39,
42. Asma, B. M. Determination of pollen viability, germination ratios and morphology of eight apricot genotypes. Afr. J. Biotechnol. 7, 4269–4273 (2008).
43. Dafni, A. & Firmage, D. Pollen viability and longevity: practical, ecological and evolutionary implications. Plant Sys. Evol 222, 113–132,

44. Rodriguez-Riano, T. & Dafni, A. A new procedure to asses pollen viability. Sex. Plant Reprod. 12, 241–244,
45. Skibbea, D. S. & Schnable, P. S. Male sterility in maize. Maydica 50, 367–376 (2005).
46. Gracen, V. E. & Grogan, C. O. Diversity and suitability for hybrid production of different sources of cytoplasmic male sterility in maize. Agron. J. 66, 654–657,
47. Raj, K. G. & Virmani, S. S. Genetic of fertility restoration of 'WA' type cytoplasmic male sterility in rice. Crop Sci. 28, 787–792,
48. Luo, D. et al. A detrimental mitochondrial-nuclear interaction causes cytoplasmic male sterility in rice. Nat. Genet. 45, 573–577,
49. Maan, S. S. & Lucken, K. A. Interacting male sterility-male fertility restoration systems for hybrid wheat research. Crop Sci. 12, 360–364,
50. Sun, J., Liu, J. & Zhang, J. A review on research and utilization of hybrid vigor of cotton. Acta Gossypii Sinica 3(1994).
51. Jean, M., Brown, G. G. & Landry, B. S. Genetic mapping of nuclear fertility restorer genes for the 'Polima' cytoplasmic male sterility n canola (Brassica napusL.) using DNA markers. Theor. Appl. Genet. 95, 321–328,
52. Schnable, P. S. & Wise, R. P. The molecular basis of cytoplasmic male sterility and fertility restoration. Trends Plant Sci. 3, 175–180,
53. Huang, M., Xu, Y., Kanjana, K., Sun, X. & Zhang, J. Flower-induction of Lemna gibbaSH0204 by salicylic acid. Plant Physiol. J. 51, 559–565 (2015).
54. Ziegler, P., Adelmann, K., Zimmer, S., Schmidt, C. & Appenroth, K. J. Relative in vitrogrowth rates of duckweeds (Lemnaceae) – the most rapidly growing higher plants. Plant Biol. 17, 33–41,
55. Wang, Z. Y., Ge, Y., Scott, M. & Spangenberg, G. Viability and longevity of pollen from transgenic and nontransgenic tall fescue (Festuca arundinacea)(Poaceae) plants. Am. J. Bot. 91, 523–530,
56. Xue, H. et al. Genetic diversity and geographic differentiation analysis of duckweed using inter-simple sequence repeat markers. Mol. Biol. Rep. 39, 547–554,

CHAPTER 9

Differential Physiological and Biochemical Responses Under Variable Culture Conditions in Micro-Propagated *Solenostemon Scutellarioides*: An Important Ornamental Plant

Ranabir SAHU, and Saikat DEWANJEE*

Advanced Pharmacognosy Research Laboratory, Department of Pharmaceutical TechnologyJadavpur UniversityKolkataIndia

ABSTRACT

Solenostemon scutellarioides is a commercially important ornamental plant. In present study, it was attempted to establish a protocol for high frequency *in vitro* regeneration of *S. scutellarioides*. Caulogenesis was found to be significant in solid MS medium supplemented with 0.5 mg dm^{-3} BAP. Combination of GA_3 (0.5 mg dm^{-3}) and BAP (0.5 mg dm^{-3}) induced high frequency regeneration coupled with higher plant height. The plantlets exposed to IBA (1 mg dm^{-3}) exhibited significant root development in terms of first appearance of root and number of roots per shoot. To improve its commercial acceptability in terms of plant architecture and foliage colouration, two weeks old *in vitro* grown plantlets were exposed to different culture conditions namely MS strength, sucrose concentration, pH and light. A variable plant types with diverse ornamental traits were developed under different culture conditions.

Keywords: anthocyanin chlorophyll *in vitro* micropropagation *Solenostemon scutellarioides*

INTRODUCTION

Solenostemon scutellarioides (L.) Codd (syn. Coleus blumei Benth.), commonly known as Coleus or Painted Nettle, is an important ornamental plant valued notably for its vibrant colourful foliage.1 In the last decade, this ornamental plant has gained popularity in the floriculture industry with the introduction of a large number of vegetatively propagated varieties with novel foliage colours. Besides novel foliage colours, opportunities exist to increase their commercial value by manipulating plant architecture. Several hundred commercial cultivars are characterized by their leaf colors, leaf shape, plant architecture and flowering characteristics. The desirable characteristics in S. scutellarioides include improved plant architecture and foliage colour quality. Bright colourful foliages improve its artistic appeal2 and faded colours decrease its product value.3 Considering these traits, commercial Coleus breeding was initiated to develop new cultivars through traditional breeding methods for improved commercial variety. The commercial cultivation of S. scutellarioides is primarily done by cutting method to regenerate similar novel variety. However, to meet with increasing commercial demand high frequency regeneration is desirable and employment of plant tissue culture may serve as an alternative source of S. scutellarioides for commercial purpose.

Plant architecture in S. scutellarioides mainly includes the branching pattern of shoots and leaf morphology. Increasing branching with improved leaf morphology is the main criteria for the selection of better commercial variety. Colouration of different organs such as flower, fruit and leaf is due to the accumulation of chlorophyll, betalains, carotenoids, flavnoids, anthocyanins etc,4 with the latter as the major pigment that impart the red and purple colours observed in S. scutellarioides leaves.5 The attractive colouration of S. scutellarioides leaves is affected by the combination of anthocyanin and chlorophyll content, which is responsible for green colouration.6 Foliage colour may be dependent on factors such as light and pH, analogously to what has been observed in flowers.7,8 Efforts to manipulate plant architecture and foliage colouration to improve commercial acceptability and manipulation of culture condition would serve productive. Therefore the objective of the present work is to establish a protocol for rapid clonal multiplication from limited donor plants and developing different commercial plant types under different culture conditions. The each type with different architecture and foliage colour would serve as the source of explants and mass multiplication achieved by using rapid clonal multiplication protocol.

RESULTS AND DISCUSSION

Nodal segments containing axillary buds have the potential to develop complete plantlets in a nutrient medium containing suitable combinations and concentrations of growth regulators.

To optimize high frequency regeneration, surface sterilized nodal segments of S. scutellarioides were cultured in MS media containing 3% sucrose and pH 5.6. After one week, explants showing axillary buds were transferred into Murashige and Skoog (MS)9 media supplemented with different combinations of phytohormones for organogenesis. Cytokinins are known to stimulate plant cell division and participate in the release of lateral bud dormancy, induction of adventitious bud formation and in growth of lateral buds.10 Different concentrations (0.5 and 1.0 mg dm−3) of cytokinins namely 6-benzylaminopurine (BAP) and kinetin (Kn) were used for the shoot regeneration (caulogenesis). Shoot regeneration response was expressed as number of regenerated shoots per explant and plant height (Figure 1 (a–c), 2a). No detrimental effect was noticed in terms of plant survival with cytokinin treatment. The overall effect of cytokinins on plant height was not promising, which resulted dwarf plantlets with reduced leaf size. The only positive effect was observed in number of shoot generation per explant. BAP (0.5 mg dm−3) was observed to be most beneficial for high frequency regeneration of shoots. No significant difference was observed in terms of plant stature and foliage colour (green with brownish shade) with BAP and Kn treatment. The positive effect of BAP at low concentration was also observed in Coleus forskohlii11 and Plumbago indica.12

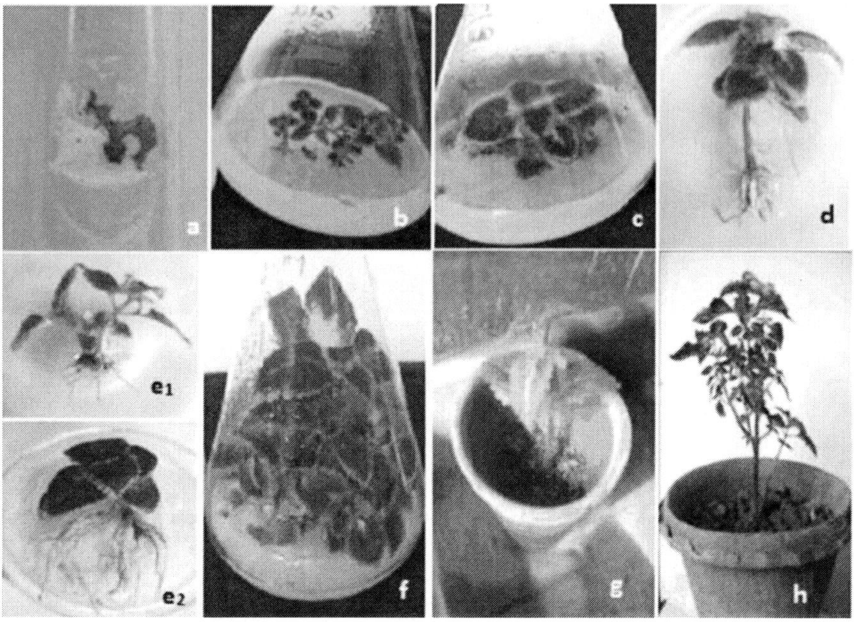

Figure 1. Different stages of micropropagation of S. scutellarioides (a–h): (a) sprouting of shoot buds; (b) initiation of shoot regeneration; (c) emergence of multiple shoot development in BAP (0.5 mg dm−3); (d) enhancement of shoot height with good survival rate in GA3 + BAP (0.5 mg dm−3 each); (e1) initiation of roots at the base of single micro-shoot; (e2) better rooting in media supplemented with IBA (1.0 mg dm−3) at the same period of culture (e1)'; (f) well established in vitro culture; (g) a hardened plant in glass house; (h) micropropagated plant growing under ex vitro environment

Gibberellic acid (GA3) is best known for its role in elongation of axial organs, such as stems, petioles and inflorescences.13 The effect of GA3 on shoot elongation is due to the increase in soluble carbohydrates induced by GA3 and available for metabolic processes.14 In present study, GA3 exhibited significant improvement in height especially to the plants previously treated with BAP (dwarf in nature) (Figure 2b). The negative effect of GA3 was observed on survival percentage after one week, particularly to those plants were exposed with higher concentration of GA3. Severe chlorosis was observed in the plants treated with GA3 (1.0 mg dm−3). But the deleterious effect of GA3 was minimized when it was used in combination with BAP. Combination of GA3 and BAP at low concentration exhibited high frequency regeneration coupled with higher plant height. Combination of GA3 (0.5 mg dm−3) and BAP (0.5 mg dm−3) exhibited best result with almost 1.4 and 1.3 fold increase in shoot numbers and plant height respectively than control (Figure 2b). Plant stature was normal as in control but some changes in plant colour were observed. A pinkish colouration was developed on the upper surface of leaf especially near to mid vein. Leaf area was broader than control and cytokinins treated plant. Leaf margin was serrated which was not common in other treatment. The positive effect of GA3 and BAP in combination was also reported in Ocimum basilicum. 15

Auxins exert a strong influence in root formation.16 For root generation (rhizogenesis), in vitro regenerated and elongated shoots were transferred into MS medium supplemented with different concentrations (0.5, 1.0 and 1.5 mg dm−3) of indole- 3-acetic acid (IAA) and indole-3-butyric acid (IBA). In course of culture, roots appeared from the base of shoots. The first appearance of root and number of roots per shoot were recorded (Figure 1 (e1, e2), 2c). Both IAA and IBA showed positive effect for early rooting as well increase number of roots per shoot. Plantlets treated with IBA (1 mg dm−1) exhibited best result as roots appeared earliest with almost 2.3 fold increase in number of roots per shoot as compared with control culture. Auxins treatment did not cause any significant morphological changes namely plant architecture and foliage colour. The success of IBA in promoting root induction has been reported in Ceropegia candelabrum17 and Hemidesmus indicus. 18

To study the effect of different external stimuli, two weeks old in vitro grown shoots were exposed with different culture conditions. The regenerated plants maintained in different culture conditions exhibited variable morpho-chemical traits to develop different plant lines with respect to architecture and foliage colouration (Figures 3–6). To study the effect of MS strength, the growth and regeneration potential was found to be improving with MS strength from 1/4 to 1 (Figure 3) but detrimental effect was observed in the cultures treated with 2MS strength. The foliage colours were also significantly varied. The culture with 1/2 and 1 MS strength exhibited beautiful foliage pattern with attractive colouration due to better anthocyanins concentration and selected as developed plant types. The effect of pH of culture media on plant architecture and leaf pattern largely varied (Figure 4). pH did not show significant variation in regeneration potential but it imparted significant difference in leaf morphology and colouration. The plants exhibited diverse attractive foliage colouration between pH 3.6–5.6. The effect of pH on both leaf morphology and foliage colour was found to be optimum at pH 5.6. Higher pH at 7.6 was found to be detrimental for the growth of S. scutellarioides. The plant exposed to pH 4.6 and 5.6 showed highest chlorophyll and anthocyanin

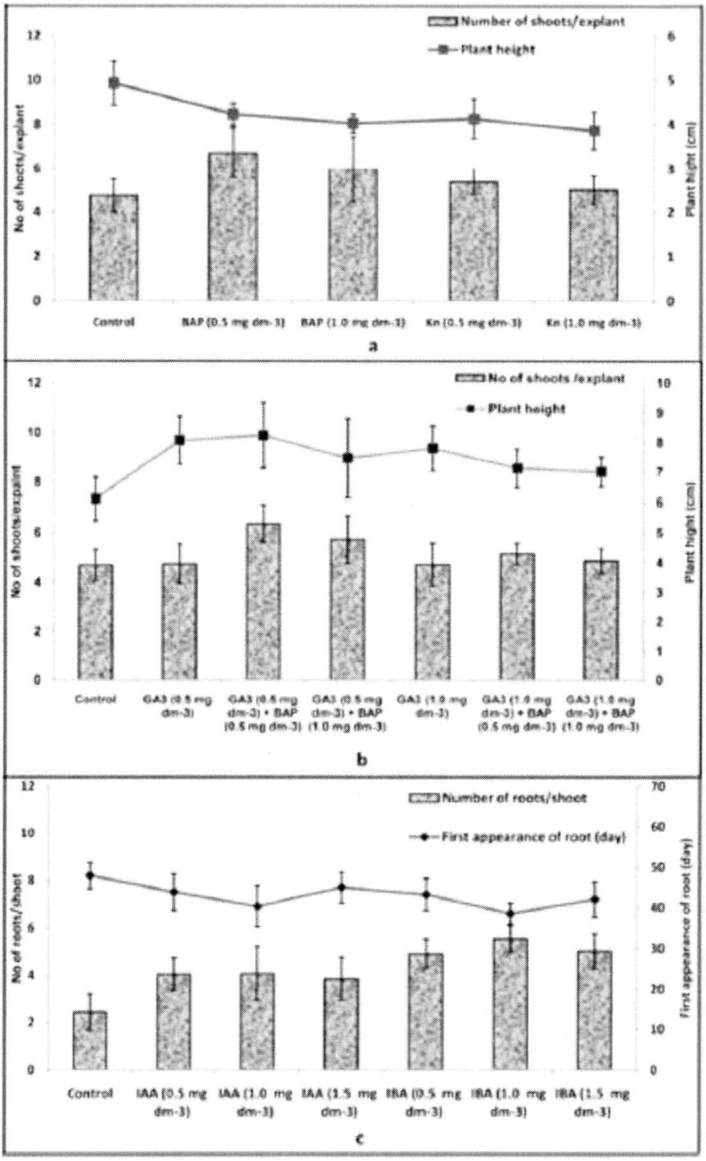

Figure 2. (a) effect of cytokinins on shoot regeneration; (b) effect of GA3 on shoot elongation; (c) effect of auxins on root regeneration from nodal explants of S. scutellarioides

content respectively. Amongst the selection, the plants regenerated in pH 3.6, 4.6 and 5.6 were chosen as developed elite types. The effect of sucrose concentration in MS media on plant architecture and foliage pattern was studied (Figure 5). Sucrose concentration did not have significant effect in shoot regeneration per explant but the height improvement was found to be optimum in culture treated with 3% sucrose. Foliage morphology and colouration was found to better in culture treated with 3% sucrose. Chlorophyll concentration in leaves gradually decreased with increasing sucrose concentration. Though sucrose concentrations gave diverse foliage colouration but for the selection of plant type, plants treated with 3% and 4% sucrose were chosen due to better leaf surface and healthy leaf structure. The effect of light on growth and regeneration potential of S. scutellarioides was gradually decreased from white, yellow, green, blue and black exposure (Figure 6). Leaf morphology remained almost similar in white, yellow and green light treatment. Detrimental effect was observed in absence of light. It also significantly reduced ornamental aspect of S. scutellarioides leaves. Chlorophyll and anthocyanin concentrations also largely varied. Green light improved chlorophyll content while white light showed higher anthocyanin content in leaves. Exposure with yellow light exhibited an intermediate concentration of both pigments. Exposure with blue light significantly reduced anthocyanins content.

The high diversity of colour in foliage of S. scutellarioides has been shown to be due variation of anthocyanins and chlorophyll ratio. Anthocyanins are key pigment for attractive pink to purple colouration in S. scutellarioides. Anthocyanins are water soluble pigment having high ecophysiological significance in plant.19 Several hypotheses described that the role of anthocyanins in osmotic regulation to drought and frost stress, antioxidant defense and protection from different light exposure. Based on the hypothesis to develop of new varieties of S. scutellarioides with improved colour quality, the shoots in vitro derived were exposed in vitro with various factors namely MS strength, sucrose concentration, pH and light. In this study, it was observed that selected factors had significant impact on foliage colour development and leaf architecture during developmental stage and the develop plants carried these traits up to ex vitro establishment. Therefore the developed variable plant types with diverse ornamental traits would serve as a secondary source of S. scutellarioides for commercial purpose.

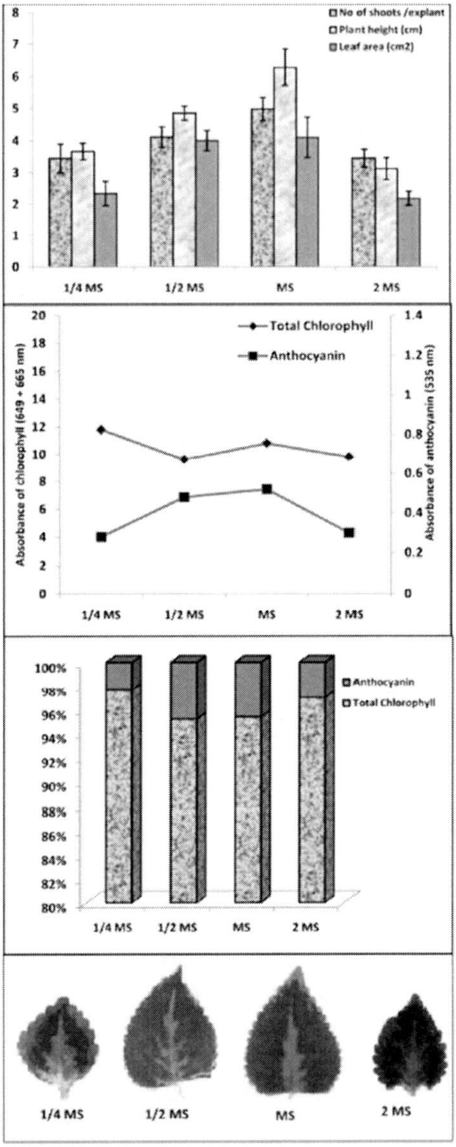

Figure 3. Effect of MS strengths on plant architecture and foliage colour development of S. scutellarioides

EXPERIMENTAL SECTION

Collection of Plant Material. S. scutellarioides L. plants were collected from Medicinal Plant Garden of Narendrapur Ramkrishna Mission, South 24 Parganas, India. The plant was authenticated by the Taxonomists of Botanical Survey of India, Shibpur, India. A voucher specimen JU/PT/PC/01/06 was deposited at our laboratory for future reference. The plants were planted in the medicinal plant garden at the Department of Pharmaceutical Technology, Jadavpur University, India.

Preparation and Sterilization of Explants. Nodal segments (0.4 ± 0.1 cm) with internodes were excised from young branches of the plants. The explants containing two buds with oppositely placed axils were used for the experiment. The explants were first washed with running tap water for removing adhering dust particles and then immersed in 1% Teepol detergent (Glaxo India Pvt. Limited, India) for 5 min with continuous shaking in a magnetic stirrer. After gentle washing with sterile distilled water for 4-5 times, the clean segments were surface sterilized with 0.1% mercuric chloride for 1 min, followed by 3-4 times washing with autoclaved distilled water In Vitro Establishment. After surface sterilization, each sterile explant was vertically implanted into single culture tube 150 mm, Borosil, India) contained 25 cm3×(25 sterilized solid MS media. The media was fortified with 3% sucrose (Hi Media, India) gelled with 0.8% Bacto agar (Hi-media, India). The pH was adjusted to 5.6 before autoclaving at 121 ºC for 20 minutes. All cultures were maintained at 22 ± 2 °C and 16 h photoperiod (light intensity 100 μmol m−2 s−1 provided by cool white fluorescent tubes, Philips, India). For adventitious bud induction, 1 week old explants were sub-cultured into 250 cm conical flasks containing 50 cm3 solid MS medium supplemented with different concentrations of two cytokinins (BAP and Kn). After two weeks, the conical flasks exhibited significant caulogenesis were sub-cultured in media containing GA3 singly or in combination with BAP at different concentrations for better shoot development. To study the effect of auxins on rhizogenesis, the healthy in vitro grown plants were transferred in solid MS medium supplemented with IAA and IBA at different concentrations. The best resulting media for rooting was identified on the basis of first appearance of root and number of roots per shoot on day 14.

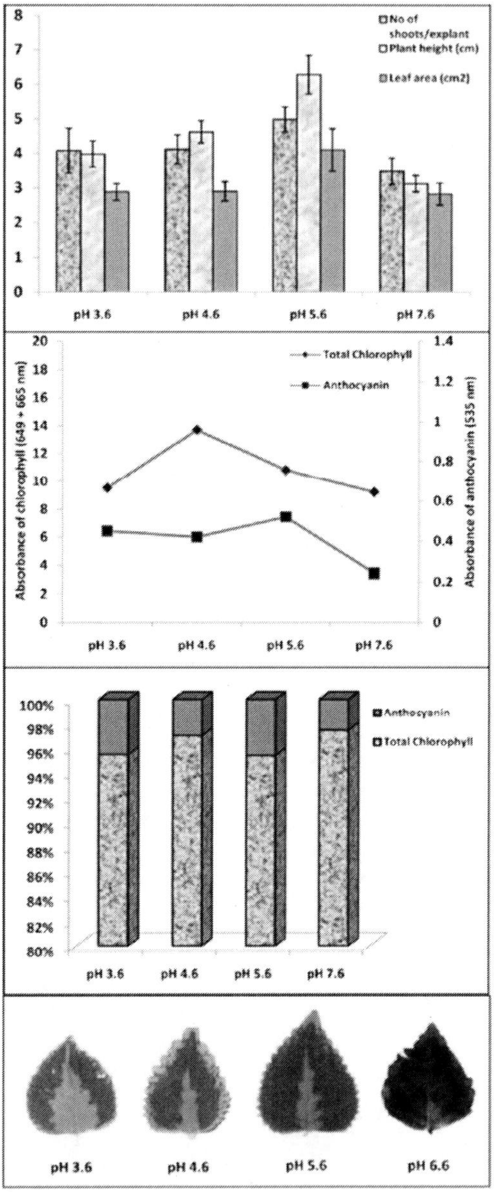

Figure 4. Effect of pH on plant architecture and foliage colour development of S. scutellarioides

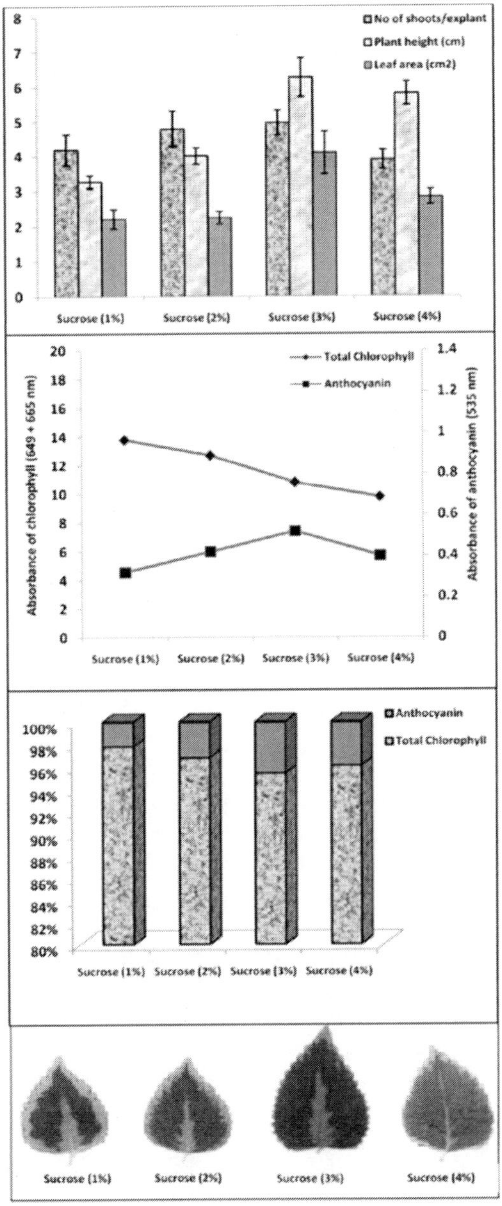

Figure 5. Effect of sucrose concentrations on plant architecture and foliage colour development of S. scutellarioides

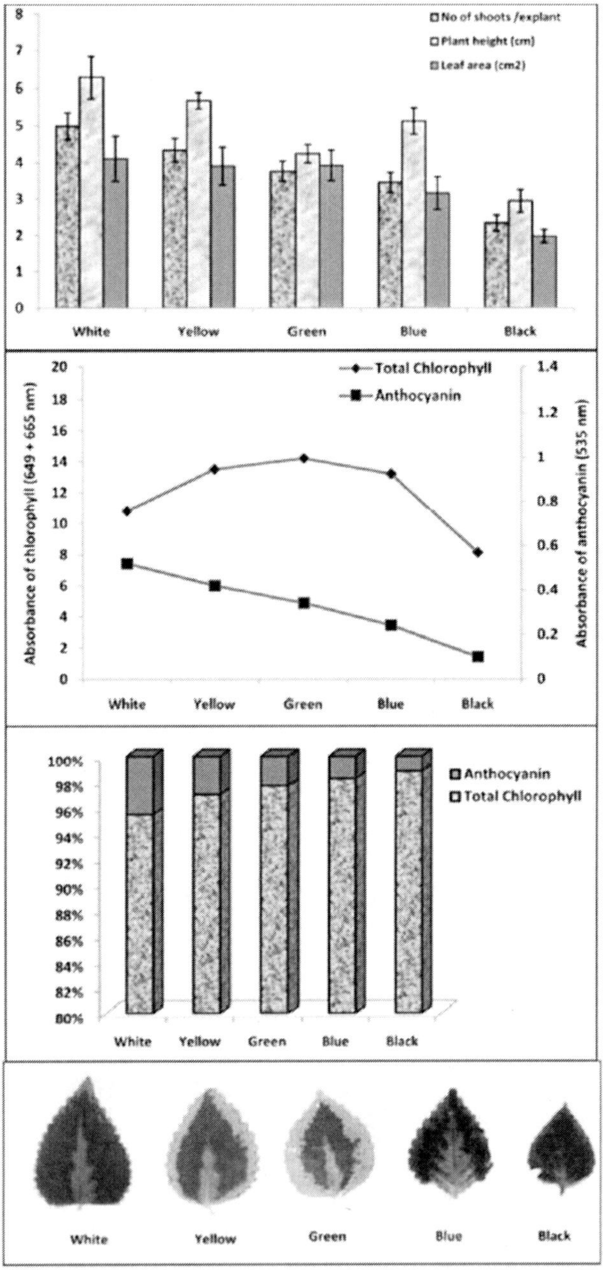

Figure 6. Effect of light on plant architecture and foliage colour development of S. scutellarioides

Ex Vitro Establishment. After two weeks, healthy plantlets with sufficient roots were removed from in vitro condition for hardening. The adhered agar was removed by repeated washing with autoclaved distilled water and the plantlets were transferred to small pots (1 for 1 plantlet), filled with autoclaved soilrite and covered with transparent polythene bags under tissue culture environment. After one week, the plantlets resumed growth were again hardened with autoclaved soilrite-soil mixture (1:1) under same condition for another week. The well hardened plants were transferred to glass house in the large earthen pots for full growth of plants under normal environmental condition for the acclimatization. After two weeks the survived plants were finally transferred in the field.

Manipulation of Media and Culture Conditions to Manipulate Plant Architecture and Foliage Color. To study the effect of different culture conditions two weeks old in vitro grown shoots were exposed with various factors namely MS strength, sucrose concentration, pH and light. One set of culture in MS with 3% sucrose, pH 5.6, under white light was maintained throughout experiment as control. Data were recorded as number of shoots per explant, plant height, leaf surface and foliage chlorophyll and anthocyanin content.

Total anthocyanins were extracted as previously described20 and absorbance was determined by spectrophotometry at 535 nm. Anthocyanin content was calculated according to the method of Murray and Hackett21 with little modification.4

For chlorophyll determination, collected leaves were frozen in liquid nitrogen. The samples were ground and chlorophyll was extracted. The absorbance was determined at 649 nm and 665 nm. The total chlorophyll a/b concentration was calculated as previously described.22 The ex vitro establishment of different plant lines were done as similar described in previous section.

Statistical Analysis. Each set of experiment consisted of 10 replicates and represented by 10 individual conical flasks. Each experiment was repeated thrice and the data were analyzed statistically using SPSS software (10.0.5, 1999, SPSS Inc).Variability in data was expressed as the mean ± standard deviation.

Electronic Supplementary

Material Supplementary material is available in the online version of this article at http://dx.doi.org/ 10.1007/s13659-012-0035-y and is accessible for authorized users.

ACKNOWLEDGMENTS

The authors are thankful to University Grants Commission, New Delhi, India for providing fund to Mr Ranabir Sahu through Rajiv Gandhi National Fellowship, 2011. We are also thankful to Dr. Md. Nasim Ali, Ramkrishna Mission Vivekananda University, Ramkrishna Mission Ashrama Narendrapur, India for providing plant material, S. scutellarioides, for this study.

REFERENCES

1. Garcia, J. F.; O'Neil, R. J. *Biol. Control***2000**, *18*, 225–234.
2. Cantazaro, C.; Bhatti, S.; Muhammad, S.; Abdullah, S. *Hortscience***2005**, *40*, 993–1147.
3. Oren-Shamir, M.; Nissim-Levi, A.; Ovadia, R.; Kagan, S.; Shaked, S. *Acta Horticult.***2003**, *624*, 171–176.
4. Nguyen, P.; Cin, V. D. *Plant Physiol. Bioch.***2009**, *47*, 934–945.
5. Lebowitz, R. *Plant Breeding Review***1985**, *3*, 343–360.
6. Rife, D. C. *J. Hered.***1948**, *39*, 85–91
7. Yoshida, K.; Kawachi, M.; Mori, M.; Maeshima, M.; Kondo, M.; Nishimura, M.; Kondo, T. *Plant Cell Physiol.***2005**, *46*, 407–415.
8. Mol, J.; Grotewold, E.; Koes, R. *Trends Plant Sci.***1998**, *3*, 212–217.
9. Murashige, T.; Skoog, F. *Physiol. Plant.***1962**, *15*, 473–497.
10. Coello, C. Y.; Miceli, C. L.; Orantes, C.; Dendooven, L.; Gutiérrez, F. A. *Gayana Bot.***2010**, *67*, 19–26.
11. Dube, P.; Gangopadhyay, M.; Dewanjee, S.; Ali, M. N. *Indian J. Biotechnol.***2011**, *10*, 228–231.
12. Bhattacharya, R.; Ray, A.; Gangopadhyay, M.; Bhattacharya, S. *Plant Cell Biotechnol. Mol. Biol.***2008**, *8*, 39–46.
13. DeMason, D. A. *Planta***2005**, *222*, 151–166.
14. Bialecka, B.; Kepcznski. J. *Plant Growth Regul.***2007**, *51*, 21–31.
15. Sahoo, Y.; Pattnaik, S. K.; Chand, P. K. *In Vitro Cell. Dev. Biol. Plant***1997**, *33*, 293–296.
16. Gaspar, T. H.; Kevers, C.; Penel, C.; Greppin, H.; Reid D. M.; Thorpe, T. A. *In Vitro Cell. Dev. Biol. Plant***1996**, *32*, 272–289.
17. Beena, M. R.; Martin, K. P.; Kirti, P. B.; Hariharan, M. *Plant Cell Tiss. Org.***2003**, *72*, 285–289.
18. Sreekumar S.; Seeni S.; Pushpangadan P. *Plant Cell Tiss. Org.***2000**, *62*, 211–218
19. Close, D. C.; Beadle, C. I. *The Bot. Rev.***2003**, *69*, 149–161.
20. Singh, A.; Selvi, M.; Sharma, R. *J. Exp. Bot.***1999**, *50*, 1619–1625.
21. Murray, J. R.; Hackett, W. P. *Plant Physiol.***1991**, *97*, 343–351.
22. Goncalves, J.; Antonio, R.; Vieira, M.; Vieira, G. *Rev. Bras. Fisiol. Veg.***2001**, *13*, 149–157.

CHAPTER 10

Flower Induction, Microscope-Aided Cross-Pollination, and Seed Production in the Duckweed *Lemna Gibba* with Discovery of a Male-Sterile Clone

Lili Fu[1], Meng Huang[1], Bingying Han[1], Xuepiao Sun[1], K. Sowjanya Sree[2], Klaus-J. Appenroth[3] & Jiaming Zhang[1]

[1]*Institute of Tropical Bioscience and Biotechnology, MOA Key Laboratory of Tropical Crops Biology and Genetic Resources; Hainan Bioenergy Center, CATAS, Haikou, Hainan Province, 571101, China.*

[2]*Department of Environmental Science, Central University of Kerala, RSTC, Padanakkad, 671314, Kerala, India. 3 Institute of Plant Physiology, University of Jena, Dornburger Str. 159, 07743, Jena, Germany.*

ABSTRACT

Duckweed species have a great potential to develop into fast-growing crops for water remediation and bioenergy production. Seed production and utilization of hybrid vigour are essential steps in this process. However, even in the extensively-studied duckweed species, *Lemna gibba*, flower primordia were often aborted prior to maturation. Salicylic acid (SA) and agar solidification of the medium promoted flower maturation and resulted in high flowering rates in *L. gibba* 7741 and 5504. Artificial cross-pollination between individuals of *L. gibba* 7741 yielded seeds at high frequencies unlike that in *L. gibba* 5504. In contrast to clone 7741, the anthers of 5504 did not dehisce upon maturation, its artificially released pollen grains had pineapple-like exine with tilted spines. These pollens were not stained by 2,5-diphenylmonotetrazoliumbromide (MTT) and failed to germinate. Therefore, clone 5504 is male sterile and has potential application with respect to hybrid vigour. Moreover, pollination of flowers of 5504 with 7741 pollen grains resulted in intraspecific hybrid seeds, which was confirmed by inter-simple sequence repeat (ISSR) markers. These hybrid seeds germinated at a high frequency, forming new clones.

INTRODUCTION

Lemnaceae, commonly known as duckweeds, is a family of small aquatic monocotyledonous plants that comprise five genera and 37 species[1,2]. They represent the smallest flowering plants with reduced morphology[3]. Duckweeds can rapidly assimilate nutrients such as nitrogen and phosphate from municipal, industrial, and agricultural wastewaters and have shown great value in wastewater remediation[4,5,6,7,8,9]. The biomass of duckweeds may contain high starch content and could be used to produce bioethanol and biobutanol[10,11,12,13,14]. Some duckweed clones contain high protein content, and have been used as high quality livestock feed[15,16,17]. Even the duckweed biomass accumulated on wastewaters may be used for feed production[18,19,20].

Lemna gibba is one of the duckweed species that has been widely used in research and applications. *L. gibba* is very efficient in removal of organic and inorganic nutrients from domestic and industrial wastewater [21,22,23,24], as well as heavy metals such as lead and uranium[25]. *L. gibba* was also found to be efficient in removal of harmful organisms such as *Giardia cysts*, *Cryptosporidium oocysts*, faecal coliforms and coliphage from wastewater[26]. It can also be used in bioassays of toxicity of industrial waste[21].

The potential of duckweeds in commercial applications has brought forward a need for domestication and the breeding of elite varieties serving different purposes. Seed production and application of hybrid vigour have an enormous importance in making duckweed a future crop plant. However, duckweeds reproduce mainly vegetatively. In this process, mother fronds form genetically identical daughter fronds that are therefore called a clone. Plants collected from different places are in most cases genetically different from one another and form different clones. Flowering is a rare event in nature and in the laboratory cultures in most duckweed species27,28. Natural flowers are morphologically identical to those induced artificially (Figure 1). Artificial induction of flowering and seed production in duckweeds is therefore needed for future efforts to breed new duckweed varieties by cross-pollination to improve their efficiency.

Figure 1.*L. gibba* plant with natural flowers. Red arrows indicate flowers; M, mother frond; D, daughter frond; G, granddaughter frond.

Many efforts have been made to induce flowering in different duckweed species. Salicylic acid (SA) is one of the most important plant hormones used in duckweed flower induction. SA induced flower formation in *L. gibba* G3 and *L. paucicostata* LP6 (valid term *L. aequinoctialis*), and the continuous presence of SA was found to be essential to obtain maximum flowering29. SA induced flowering in *L. gibba* G3 in coordination with day length30,31. Flower promotion was obtained with day lengths of more than 9 h, but never with less than 8 h. Salicylic acid treatment caused a shift in the critical day length of about 2 hours31. Benzoic acid is another chemical that is effective in flower induction. When *L. gibba* G3 was grown on 1/10 M medium or 1/2 H medium under long day conditions, addition of benzoic acid to either media led to a flowering response, which was inhibited by the plant hormones IAA, GA3, ABA, and zeatin[32,33].

Analysis of flower induction in previous reports was based on dissecting the budding pockets of *L. gibba*. Fronds with flower primordia, i.e. immature flowers, visible under a dissecting microscope were usually considered as flowering[31,32,34], whereas no data on induction of mature flowers are available up to date to the best of our knowledge. However, only mature flowers can be used for cross-pollination for the purpose of breeding. In this study, we present efficient methods to induce the formation of mature flowers and production of intraspecific hybrid seeds in *L. gibba*. For that purpose, two geographically distinct *L. gibba*, clones 7741 (origin: Italy, better known by the older term G3) and 5504 (origin: China), and one clone of *L. minor*, clone 7210 (origin: Japan) were investigated. The genetic background of these clones was investigated by barcoding according to the described methods[35,36,37](Supplementary material, Figure S1). From clone 7741, seeds were produced by artificial self-pollination. Moreover, intraspecific hybrid seed production by pollen transfer between different clones of *L. gibba* under a microscope was successfully carried out for the first time. This strategy will be used in breeding programmes in the future to make duckweed a crop plant. Moreover, the analysis of pollen viability and cross-pollination led to the identification of a male-sterile line of *L. gibba* which is the first report in Lemnaceae family.

RESULTS

Flower maturation in *L. gibba* is limited in conventional culture media

Three commonly used duckweed media including liquid and/or semi-solid media H, E, and MH were used to study their influence on the rate of flowering. Two clones of *L. gibba*, 7741 and 5504, and one clone of *L. minor*, 7210 were incubated on these media under 16 h light period according to the previous reports for *L. gibba* G3[31,38]. The fronds of all clones grew well on all media (Figure 2a, controls). Immature flowers were found in more than 50% of the colonies of *L. gibba* 5504 and 7741, but no flower primordia could be observed in *L. minor* 7210 (data not shown). Mature flowers were hardly induced in any of the clones (Table 1, Figure 2b,c), the highest rate of mature flowers obtained in solidified MH medium was approximately 2% for both of the *L. gibba* clones. First, we investigated photoperiod as a possible reason for the low percentage of mature flower formation. The clones were cultured under 8–16 h photoperiods. Immature flowers were induced at high frequencies in *L. gibba* 7741 and 5504 but not in *L. minor* 7210 under photoperiod ≥10 h (data not shown). However, most of these flower primordia in *L. gibba* were aborted and only a very low number of mature flowers were developed when grown in MH medium. Mature flowers were not observed at all in H and E media (Table 1). Flower primordia often co-existed with frond primordia in the same budding pocket (Figure 2c).

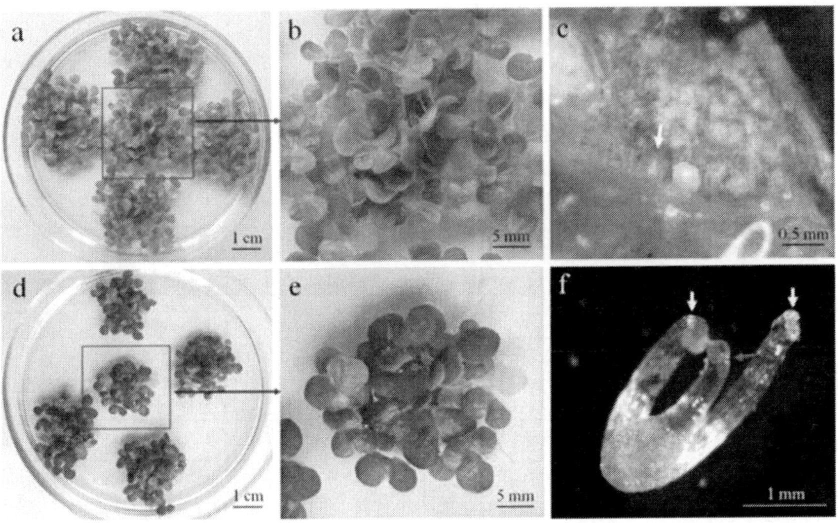

Figure 2. Comparison of *L. gibba* clone 7741 plants grown in media without ((**a**–**c**); upper row as a control) and with ((**d**–**f**); bottom row) salicylic acid (SA). (**a**) Control fronds grown in medium without SA showing high amount of biomass 15 days after inoculation. (**b**) Magnification of the *red-squared* region in (**a**) showing the absence of flowers. (**c**) Dissected budding pocket of a frond growing on SA-free medium; the white arrow indicates frond primordium, and the red arrow indicates flower primordium. (**d**) Fronds growing on medium supplemented with 20 μM SA showing lower amount of biomass after 15 days. (**e**) Magnification of the *red-squared* region in (**d**), showing flowers (red arrows); (**f**) Mature flower dissected from a budding pocket with pistil (red arrow), and two stamen (white arrows; anthers have shed).

SA promotes flower maturation

When SA (e.g. 20 μM) was supplemented to E and MH media, the vegetative growth of all three clones, regardless of flower formation, slowed down although at different concentrations (cf. Figure 2a,d; Figure 3A). Mature flowers with one pistil and/or two stamens bending upward were observed under stereomicroscope (for *L. gibba* 7741 see Figure 2e,f). Salicylic acid (SA) in MH medium resulted in higher flower induction rates than in E medium (Table 1). *Lemna gibba* 7741 had flowering rates of 51% and 74% in E + SA and MH + SA, respectively, while clone 5504 had flowering rates of 6.5% and 54%, respectively (Table 1). Even in the presence of SA, *L. minor* 7210 did not show any flower development.

Table 1: Influence of different basic media and salicylic acid on the rate of mature flower formation.

Clone	Nutrient medium	Time to first flower emergence (d)	Flowering rate (%)
Lemna gibba 7741	E	n/a	0
	H	n/a	0
	MH	20	1.7 ± 0.5
	E + 20 μM SA	10	51.0 ± 1.1
	MH + 20 μM SA	12	73.5 ± 3.5
Lemna gibba 5504	E	n/a	0
	H	n/a	0
	MH	20	2.3 ± 0.8
	E + 20 μM SA	14	6.5 ± 3.0
	MH + 20 μM SA	12	54.1 ± 4.2
Lemna minor 7210	E	n/a	0
	H	n/a	0
	MH	n/a	0
	E + 20 μM SA	n/a	0
	MH + 20 μM SA	n/a	0

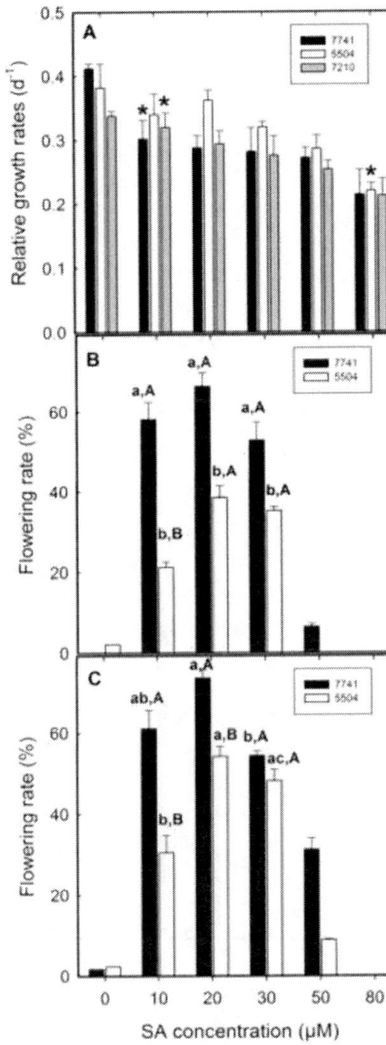

Figure 3. Dose-response effects of salicylic acid (SA) on growth and flowering rates. (**A**) Inhibitory effect of SA on the growth of *L. gibba* clones 7741 and 5504 and *L. minor* clone 7210 in semi-solid medium. Asterisks indicate the lowest concentration that significantly inhibited the growth of the three clones (students's t-test, $\alpha < 0.05$). (**B**) Effects of SA concentration in liquid medium on the flowering rates of *L. gibba* 7741 and 5504. (**C**) Effects of SA on solidified medium on the flowering rates of *L. gibba* 7741 and 5504. Note: Data are means of three independent experiments and are presented as mean ± standard error. Significance of differences in (**B**) and (**C**), were tested by one-way ANOVA ($\alpha < 0.05$); Lower case letters mark significance of the effect of different SA concentrations, capical letters evaluate the differences in the response of both *L. gibba* clones. The results on *L. minor* 7210 in (**B**) and (**C**) were not presented because there was no stimulating effect.

The first mature flowers were observed approximately 10–14 days after application of SA (Table 1, Figure 4). The number of newly developed mature flowers scored daily, increased steadily in the following days until day 20, and declined thereafter in the two *L. gibba* clones. Approximately 70% of total mature flowers were obtained between day 17 and 21, with a peak around day 20 (Figure 4). Approximately 17 and 31 flowers per plate on an average were obtained at day 20 in *L. gibba* 5504 and 7741, respectively. The flowering rates in the following experiments were presented always at day 20.

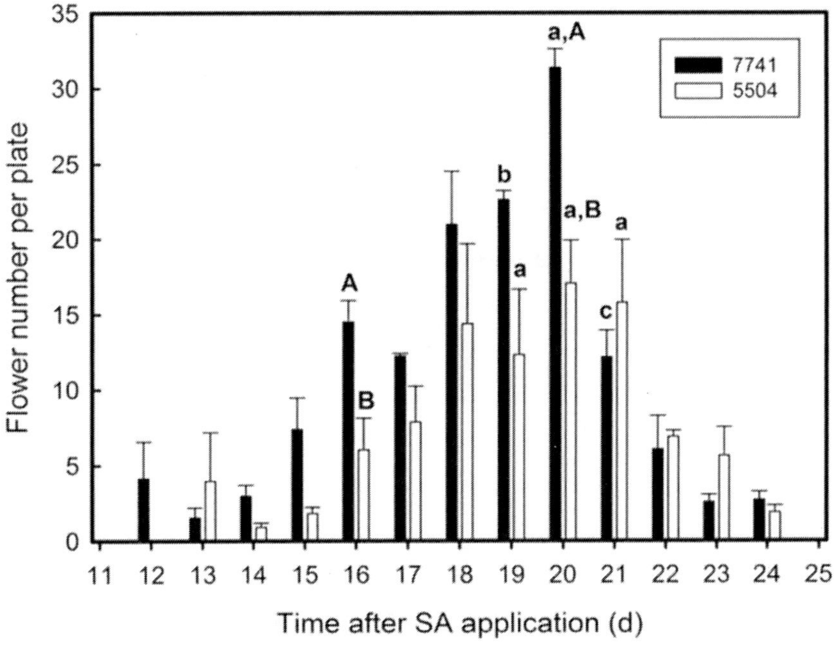

Figure 4. Temporal distribution of the number of flowering fronds. The fronds were cultured in 1/2 MH medium supplemented with 20 μM salicylic acid. Significance of differences from day 15 to day 21 after SA application were evaluated by ANOVA at a significance level of 0.05. Lower case letters indicate differences in the flower number over time, capital letters mark differences between *L. gibba* clones 7741 and 5504.

Effect of SA concentration and agar solidification on flowering rate

Flowering rates were very low without SA in the nutrient medium (Table 1). When the lowest tested concentration of SA (10 μM) was supplemented to the liquid medium, the flowering rates already abruptly increased (Figure 3B). The optimum SA concentration was 20 μM, although hardly any significant difference could be detected between 10 and 30 μM (in clone 5504 the rate was lower at 10 μM). At 20 μM SA flowering rates were 66% and 38% for *L. gibba* 7741 and 5504, respectively. When the SA concentration was increased further, flowering rates decreased (Figure 3B). Almost no flowers were induced when the SA concentration was increased to 80 μM and frond death rate increased to approximately 70% (data not shown).

Semi-solid medium has not been tested previously for flower induction in duckweeds. The general effect of SA on the flowering rates in this medium was similar to that in the liquid medium, and the optimum SA concentration was also 20 μM (Figure 3C). However, compared to the liquid media, the flowering rates in the semi-solid media were slightly higher in general and significantly higher at 20, 30 and 50 μM SA in clone 5504 (Figure 3B,C). The highest flowering rate in the semi-solid media reached 74% and 54% for clones 7741 and 5504 at 20 μM SA, respectively.

Difference in emergence of flowers and pollen characteristics in *Lemna gibba* 7741 and 5504

The stamens of *L. gibba* 7741 usually appeared ahead of the pistil (Figure 5a), and the anthers released pollen grains rapidly after the appearance of pistil (Figure 5b). Each anther of this clone contained approximately 210 pollen grains. In contrast, in *L. gibba* 5504 the pistil emerged first (Figure 5c), and the stamens could be observed only by dissecting the budding pocket at this moment (Figure 5d). A dew droplet was often observed on top of the pistil before stamens emerged (Figure 5e), which was suggested to help in pollination. Moreover, the anthers of *L. gibba* 5504 were never observed to release pollen grains until they withered (Figure 5f). Their pollen grains were released artificially with a dissecting needle, having a smaller number of approximately 180 grains per anther.

Scanning electron microscopy indicated that the slits of the mature anther of *L. gibba* 5504 remained closed (Figure 6a), while that in the anthers of *L. gibba* 7741 was widely open (Figure 6b). The artificially released pollen grains of *L. gibba* 5504 had pineapple-like exine with tilted spines of 0.6 to

1.3 μm in length (Figure 6c). The sizes of the pollen grains varied in diameter from 15 to 21 μm. In contrast, the pollen grains of *L. gibba* 7741 had smooth exine with uniformly and vertically distributed spines (Figure 6d). The sizes of 7741 pollen grains were relatively uniform with a diameter between 20 to 22 μm.

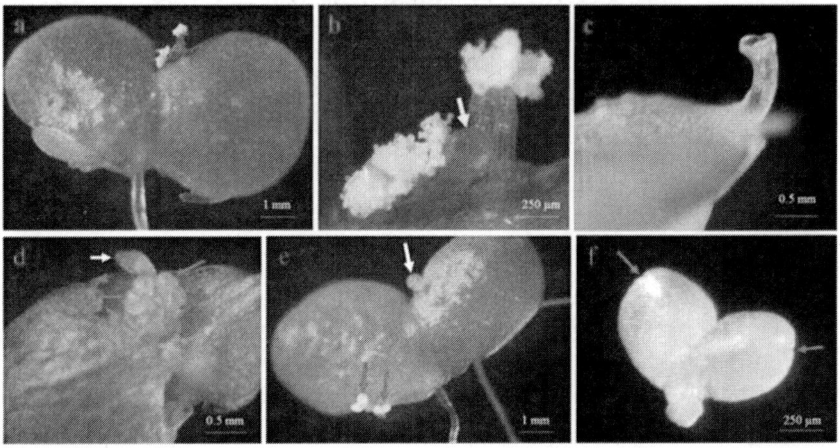

Figure 5. Flowering of *L. gibba* clones 7741 (**a**,**b**) and 5504 (**c–f**) under the influence of salicylic acid. (**a**) Flowering colony of clone 7741. (**b**) Magnification of the flower in (**a**), showing the stigma (yellow arrow) and pollens on the stamens. (**c**) Pistil emerged from the budding pocket of clone 5504 bending upward. (**d**) the same frond in (**c**) dissected from the ventral side to show the two stamens in the budding pocket (red arrows), the white arrow indicates the pistil. (**e**) Colony of 5504 with two stamens (red arrows) in one frond after the pistil withered, and a liquid droplet secreted by the pistil in another frond (white arrow). (**f**) Anther of clone 5504 that did not dihisce before withering (red arrows indicate the closed slits).

Pollen viability test by chemical staining and germination

Several chemical staining methods were tested to distinguish viable and non-viable pollen grains in duckweeds. The MTT method was most effective, staining viable pollen grains of *L. gibba* 7741 violet-purple whereas dead pollen grains (killed by heating) lacking almost any stain (Supplementary Figure S2, Supporting Information). Pollen grains of *L. gibba* 5504 and 7741 were then stained in parallel experiments (Figure 7a,b). Most of the pollens of clone 7741 were stained dark violet (approximately 67 ± 3%; Figure 7a), and were considered viable.

Approximately 20% of the pollen grains of clone 7741 were not stained and were considered non-viable. The rest of the pollen grains (approximately 10%) were stained pale violet, and their viability was therefore uncertain. In contrast, none of the pollen grains of clone 5504 were stained dark violet (Figure 7b).

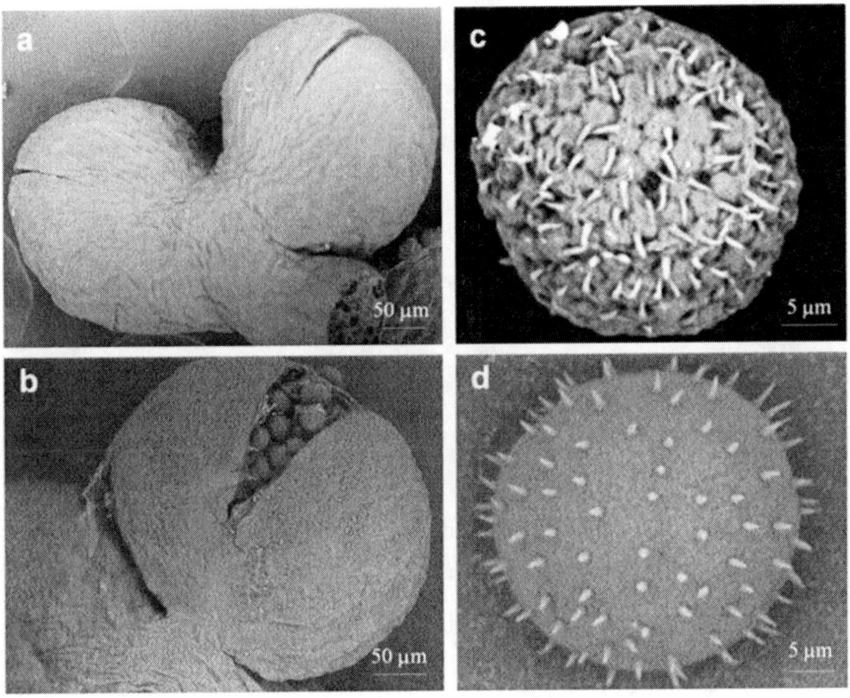

Figure 6. Scanning electron micrographs of anthers and pollen grains. (**a**) Anther of *L. gibba* 5504 showing the closed slit. (**b**) Anther of *L. gibba* 7741 with opened slit. (**c**) Pollen grain of clone 5504 showing pineapple-like exine and tilted spines. (**d**) Pollen grain of clone 7741 showing the smooth exine and uniformly distributed spines.

The viability of pollens was further tested by germination experiments. Some of the pollen grains of clone 7741 germinated within five minutes, and germination frequency was approximately 40% as counted 3 h later (Figure 7c), whereas none of the pollen grains of clone 5504 germinated even if the incubation time was prolonged to 24 h (Figure 7d). Therefore, we concluded from both experiments (Figure 7b,d) that pollen of clone 5504 was sterile.

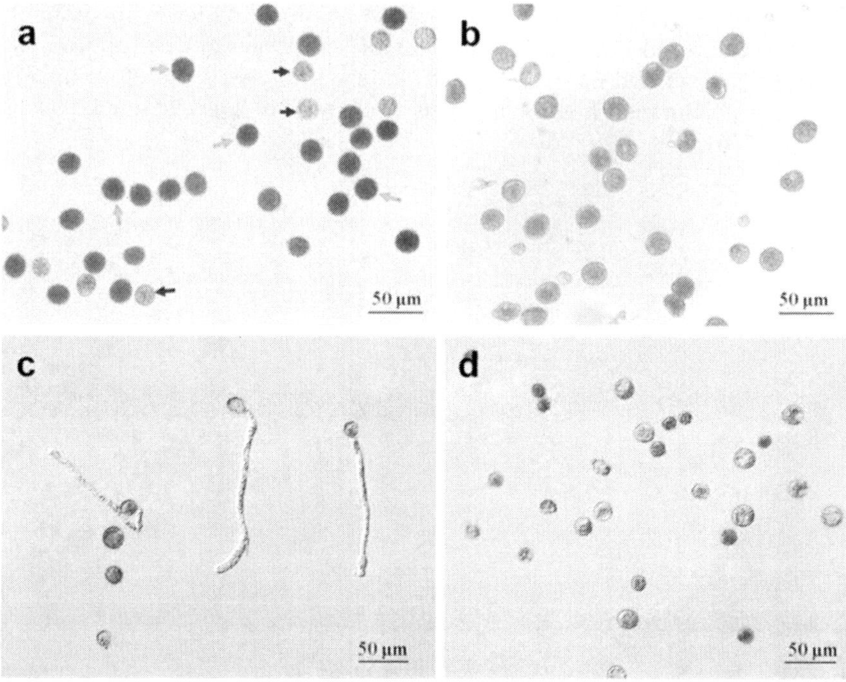

Figure 7. Viability test of pollen grains from *L. gibba* 7741 and 5504. (**a**) Pollen grains from clone 7741 stained with MTT solution. (**b**) Pollen grains from clone 5504 stained with MTT solution. (**c**) Germinating pollen grains of clone 7741. (**d**) Lack of germination of pollen grains of clone 5504 after 16 h of incubation under germination conditions. The green and the blue arrows indicate the viable and the non-viable pollen grains, respectively.

Seed formation following pollination

Natural pollination of flowers of *L. gibba* in the culture plates rarely produced seeds. Artificial pollination between individuals of *L. gibba* 7741, however, produced seeds (Figure 8a) with a setting rate of approximately 20–30%. Two or three seeds were most often observed enclosed in the thin-layer pericarp of the fruit (Figure 8b), one or four seeds were observed occasionally in one fruit. The mature seeds were white and oval with a purple operculum to the posterior end (Figure 8c). Scanning electron microscopy showed that the seed coat was covered with longitudinal-ribs (Figure 8d).

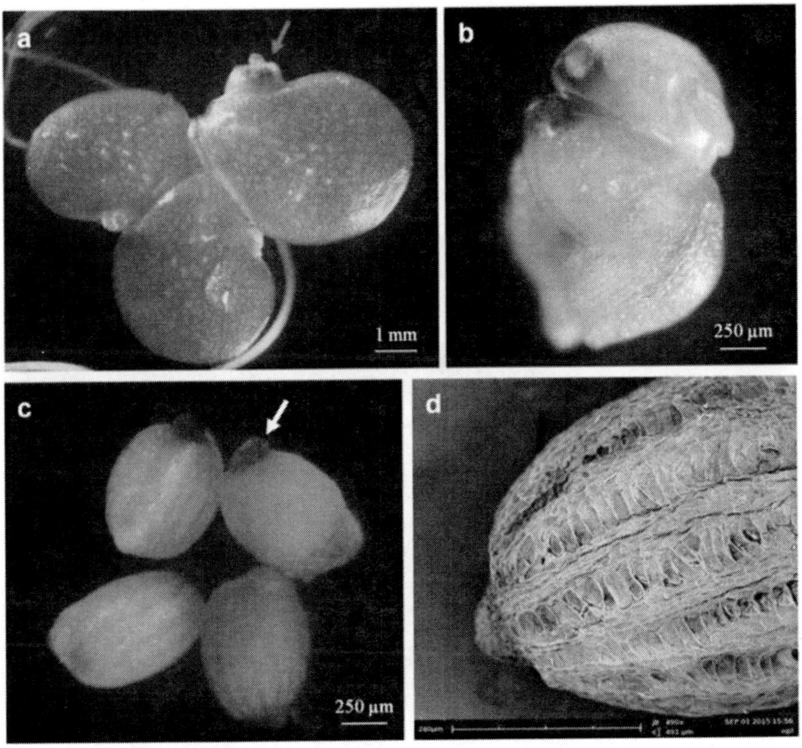

Figure 8. Seeds of *L. gibba* 7741. (**a**) Colony with an immature fruit (red arrow). (**b**) Immature fruit with two seeds. (**c**) Mature seeds with purple operculum (white arrow). (**d**) Scanning electron graph of a seed showing the longitudinal-ribs.

In contrast to clone 7741, no seeds were obtained from clone 5504 after natural and artificial pollination between individuals of this clone.

Artificial cross-pollination between *L. gibba* clones and validation of hybrids

Artificial cross-pollination of the stigmas of clone 7741 using pollen grains of clone 5504 did not result in any seed formation. However, artificial pollination of the stigmas of clone 5504 using the pollen grains of clone 7741 produced intraspecific hybrid seeds. These seeds were morphologically identical to those resulted from self-pollination of clone 7741 (Figure 9a). Also, the hybrid seeds germinated normally at the same rate as those formed after self-fertilization: seeds from clone 7741 alone germinated to 100% (100 seeds tested), and hybrid seeds germinated to

97% (35 seeds tested). Hybrid colonies were formed after germination as shown in Figure 9. The initial hybrid fronds were small compared to their parents (Figure 9b); however, the subsequent fronds grew larger gradually until they reached the size of their parents.

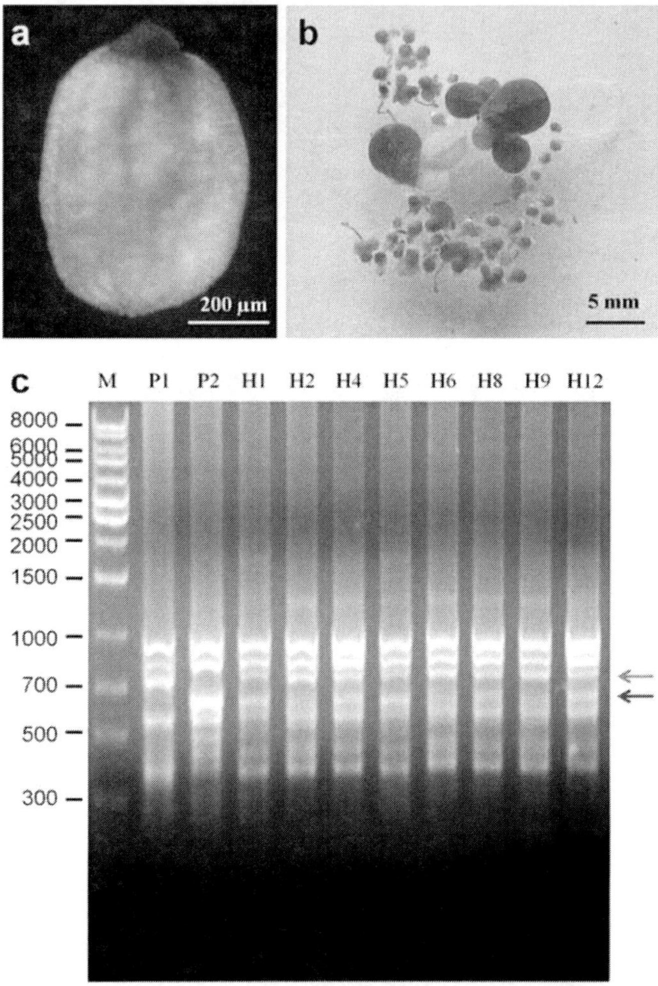

Figure 9. Germination of hybrid seeds and ISSR validation. (**a**) A typical hybrid seed from crossing between clones 5504 and 7741. (**b**) Hybrid clones five days after germination, their female (F) and male (M) parents are used as size controls. (**c**) ISSR validation of the hybrid clones; M, molecular weight marker (Axygen 1 kb ladder); P1, male parent 7741; P2, female parent 5504; H1 through to H12, hybrid clones; the green arrow indicates 7741-specific band, the red arrow indicates 5504-specific band.

In order to prove that indeed hybrids were formed between the paternal clone 7741 and the maternal clone 5504, the parental and hybrid clones were investigated by ISSR by testing 50 primers. Primer B825 showed parent-specific bands and was used to test the assumed hybrid clones (Figure 9c). Results indicated that both parental bands could be clearly detected in the hybrid lines, therefore, the hybrid lines inherited the genetic properties of both parents (Figure 9c).

DISCUSSION

Flower induction in *Lemna gibba* but not in *Lemna minor*

In order to develop a method for efficient flower induction, which serves the purpose of cross-pollination, we measured the flowering rate as the number of colonies with mature flowers divided by the number of total colonies. The previously used number of immature flowers30, 38, 39 is not useful for subsequent pollination and seed formation experiments. We successfully induced mature flowers at high frequencies for *L. gibba* clones 7741 and 5504, respectively. No flower induction was observed, under any of the conditions tested, in *L. minor* 7210, which suggested that different duckweed species responded differently to SA induction and may need different methods to induce flowering, e.g., in the case of *L. minor*, ethylenediamine-N,N'-bis (2-hydroxyphenylacetic acid) (EDDHA) is a successful flower-inducer40. Although no flowering was induced in *L. minor* 7210 by SA, its growth rate was reduced significantly, which indicated that growth inhibition was not a result of flowering. Flowering and vegetative growth are not always competitive in duckweeds, for example, *Wolffia microscopica* was reported to grow very rapidly during continuous flowering season41.

In the previous reports, authors had considered immature flowers to quantify flower induction30, 38, 39. However, this parameter is not useful for subsequent pollination and seed formation experiments as most of the immature flowers were aborted during subsequent development. In the present paper a method was developed to induce the formation of mature flowers at high rates, calculated as the number of colonies with mature flowers divided by the number of total colonies. Without SA in the MH or E media, flower primordia (i.e. immature flowers) were induced in *L. gibba* 7741 and 5504 at high frequencies. However, mature flowers were obtained only at very low percentages (ca. 2% in MH medium). Supplementation of media with SA increased the rates of formation of mature flowers dramatically, for both *L. gibba* 7741 and 5504. The results in semi-solid MH medium with SA were even slightly higher than in liquid medium. Formation of high percentage of mature flowers is a prerequisite for the production of seeds.

It was reported that ammonium in the Hutner's medium inhibited the flower primordia formation in *L. gibba* G3[30,38]. This inhibition effect could be reversed by addition of SA and hexacyanoferrate (III) to the medium[30]. However, we obtained high percentage of flower primordia in *L. gibba* 7741 and 5504 even in the ammonium containing MH medium, which was more than 50%, respectively. Therefore, ammonium may not be an inhibition factor for flower primordial differentiation at least in clones 7741 and 5504.

Moreover, the flowering rates presented here were based on the mature flower counts at the end of the experiment (20 days), when flower percentage was at the climax (Figure 3), ignoring the flowers matured before the end of the experiments. Therefore, the actual flowering rates were even higher than what are presented here. We concluded that the induction of mature flowers with SA is a useful technique for later cross-pollination and breeding.

Evaluation of pollen viability

In order to produce seeds, pollen must be viable. Chemical staining is a widely used method to examine pollen viability[42,43,44]. We used several staining reagents to test the viability of *L. gibba* pollen grains and found that the MTT staining was the most effective (Supplementary Figure S2). Heat-killed pollens were recommended previously as a negative control[44]. We adopted this method to prepare dead pollens. However, the spines or echinates were removed from the exine by the heat-killing process (Supplimentary material, Figure S2), which resulted in varied morphology, and therefore need to be improved in the future. Regardless of this, pollens of *L. gibba* 7741 clearly showed viability as tested by staining and pollen germination. This was not the case for pollens of *L. gibba* 5504 (Figure 7). Thus, vital pollens were produced from mature flowers in clone 7741 but not in clone 5504.

Hybrid vigor and male sterility

It has been shown in the present paper that the success of spontaneous self-fertilization can be greatly improved by artificial self-fertilization and that even intraspecific cross-fertilization between different clones of *L. gibba* is possible to succeed with high rate. Artificial pollen transfer from clone 7741 to stigmas of clone 5504 resulted in fertilizaiton and formation of seeds.

These seeds germinated normally and formed new vegetative plants. To prove that these plants were intraspecific hybrids formed by clone 7741 as pollen-donor and 5504 as pollen acceptor, parent plants and seedlings were analyzed by ISSR. The presence of parent-specific bands in the seedlings demonstrated that these plants are indeed hybrids. Artificial transfer of pollens from clone 5504 to stigmas of clone 7741 did not result in fertilization. We conclude that *L. gibba* 5504 is male-sterile. To the best of our knowledge, this is the first report about male-sterility in the duckweed family. Our conclusion is based on the following results: First, the emergence of the stamens of clone 5504 lagged behind the pistil. Second, the anthers of clone 5504 were not ruptured and pollen grains were not released. Third, the pollen grains of clone 5504 had an underdeveloped and pineapple-like exine with tilted spines. Fourth, they were not stained by MTT, and finally, these pollens did not germinate, and did not result in seeds by pollinating on stigmas of both clones 5504 and 7741.

Male sterility is one of the bases in commercialization of hybrid vigour and has been widely used in hybrid seed production of maize45,46, rice47,48, wheat49, cotton50, oilseed rape51, and many other crops52. The discovery of a male-sterile line of duckweed may be useful in future application of hybrid vigour in different applications of duckweed.

METHODS

Plant materials and culture media

Three axenic clones, *Lemna gibba* L. 5504, 7741 and *L. minor* 7210, were used. Clone 5504 was isolated from the Innovation River in Pudong, Shanghai, China, with an original accession number SH020453 and maintained at the Biological Collection Centre of the Institute of Tropical Bioscience and Biotechnology, Chinese Academy of Tropical Agricultural Sciences, as well as at the Rutgers Duckweed Stock Cooperation at Rutgers University, New Jersey, USA under international accession number 5504. Clone 7741 was a standard *L. gibba* clone derived from clone G335, the later was originally isolated from Sicilia, Siracusa, Italy by R. Kandeler, Vienna, Austria. Clone 7210 was identified as *L. minor* (Supporting Information, Figure S1). It was originally isolated in Japan, maintained in Landolt's collection, Zurich, Switzerland and obtained from the Rutgers Duckweed Stock Cooperative. The identities of these clones were verified by phylogenetic analysis based on their *atpF-atpH* intergenic sequences and their *rps16* intron sequences as suggested by Borisjuk *et al.*36 (Supporting Information, Figure S1).

The components in the three basic nutrient media used in this study are provided in Table S1 (Supporting Information). All nutrient media were supplemented with 29 mM sucrose with or without 6 g L^{-1} agar (Biotechnology grade, Beijing Solarbiao Science & Technology co., Ltd, Beijing, China) for solidification. Media were autoclaved and the SA stock solution was filter sterilized (0.22 µm filter units; Millipore, USA).

Observation of flower primordium and determination of critical day length

To observe flower induction and to determine critical day length, fronds were cultured in MH medium under 8 h light/16 h darkness for 20 days, then transferred to fresh MH or E-medium and continued to culture under 8, 9, 10, 12, 14, and 16 h photoperiod for 20–30 days. The budding pockets were dissected and observed under a dissecting microscope.

Effects of SA on flower induction

Three-frond colonies of the above mentioned clones were cultured in three replicates on semi-solid MH medium supplemented with 29 mM sucrose and 20 µM SA, and cultivated under 16 h photoperiod at 40 $\mu mol \cdot m^{-2} \cdot s^{-1}$ photosynthetic active radiation (PAR) at 25 ± 1 °C. Each replicate contained 10 plates, and each plate contained five colonies. In contrast to previous reports, only fronds with mature flowers were considered as flowering, which is in accordance with the purpose of the investigation. Mature flowers were counted every day from their first emergence until day 24.

To study the effect of the SA concentration on the flowering rate, three-frond colonies were inoculated with threefold replicates in the MH medium supplemented with different concentrations of SA (0, 10, 20, 30, 50, and 80 µM) and cultivated under 16 h daily photoperiod at 40 $\mu mol \cdot m^{-2} \cdot s^{-1}$ PAR at 25 ± 1 °C. Flowering rates were calculated 20 days after inoculation. The flowering rate was defined as the total number of plants (frond colonies) with mature flowers at the time of counting divided by the total number of plants inspected.

Effects of SA on frond growth rate

Three-frond colonies were inoculated in three replicates on the basal medium supplemented with different concentrations of SA (0, 10, 20, 30, 50, and 80 µM) and incubated under 16 h day photoperiod at

40 μmol · m^{-2} · s^{-1} PAR at 25 ± 1 °C. Frond number of each sample was counted seven days after inoculation. The relative growth rate (RGR) was calculated as described previously54: RGR = ($\ln N_2$ − $\ln N_1$)/t, where N_1represents the initial frond number, N_2 represents the final frond number, and t represents the days of incubation.

Estimation of pollen grains in the anthers

To estimate the quantity of pollen grains in the anthers, 30 mature anthers from randomly chosen colonies were put in three 1.5 mL centrifuge tubes with 10 anthers in each tube and broken with a vaccination needle. The broken anthers were then treated with 0.5 mL 2.5% pectinase (Y23, Yakult Pharmaceutical Industry Co., Ltd., Tokyo, Japan) solution for 4 h at 28 °C, followed by dilution with 0.5 mL 2.5% sucrose solution. The centrifuge tubes were then vortexed for 10 sec. The resulting pollen suspension was spread on glass slides in 5 μL droplets, and the pollen grains in each droplet were counted under a light microscope (Axioskop 40, Zeiss, Jena, Germany) with triplicates for each sample. Pollen grain quantity in each anther was calculated as the average number of pollen grains in each 5 μL droplet × the total volume of pollen suspension in one tube (1000 μL)/the volume of each droplet (5 μL)/the number of anthers in each tube (10).

Pollen viability tests by chemical staining

Four different staining methods were tested for their effectiveness to distinguish between viable and non-viable pollen (Trognitz 1991; Khatun & Flowers 1995). As control, pollens incubated at 80 °C for 2 h were considered dead55. The staining reagents included: (1) 2,3,5-triphenyl tetrazolium chloride (TTC), (2) 2,5-diphenyl monotetrazolium bromide (MTT), (3) 5-bromo-4-chloro-3-indoyle-β-galactoside (X-gal) (Rodriguez-Riano & Dafni 2000), and (4) fluorescein diacetate (FDA) (Kearns & Inouye 1993).

The MTT reagent was finally selected. Mature anthers were placed on a glass slide and incubated with 5 μL MTT solution. The working solutions of MTT contained 1% staining reagent with 5% sucrose that was added to the anthers. Pollen grains were released into the solution by breaking the anthers with a vaccination needle. After incubating for 10 min, the pollen grains were observed under a light microscope (Axioskop 40, Zeiss, Jena, Germany), and the viability was evaluated by the colour of the pollen. Ten anthers were examined per clone. The pollen viability is given as percentage of the number of viable pollen grains to the total number of pollen grains observed.

Pollen viability test by germination

Germination medium (MH medium supplemented with 1% sucrose) was spread on glass slides to a thin layer. Pollen grains were then inoculated by scratching the anthers on the surface of the medium. The germination of the pollen grains was observed under microscope at intervals of five minutes.

Scanning electron microscopy

Mature anthers were fixed in formaldehyde-acetic acid-ethanol solution (FAA, 100 ml solution contains 5 ml 40% formaldehyde, 5 ml glacial acetic acid, and 90 ml 70% ethanol) at 4 °C overnight, and then dehydrated in a series of tert-butyl alcohol (30%, 50%, 70%, 80%, and 90%) for 15 min at each concentration. The anthers were dehydrated with 100% tert-butyl alcohol three times for 1 hour. The dehydrated anthers were then freeze-dried and inspected with a scanning electron microscope (FEI Phenom, Phenom Wold, Eindhoven, Netherlands). Pollen grains were released from the dehydrated anthers with a vaccination needle and suspended in tert-butyl alcohol. Aliquots of the pollen suspension were then spread over glass slides, adhered to electron microscopy sample tables (FEI Phenom), and observed with a scanning electron microscope (FEI Phenom). Twenty pollen grains from each clone were randomly chosen to measure their size and to study their morphology.

Artificial cross-pollination under stereomicroscope

Flowering fronds of *L. gibba* 7741 and 5504 were transferred to fresh semi-solid MH medium supplemented with 29 mM sucrose and were incubated at 40 μmol . m^{-2} . s^{-1} PAR and 25 ± 1 °C for 24 h. Artificial pollination was then performed in the tissue culture hood under a stereomicroscope. Pollen grains of clone 7741 were pollinated on the stigma of clone 5504 under a stereomicroscope by moving the whole colony with tweezers towards the stigma. The pollinated fronds were then transferred to fresh plates containing MH medium supplemented with 29 mM sucrose and 20 μM SA and incubated under 16 h daily photoperiod of 40 μmol · m^{-2} · s^{-1} PAR at 25 ± 1 °C. Seeds were harvested after 20 days. Similar experiments were carried out by transferring pollens of clone 7741 on to the stigma of clone

5504 in order to test the possibility for intraspecific cross-pollination between different clones.

Seed germination and validation of hybrid formation

Fresh seeds were punctured on the seed coat with a scalpel, and were incubated in MH medium under 16 h daily photoperiod of 40 μmol · m^{-2} · s^{-1} PAR at 25 ± 1 °C. Each line that was vegetatively propagated from one seed was considered as a new clone.

DNA was isolated using a genomic DNA extraction kit (Tiangen Biotech Co. LTD, Beijing, China) according to the reccommended protocol. Fifty primers that were selected from the universal inter-simple sequence repeat (ISSR) primers56 were synthesized and used to distinguish genetic differences between parental clones 7741 and 5504. One primer (B825, 5′ ACACACACACACACACT 3′) showed clone-specific bands and was used to validate the genetic background of the hybrid clones. The PCR reactions were performed at 94 °C for 4 min, followed by 30 cycles of 94 °C for 30 sec, 48 °C for 30 sec, and 72 °C for 1 min. The PCR products were separated on 1.5% agarose gel, stained with Goldview Nucleic Acid Gel Stain (Solarbio, Beijing, China), and visualized using the gel imager Tanon 410 (Tanon, Shanghai, China).

ADDITIONAL INFORMATION

Publisher's note: Springer Nature remains neutral with regard to jurisdictional claims in published maps and institutional affiliations.

REFERENCES

1. Sree, K. S., Bog, M. & Appenroth, K.-J. Taxonomy of duckweeds (Lemnaceae), potential new crop plants. Emir. J. Food Agric. 28, 291–302,
2. Appenroth, K.-J., Borisjuk, N. & Lam, E. Telling duckweed apart: Genotyping technologies for the Lemnaceae. Chin. J. Appl. Environ. Biol. 19, 1–10,
3. Landolt, E. Biosystematic investigations in the family of duckweeds (Lemnaceae) Vol. 2: The family of Lemnaceae: a monographic study. – Morphology, karyology, ecology, geographic distribution, nomenclature, descriptions., (Eidgenössische Technische Hochschule Zürich, 1986).
4. Ziegler, P., Sree, K. S. & Appenroth, K. J. The use of duckweed in relation to water remediation. Desalination and Water Tretament 63, 327–342,
5. Ziegler, P., Sree, K. S. & Appenroth, K. J. Duckweeds for water remediation and toxicity testing. Toxicol. Environ. Chem. 98, 1127–1154,

6. El-Shafai, S. A., El-Gohary, F. A., Nasr, F. A., van der Steen, N. P. & Gijzen, H. J. Nutrient recovery from domestic wastewater using a UASB-duckweed ponds system. Biores. Technol. 98, 798–807,
7. Ge, X., Zhang, N., Phillips, G. C. & Xu, J. Growing Lemna minorin agricultural wastewater and converting the duckweed biomass to ethanol. Biores. Technol. 124, 485–488,
8. Xu, J. & Shen, G. Growing duckweed in swine wastewater for nutrient recovery and biomass production. Biores. Technol. 102, 848–853,
9. Zhao, Y. et al. Pilot-scale comparison of four duckweed strains from different genera for potential application in nutrient recovery from wastewater and valuable biomass production. Plant Biol. 17, 82–90,
10. Li, X., Jin, Y., Gao, X., Zhang, G. & Zhao, H. Fermentation method of high ratios of biobutanol with Landoltia punctata. China Brewing 31, 85–88 (2011).
11. Qian, C. et al. Improving production of bioethanol from duckweed (Landoltia punctata) by pectinase pretreatment. Energies 5, 3019–3032,
12. Tao, X. et al. Comparative transcriptome analysis to investigate the high starch accumulation of duckweed (Landoltia punctata) under nutrient starvation. Biotechnol. Biofuels 6, 72,
13. Liu, Y. et al. Uniconazole-induced starch accumulation in the bioenergy crop duckweed (Landoltia punctata) I: transcriptome analysis of the effects of uniconazole on chlorophyll and endogenous hormone biosynthesis. Biotechnol. Biofuels 8, 57,
14. Liu, Y. et al. Uniconazole-induced starch accumulation in the bioenergy crop duckweed (Landoltia punctata) II: transcriptome alterations of pathways involved in carbohydrate metabolism and endogenous hormone crosstalk. Biotechnol. Biofuels 8, 64,
15. Anderson, K. E., Lowman, Z., Stomp, A. M. & Chang, J. Duckweed as a feed ingredient in laying hen diets and its effect on egg production and composition. Intern. J. Poultry Sci. 10, 4–7,
16. Leng, R. A., Stambolie, J. H. & Bell, R. Duckweed - a potential high-protein feed resource for domestic animals and fish. Livestock Res. Rural. Dev. 7, Article 5 (1995).
17. Bairagi, A., Sarkar Ghosh, K., Sen, S. K. & Ray, A. K. Duckweed (Lemna polyrhiza) leaf meal as a source of feedstuff in formulated diets for rohu (Labeo rohitaHam.) fingerlings after fermentation with a fish intestinal bacterium. Biores. Technol. 85, 17–24,
18. Iqbal, S. Duckweed aquaculture: potentials, possibilities and limitations for combined wastewater treatment and animal feed production in developing countries. SANDEC Report No. 6/99 (1999).
19. Islam, M. S. et al. Wastewater-grown duckweed may be safely used as fish feed. Can. J. Microbiol. 50, 51–56,
20. Cheng, J. J. & Stomp, A.-M. Growing duckweed to recover nutrients from wastewaters and for production of fuel ethanol and animal feed. CLEAN – Soil, Air, Water 37, 17–26,
21. Cayuela, M. L., Millner, P., Slovin, J. & Roig, A. Duckweed (Lemna gibba) growth inhibition bioassay for evaluating the toxicity of olive mill wastes before and during composting. Chemosphere 68, 1985–1991,
22. Ran, N., Agami, M. & Oron, G. A pilot study of constructed wetlands using duckweed (Lemna gibbaL.) for treatment of domestic primary effluent in Israel. Water Res. 38, 2240–2247 (2004).
23. Zimmo, O. R., Al-Sa'ed, R. M., van der Steen, N. P. & Gijzen, H. J. Process performance assessment of algae-based and duckweed-based wastewater treatment systems. Water Sci. Technol. 45, 91–101 (2002).
24. Zimmo, O. R., van der Steen, N. P. & Gijzen, H. J. Nitrogen mass balance across pilot-scale algae and duckweed-based wastewater stabilisation ponds. Water Res. 38, 913–920,

25. Mkandawire, M., Taubert, B. & Dudel, E. G. Capacity of Lemna gibbaL. (duckweed) for uranium and arsenic phytoremediation in mine tailing waters. Int. J. Phytoremediation 6, 347–362,
26. Falabi, J. A., Gerba, C. P. & Karpiscak, M. M. Giardia and Cryptosporidium removal from waste-water by a duckweed (Lemna gibbaL.) covered pond. Lett. Appl. Microbiol. 34, 384–387,
27. Landolt, E. & Kandeler, R. The Family of Lemnaceae - A Monographic Study. Vol. 4, Biosystematic Investigations in the Family of Duckweeds(Lemnaceae). Vol. 4 (Veröffentlichungen des Geobotanischen Institutes der ETH, 1987).
28. Pieterse, A. H. Is flowering in Lemnaceae stress-induced? A review. Aquat. Bot. 104, 1–4,
29. Khurana, J. P. & Cleland, C. F. Role of salicylic acid and benzoic acid in flowering of a photoperiod-insensitive strain, Lemna paucicostataLP6. Plant Physiol. 100, 1541–1546,
30. Tanaka, O. & Cleland, C. F. Comparison of the ability of salicylic acid and ferricyanide to induce flowering in the long-day plant, Lemna gibbaG3. Plant Physiol. 65, 1058–1061,
31. Cleland, C. F. & Tanaka, O. Effect of daylength on the ability of salicylic acid to induce flowering in the long-day plant Lemna gibbaG3 and the short-day plant Lemna paucicostata6746. Plant Physiol. 64, 421–424,
32. Cleland, C. F. & Briggs, W. S. Gibberellin and CCC effects on flowering and growth in the long-day plant Lemna gibbaG3. Plant Physiol. 44, 503–507,
33. Fujioka, S. et al. The role of benzoic acid and plant hormones in flowering of Lemna gibbaG3. Plant Cell Physiol. 26, 655–659, d
34. Tanaka, O. et al. Flower-Inducing Activity of Lysine in Lemna paucicostata6746. Plant Cell Physiol. 38, 124–128,
35. Bog, M. et al. Genetic structure of the genus LemnaL. (Lemnaceae) as revealed by amplified fragment length polymorphism. Planta 232, 609–619,
36. Borisjuk, N. et al. Assessment, validation and deployment strategy of a two-barcode protocol for facile genotyping of duckweed species. Plant Biol. 17, 42–49,
37. Xu, Y. et al. Species distribution, genetic diversity and barcoding in the duckweed family (Lemnaceae). Hydrobiologia 743, 75–87,
38. Cleland, C. F. & Briggs, W. R. Flowering responses of the long-day plant Lemna gibbaG3. Plant Physiol. 42, 1553–1561,
39. Cleland, C. F. & Ajami, A. Identification of the flower-inducing factor isolated from aphid honeydew as being salicylic acid. Plant Physiol. 54, 904–906,
40. Krajncic, B. & Nemec, J. Mechanisms of EDDHA effects on the promotion of floral induction in the long-day plant Lemna minor(L.). J. Plant Physiol. 160, 143–151,
41. Sree, K. S. et al. The duckweed Wolffia microscopica: A unique aquatic monocot. Flora 210, 31–39,
42. Asma, B. M. Determination of pollen viability, germination ratios and morphology of eight apricot genotypes. Afr. J. Biotechnol. 7, 4269–4273 (2008).
43. Dafni, A. & Firmage, D. Pollen viability and longevity: practical, ecological and evolutionary implications. Plant Sys. Evol 222, 113–132,
44. Rodriguez-Riano, T. & Dafni, A. A new procedure to asses pollen viability. Sex. Plant Reprod. 12, 241–244,
45. Skibbea, D. S. & Schnable, P. S. Male sterility in maize. Maydica 50, 367–376 (2005).
46. Gracen, V. E. & Grogan, C. O. Diversity and suitability for hybrid production of different sources of cytoplasmic male sterility in maize. Agron. J. 66, 654–657,
47. Raj, K. G. & Virmani, S. S. Genetic of fertility restoration of 'WA' type cytoplasmic male sterility in rice. Crop Sci. 28, 787–792,
48. Luo, D. et al. A detrimental mitochondrial-nuclear interaction causes cytoplasmic male sterility in rice. Nat. Genet. 45, 573–577,
49. Maan, S. S. & Lucken, K. A. Interacting male sterility-male fertility restoration systems for hybrid wheat research. Crop Sci. 12, 360–364,

50. Sun, J., Liu, J. & Zhang, J. A review on research and utilization of hybrid vigor of cotton. Acta Gossypii Sinica 3(1994).
51. Jean, M., Brown, G. G. & Landry, B. S. Genetic mapping of nuclear fertility restorer genes for the 'Polima' cytoplasmic male sterility n canola (Brassica napusL.) using DNA markers. Theor. Appl. Genet. 95, 321–328,
52. Schnable, P. S. & Wise, R. P. The molecular basis of cytoplasmic male sterility and fertility restoration. Trends Plant Sci. 3, 175–180,
53. Huang, M., Xu, Y., Kanjana, K., Sun, X. & Zhang, J. Flower-induction of Lemna gibbaSH0204 by salicylic acid. Plant Physiol. J. 51, 559–565 (2015).
54. Ziegler, P., Adelmann, K., Zimmer, S., Schmidt, C. & Appenroth, K. J. Relative in vitrogrowth rates of duckweeds (Lemnaceae) – the most rapidly growing higher plants. Plant Biol. 17, 33–41,
55. Wang, Z. Y., Ge, Y., Scott, M. & Spangenberg, G. Viability and longevity of pollen from transgenic and nontransgenic tall fescue (Festuca arundinacea)(Poaceae) plants. Am. J. Bot. 91, 523–530,
56. Xue, H. et al. Genetic diversity and geographic differentiation analysis of duckweed using inter-simple sequence repeat markers. Mol. Biol. Rep. 39, 547–554,

CHAPTER 11

Additive QTLs on three chromosomes control flowering time in woodland strawberry (*Fragaria vesca* L.)

Samia Samad[1,2,7], Takeshi Kurokura[3,7], Elli Koskela[1] , Tuomas Toivainen[1] , Vipul Patel[4] , Katriina Mouhu[1] , Daniel James Sargent[2,5] and Timo Hytönen[1,6]

[1] *Department of Agricultural Sciences, Viikki Plant Science Centre, University of Helsinki, 00014 Helsinki, Finland;*

[2] *Fondazione Edmund Mach, Research and Innovation Centre, San Michele All'adige, 38010 TN, Italy;*

[3] *Faculty of Agriculture, Utsunomiya University, Tochigi, 321-8505, Japan;*

[4] *Department of Plant Developmental Biology, Max Planck Institute for Plant Breeding Research, 50829 Cologne, Germany;*

[5] *Driscoll's Genetics Limited, East Malling Enterprise Centre, East Malling, Kent ME19 6BJ, UK and*

[6] *Department of Biosciences, Viikki Plant Science Centre, University of Helsinki, 00014 Helsinki, Finland.*

[7] *These authors contributed equally to this study*

ABSTRACT

Flowering time is an important trait that affects survival, reproduction and yield in both wild and cultivated plants. Therefore, many studies have focused on the identification of flowering time quantitative trait locus (QTLs) in different crops, and molecular control of this trait has been extensively investigated in model species. Here we report the mapping of QTLs for flowering time and vegetative traits in a large woodland strawberry mapping population that was phenotyped both under field conditions and in a greenhouse after flower induction in the field. The greenhouse experiment revealed additive QTLs in three linkage groups (LG), two on both LG4 and LG7, and one on LG6 that explain about half of the flowering time variance in the population. Three of the QTLs were newly identified in this study, and one co-localized with the previously characterized *FvTFL1* gene. An additional strong QTL corresponding to previously mapped *PFRU* was detected in both field and greenhouse experiments indicating that gene(s) in this locus can control the timing of flowering in different environments in addition to the duration of flowering and axillary bud differentiation to runners and branch crowns. Several putative flowering time genes were identified in these QTL regions that await functional validation. Our results indicate that a few major QTLs may control flowering time and axillary bud differentiation in strawberries. We suggest that the identification of causal genes in the diploid strawberry may enable fine tuning of flowering time and vegetative growth in the closely related octoploid cultivated strawberry.

INTRODUCTION

Synchronizing floral development with local climatic conditions is important for crop and wild plants because of its significant impact on reproduction, yield and ultimately survival. The genetic control of flowering time is thus of immense significance. As a result, numerous mapping studies have focused on the identification of flowering time QTLs in different crops, and a detailed molecular understanding of flowering time regulation in model plants has facilitated the identification of causal genes.[1,2,3,4] In the cultivated strawberry, one of the most economically important berry crops throughout the world, flowering time is a major breeding target because of year-round demand for fresh berries. To extend the cropping season, breeders are developing early and late flowering cultivars as well as everbearing (EB; also called day-neutral) cultivars, which display a continuous flowering habit.[5,6,7,8]

In seasonal flowering 'Junebearing' strawberries, flower induction occurs in autumn, and the timing of the induction correlates with flowering time in the following season.[9] Both short days (SD) and cool temperatures can induce flowering. Typically, flower induction occurs independently of day-length at temperatures below ~13 °C, whereas SDs are required at intermediate temperatures, and at temperatures above 20–24 °C floral induction is inhibited.[10,11,12,13] There is however significant variation in critical temperature limits between cultivars,10,14and some cultivars possess an obligatory SD requirement for flower induction even under cool temperatures.15 In contrast to seasonal flowering strawberries, EB cultivars flower earlier under long days (LD) than SDs.16,17 Both seasonal flowering and EB habits are also found in the woodland strawberry.18

As the cultivated strawberry is a complex allo-octoploid ($2n=8x=56$), the woodland strawberry (*F. vesca*; $2n=2x=14$) and other diploid species have been used as a genetically facile surrogate system for genetic investigation. Several genetic linkage maps have been developed for diploid strawberry species. Initially saturated maps were developed using large numbers of microsatellite (SSR) and other PCR-based markers,19,20,21,22,23,24,25 and the *Fragaria* SSR-based reference linkage map was used to anchor the *F. vesca* 'Hawaii-4' genome sequence.26More recently, high-throughput genotyping has been employed using various platforms to produce dense SNP-based linkage maps that have enabled more precise sequence scaffold anchoring and orientation,27whereas the most comprehensive linkage map for diploid *Fragaria* to date is that of *F. iinumae* produced using the *Fragaria* Axiom SNP genotyping array (Axiom IStraw90, Affymetrix, Santa Clara, CA, USA), supplemented with SNPs scored using genotyping-by-sequencing (GBS).28,29 The linkage maps have been used to map a number of genetic loci in diploid strawberry species, including fruit color, runnering, leaf color, and seasonal flowering,20,30,31,32 and represent a powerful tool for the genetic dissection of traits of economic importance in the cultivated strawberry.

Previous genetic mapping studies have shown that the gene causing seasonal flowering habit in woodland strawberry is located on linkage group 6 (LG6).21,31 This gene has been recently shown to encode a strong repressor of flowering, TERMINAL FLOWER 1 (FvTFL1), and a 2-bp deletion in the first exon of *FvTFL1* causes expression of the EB phenotype in woodland strawberry.32 Furthermore, homologs of FvTFL1 have been found to repress flowering in several rosaceous species suggesting the conservation of the genetic mechanism of flowering time regulation.33,34,35,36 Only a few other flowering time genes have been functionally characterized in woodland strawberry. Using genetic transformation, recent studies have revealed how homologs of *FLOWERING LOCUS T* (*FvFT1*) and *SUPPRESSOR OF THE OVEREXPRESSION OF CONSTANS1* (*FvSOC1*) mediate photoperiod and temperature signals and how they are integrated to control flower induction.32,37,38,39 In the seasonal flowering woodland strawberry, cool temperatures below 13 °C cause photoperiod-independent downregulation of *FvTFL1* by an unknown mechanism leading to flower induction, whereas at higher temperatures, photoperiod controls flowering through *FvTFL1*. Under LD, leaf-expressed FvFT1 activates *FvSOC1* in the shoot apex leading to the upregulation of *FvTFL1* and the maintenance of a vegetative stage, whereas under SD, this *FvFT1-FvSOC1-FvTFL1* pathway is silenced and flower induction occurs. However, at temperatures above 20 °C *FvTFL1* is highly upregulated by an unknown activator that functions independently of FvSOC1, and plants remain vegetative.[37,39] This model contrasts with the mechanism in *Arabidopsis* in which both FT and SOC1 function as floral activators.[1] However, FvFT1 is a strong floral activator in EB woodland strawberries that are lacking functional *FvTFL1*.[32,38]

The genetic control of the EB habit has also been studied in the cultivated strawberry. Weebadde *et al.*[5] suggested the presence of several QTLs with significant genotype×environment interaction,[5] whereas other studies indicated that a single dominant locus causes continuous flowering.[40,41] The presence of a single QTL was supported by recent genetic mapping studies by two groups,[6,8] who found a major dominant QTL on LG4 in two independent crossing populations.[6,8] Moreover, Honjo *et al.*[42] demonstrated that the EB trait is controlled by the same gene in the EB parents of these populations.[42] In addition to the EB trait, this locus called *PFRU* also controls the production of runners, which are long shoots that enable the efficient clonal reproduction of the species.[6] Although studies indicate that different genes cause the EB phenotype in the woodland and cultivated strawberries, a recent study has shown that *TFL1* homologs encode major floral repressors in both species, and that the silencing of this gene also causes the EB habit in cultivated strawberry.[43] Moreover, the regulation of *FaTFL1* correlates with flower induction in different seasonal flowering cultivars.[43,44]

As strawberries form terminal inflorescences from the apical meristems of the crowns, the number of inflorescences, and consequently the yield potential, depends on the number of side shoots called branch crowns that are produced.[45,46,47,48] SD conditions promote the differentiation of strawberry axillary buds to branch crowns, whereas under LDs axillary buds typically differentiate to runners.[10,14,45] Changes in gibberellin biosynthesis and signaling are known to mediate the photoperiodic differentiation of axillary buds.[49] This differentiation is not understood at the molecular level, although runnerless mutants are known in woodland strawberry.[18]

Here, we aimed to identify novel genetic loci controlling flowering time and axillary bud differentiation using the large woodland strawberry F2 mapping population produced by Koskela *et al.*[32] We demonstrate that five QTLs with additive effects explain about half of the flowering time variance found in the population. One of the QTLs co-localized with the previously identified *FvTFL1* and another with *PFRU*, whereas the others were newly identified in this study. In addition, we report the mapping of two QTLs that affect the number of branch crowns and runners in our mapping population.

MATERIALS AND METHODS

Mapping population and phenotyping

Selected lines from a F2 population derived from the cross *F. vesca f. semperflorens* 'Hawaii-4' (H4)×*F. vesca* subsp. *vesca* (FV), (denoted H4×FV),32 were used in this study. The population consisted of 735 seasonal flowering plants that were planted in an experimental field at the University of Helsinki at the beginning of August 2009. In the spring 2010, flowering time was observed in the field. On the basis of these observations and previously scored genotypes for the *FvTFL1* gene,32335 seasonal flowering lines containing one or two functional *FvTFL1* alleles were selected for flowering time observations. At the beginning of August 2010, four clonal daughter plants per line were potted into 6 cm square plastic pots filled with peat moss (Kekkilä, Finland). Plants were kept under natural environmental conditions and periodically supplemented with fertilizer (NPK 17-4-25; Kekkilä) applications until 13 December 2010. Then, the plants were moved into a greenhouse and repotted into 8×8 cm plastic pots and arranged according to completely randomized design. The greenhouse temperature was set to 18 °C and plants were kept under an 18 h photoperiod illuminated with high pressure sodium lamps (Airam 400 W, Kerava, Finland) providing a light intensity of 100 μmol m^{-2} s^{-1}. Plants were automatically watered with the fertilizer solution. The date of first flower of all genotypes was recorded and flowering time was calculated as days after the anthesis of the earliest flowering F2 line. Runners were removed and counted regularly in the greenhouse, and the total number of runners produced per plant before 8 March 2011 was calculated. In addition, the number of branch crowns was observed on the same date. On the basis of the greenhouse phenotypic data, 32 F2 lines with extreme flowering time values (16 early and 16 late flowering lines) were selected for growth chamber experiment. Four clones of each F2 line were rooted in 8×8 cm plastic pots and plants were subjected to 12 h short-day treatment (AP67 LED lamps; 200 μmol m^{-2} s^{-1}; Valoya, Finland) at 11 °C for 6 weeks followed by flowering time observations in the greenhouse as described above. In the greenhouse and growth chamber experiments, mean phenotypic values of four replicates were used for QTL mapping.

DNA extraction and quantification

DNA was extracted from young leaves of the parents 'H4' and 'FV', along with the F1 and 335 F2 seedlings of the progeny, using a modified version of the CTAB method.[31] The DNA was quantified using a Nanodrop spectrophotometer (Thermo Fisher Scientific, Waltham, MA, USA) and was diluted 1:100 (5–10 ng μl^{-1}) for use in PCR. For genome resequencing an additional purification step was carried out. NaCl was added at a final concentration of 0.2 M, and DNA was precipitated by adding 2/3 volumes of cold isopropanol. After centrifugation, the pellet was washed with 100 μl of 70% ethanol and re-suspended in 50 μl of TE.

Amplification and scoring of genetic markers

A total of 73 published SSR primer pairs (Supplementary File 1) labeled on the forward primer with either 6-FAM or HEX fluorescent dyes[19,20,21,22,25,50,51,52] were screened for polymorphism in the parental and F1 lines using the 'Type-it' PCR mastermix (Qiagen, Hilden, Germany), following the PCR protocol reported in Sargent *et al.*[53] In addition, 12 SSR primer pairs were tested using the fluorescent labeling method described by Schuelke[24] (Supplementary File 1). Samples were diluted 1:75, separated by capillary electrophoresis (Foster City, CA, USA) on a genetic analyzer (ABI 3100, ABI 3130 XL or ABI 3700, Applied Biosystems), and the resultant data were collected and analyzed using GeneMapper or Peak Scanner 2 (Applied Biosystems).

SNPs were identified in the parental genome sequences and the temperature switch PCR (TSP) protocol[23] was used to develop assays for 49 SNP markers within flowering time QTL regions identified in this investigation. In brief, two sets of primers, a locus specific (LS) set and a nested-locus specific (NLS) set were designed per SNP using Primer3 v.0.40 (ref. 54) and NetPrimer (http://www.premierbiosoft.com/ netprimer /netprlaunch/netprlaunch.html), respectively. The LS primers were designed to have a Tm between 60–65 °C (optimal 63 °C) and a product size >400 bp, whereas the NLS primers were designed to have a T_m of 43–47 °C (optimal 45 °C) and were extended using a non-complementary 5′ tail region that increased the overall primer T_m to 52–55 °C (optimal 53 °C).[23] PCR was performed as described previously.[23] The primer concentrations and the numbers of PCR cycles were adjusted for each primer individually to optimize marker amplification for ease of scoring. Products were visualized over UV light following electrophoresis through a 1.5% (w/v) TAE agarose gel containing gel red (Applied Bioprobes, Rockville, MD, USA) at a constant voltage of 100 V for 2–3 h, depending on the sizes of the fragments produced. The TSP loci were named UH, followed by the chromosome number and finally a unique primer pair reference number (Supplementary File 1).

Genotyping-by-sequencing

DNA samples (30–100 ng μl^{-1}) of 188 seedlings selected to represent the full seedling population genotypically (based on the SSR and TSP data) and phenotypically, were sent to the Cornell University where genotyping-by-sequencing (GBS) was carried out according to Elshire *et al.*55 GBS libraries were sequenced using the Illumina HiSeq2500 platform (San Diego, CA, USA). GBS tags with a minimum number of three reads and unique alignment to the reference genome (https://www.rosaceae.org/species/fragaria_vesca/genome_v2.0.a1)27were identified, and from them SNPs were called by Genomic Diversity Facility in Cornell using the GBS-pipeline56 implemented in the TASSEL 3.0 software application.57 To remove genotyping errors, several filtering steps were performed for each polymorphic site using vcftools v. 0.1.12b.58 A site was accepted if genotype calling rate was over 90% per site, the minor allele frequency was >0.1, and the proportion of heterozygote genotypes fit to the Hardy–Weinberg expectations (2pq, $P<0.01$). Missing data was imputed with BEAGLE4 (ref. 59) and the TASSEL5 was used to produce ABH file with 4672 SNPs. Following an initial round of linkage mapping (described below), a second filtering and manual imputation step was performed using the rationale previously reported by Ward *et al.*60 This second imputation step removed suspected false homozygous calls that created multiple double-recombination events, inflating mapping distances and confounding local marker ordering.

Heritability, linkage mapping and QTL analysis

The broad sense heritability for flowering time and numbers of runners and branch crowns were calculated in a population of 188 individuals according the formula $H^2=\sigma^2 g/(\sigma^2 g+\sigma^2 e)$, where $\sigma^2 g$=(MS(genotype)−MS(environment))/r and $\sigma^2 e$=MS(environment). MS(genotype) and MS(environment) were obtained from analysis of variance for completely randomized design as the mean sums of squares for genotype and residual error, respectively, and r was the effective number of replicates (three to four per genotype). Estimation of these variance components were performed by REML as implemented in PROC VARCOMP procedure in

SAS/STAT software, version 9.4 (Copyright 2013, SAS Institute, Cary, NC, USA). Linkage mapping was performed using JOINMAP 4.1 (Kyazma, Wageningen, The Netherlands)[61] using maximum likelihood and the default mapping criteria. The maps presented were illustrated using MapChart 2.2.[62] The significance of association between individual markers on the H4×FV linkage map and the phenotypes was calculated with linear regression using MapQTL 6.0 (Kyazma).[63] QTLs were identified employing interval mapping with a step size of 1 cM, considering a maximum of five neighboring markers. The genome-wide log of odds (LOD) threshold was determined over 10 000 permutations and the most significant markers were then used as co-factors for restricted multiple QTL mapping (rMQM) with a step size of 1 cM. Linkage mapping and QTL analyses were performed on 335 individuals using SSR and TSP markers and on 186 individuals using all marked data.

Genome resequencing and the identification of single-nucleotide polymorphisms between the 'H4' and 'FV' genomes

The genome of the 'FV' parent was re-sequenced at DNA Sequencing and Genomics Laboratory, Institute of Biotechnology, University of Helsinki, Finland. Briefly, DNA was sheared using a Bioruptor NGS sonicator (Diagenode, Denville, NJ, USA) and the obtained fragments were end-repaired and A-tailed, and truncated Illumina Y-adapters were ligated. In a PCR step (20 cycles), full-length P5 and indexed P7 adapters were introduced using KAPA Hifi DNA Polymerase (KAPA Biosystems, Wilmington, MA, USA). The obtained libraries were purified and size selected using AMPure XP beads (Beckman Coulter, Brea, CA, USA). The obtained final libraries were paired-end (300 bp+300 bp) sequenced on a MiSeq Sequencer (Illumina, San Diego, CA, USA). Adapter sequences were removed from paired-end reads with cutadapt 1.8.1.,[64] and reads were aligned against the strawberry reference genome v2.0.a1 with bwa-mem[65] using default settings. After the alignment, samtools 1.3 (refs 66,67) was used for sorting, indexing, filtering and removing PCR duplicates. Reads with mapping quality of <25 were discarded. After filtering, average coverage was 14.8, and 91.5% of the reference genome was covered. Genetic variants including SNPs and indels were called with bcftools after calculating genotype likelihoods with samtools mpileup function. To ensure high quality variants, additional filtering was conducted with vcftools,[58] and only sites having at least 10-fold coverage and <50-fold coverage were accepted for further analysis. Genomic regions spanning flowering time QTL were identified from the genomic positions of flanking

SSR markers mapped in the full mapping progeny, within which polymorphic SNPs were identified for subsequent marker development.

Identification of candidate genes in the QTL regions

To identify candidate genes for the QTLs, predicted genes located between the markers flanking the most significant markers were BLAST searched against the RefSeq or UniRef90 protein database using the blastx algorithm in Blast2GO68 or blast+ (2.4.0). Query sequences longer than 8000 bp were restricted to the first 8000 bp for Blast2GO as this is the maximum length accepted by the software. Homologs of known flowering time genes were selected as candidate genes. Parental genetic variation in the coding regions of candidate genes were retrieved in the Illumina whole genome resequencing data using vcftools,58 and translated amino acid sequences were compared to reveal non-synonymous variation in the coding regions.

RESULTS

Segregation of flowering time and vegetative traits in H4×FV mapping population

Initial field observations revealed over one month variation in flowering time among the 735 seasonal flowering F2 lines of H4×FV mapping population in Helsinki in summer 2010 (Supplementary Figure 1). The first plants started to flower at week 20, and almost 70% of the plants flowered by week 22. However, many plants flowered significantly later, and about 5% of plants flowered at week 25 or later. In autumn 2010, 335 F2 selected lines were subjected to natural autumn conditions followed by growth observations in a greenhouse. Again, significant variation in flowering time was observed (Figure 1a). On average, these plants started to flower 39 days after the transfer to the greenhouse, and an 18-day variation in flowering time was observed. The broad sense heritability for flowering time was H^2=0.61. Greenhouse flowering data correlated positively, but weakly with field data (r=0.39). Flowering time of the 16 earliest and the 16 latest F2 lines were further tested in a growth chamber. In this experiment, over two week variation in flowering time and clear positive correlation with greenhouse flowering data (r=0.71) was observed (Supplementary Figure 2).

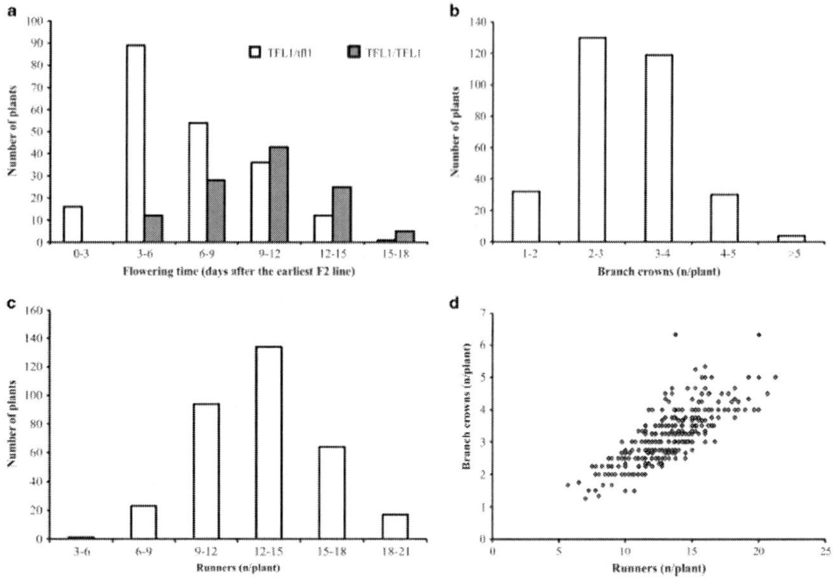

Figure 1. Segregation of flowering time and vegetative traits in H4×FV mapping population. Segregation of flowering time (**a**), the number of branch crowns (**b**) and runners (**c**), and correlation between the number of branch crowns and runners (**d**) are shown. Flowering time data is shown separately for plants homozygote or heterozygote for functional *FvTFL1* alleles (FV alleles). Four clones of each of 335 F2 plants were phenotyped and mean values were used for analyses. Standard deviations of clones of each F2 line varied between 0.0 and 4.0 for flowering time, 0.0 and 2.1 for branch crowns, and 0.5 and 7.6 for runners.

Variation was also observed in the vegetative development of the F2 lines. By the end of the experiment, most F2 lines produced 2–4 branch crowns per plant. However, a significant number of genotypes produced an average of less than two, or more than four branch crowns (Figure 1b). The total number of runners produced per genotype varied between 5 and 22 (Figure 1c). The broad sense heritability (H^2) was 0.48 and 0.41 for branch crowns and runners, respectively. The number of runners was positively correlated with the number of branch crowns (r=0.785; Figure 1d). However, no correlation was found between flowering time and the number of branch crowns or runners.

Marker development and linkage map construction

Of the 73 reported SSR and 49 TSP markers screened (Supplementary File 1), 47 and 13 markers, respectively, were polymorphic between the 'H4' and 'FV' genotypes and segregated in the progeny. Following linkage analysis on 335 F2 lines, these 60 markers coalesced into seven discrete linkage groups spanning the majority of the *Fragaria* genome (Supplementary Figure 3). To increase mapping resolution, GBS was carried out in 188 F2 seedlings. Out of 271 473 572 barcoded Illumina reads, 2 391 856 GBS tags with a minimum coverage of three were found, and 1 755 780 tags (73.4%) were uniquely aligned on the H4 reference genome. Initial SNP calling revealed 133 930 SNPs, but after Hardy–Weinberg filtering and removal of SNPs that had more than 10% missing data and a minor allele frequency below 0.1, an ABH file of 4672 SNPs was retained. Following initial mapping, imputation and filtering, 2395 loci in 186 F2 lines, comprising 53 SSR/TSP markers and 2342 GBS-derived SNP markers coalesced into a linkage map of the seven expected linkage groups that covered a total genetic distance of 558 cM (Figure 2). Plotting genetic distances against the physical positions of the markers on the *Fvb* genome sequence assembly demonstrated that the genetic map was predominantly co-linear with the genome sequence (Figure 2) and thus marker placement on the map could be considered generally reliable.

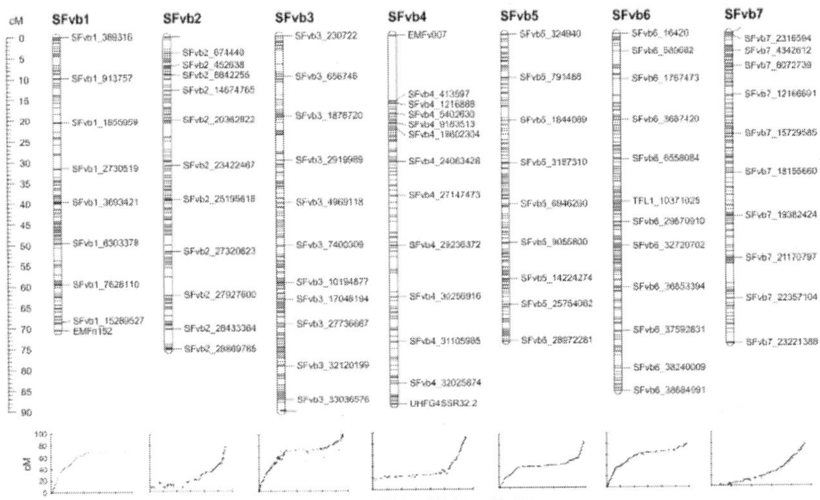

Figure 2. Linkage map of 186 F2 progeny of the H4×FV cross comprising of 2395 molecular markers. The numbers at the top denote the chromosome number and the genetic distances are in centiMorgan (cM, scale in the left). In the small panels, all markers of each chromosome are plotted according to their position in the genome v.2.0 (*x*-axis) and in the genetic map (*y*-axis).

Flowering time QTLs in greenhouse and field experiments

QTL analysis was first carried out using greenhouse flowering time data on the progeny of 335 seedlings and segregation data for 60 genetic loci. This analysis revealed four significant QTLs, one on both LG6 and LG7, and two on LG4 (Supplementary Figure 4). QTL analysis was also performed in growth chamber flowering time data using 32 F2 lines that were selected based on their extreme flowering time phenotypes in the greenhouse experiment. Using these lines, the same four QTLs with a LOD value above 3 were detected in both greenhouse and growth chamber flowering time data (Table 1). To increase the mapping resolution, we performed QTL analysis using the additional GBS based markers on selected 186 seedlings including individuals that exhibited informative recombination close to the QTL regions mapped in a larger progeny. This high-resolution mapping identified one additional QTL peak on LG7 and narrowed down the other QTL peaks (Figure 3). The peak of the LOD of the QTL on LG6 corresponded to the position of the *FvTFL1* gene previously identified by Koskela *et al.*32 (Figure 3).32 This QTL explained 17% of the observed variance; plants homozygous for 'FV' alleles of *FvTFL1* flowered later than heterozygous lines (Table 2). The peaks of the LOD for two significant loci on LG4 were associated with markers SFvb4_7853414 and bx083, both explaining about 15% of the observed variance. In both cases, the allele from the 'FV' parent promoted flowering. The peaks of the LOD of the significant QTLs on LG7 were associated with markers SFvb7_16204472 and SFvb7_21710529 that explained over 12% of the observed variance each, with the allele of the 'FV' parent promoting earlier flowering. The analysis of different haplotype combinations indicated that QTLs on all three LGs had additive effects on flowering time (Figure 4; Table 2). Plants with 'FV' markers in both LG4 and LG7 flowered first, whereas the latest plants typically had one or two 'H4' alleles in both LGs. In addition, lines containing two functional *FvTFL1* alleles tended to flower later than heterozygous lines, regardless of LG4 and LG7 marker genotypes.

Table 1. Flowering time QTLs mapped in 32 F2 lines with extreme phenotypes

Marker	*Linkage group*	*Position (cM)*	*LOD value, growth chamber*	*LOD value, greenhouse*
UHFG4M19[a]	LG4	18.91	3.8	6.42
bx083	LG4	37.70	4.1	7.42
TFL1	LG6	17.84	5.01	13.59
BFaCT44	LG7	40.59	3.57	5.59

[a]Markers at 15.91–20.17 cM showed the same LOD value. The most significant markers for each QTL, their positions and LOD values in growth chamber and greenhouse experiments are shown.

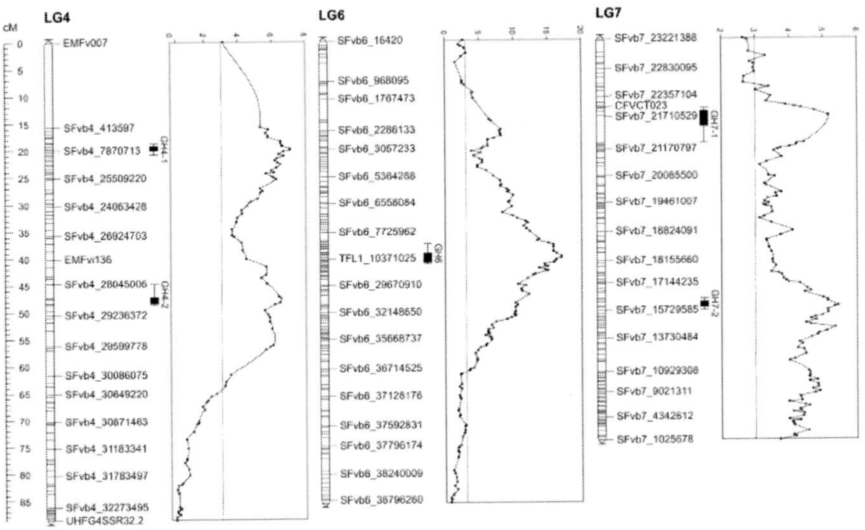

Figure 3. Flowering time QTLs in a greenhouse experiment. QTLs for flowering time were detected on chromosomes 4, 6, and 7. The genetic map (left) and LOD scores (right) are shown for each chromosome. Bars denote areas with the highest LOD scores (QTL regions). Field-grown plants were moved to the greenhouse in the middle of December 2011 for flowering time observations. Four clones of each of 186 F2 plants were phenotyped and mean values were used for the QTL analysis.

Table 2: QTLs mapped in the greenhouse experiment

QTL/ phenotype	*Marker*	*Position (bp)*	*Position (cM)*	*LOD value*	*Explained variance (%)*	*Dominance*	*Additive*
FT	SFvb4_7870713	Fvb4: 7870713	20.012	7.1	16.1	−0.137	1.807
FT	Bx083/ SFvb4_28667943	Fvb4: 28667943	47.485	6.58	15	0.427	1.784
FT	TFL1	Fvb6: 10371025	40.306	7.59	17	−4.380	1.623
FT	SFvb7_16204472	Fvb7: 16204472	24.895	5.45	12.6	0.232	1.775
FT	SFvb7_21710529	Fvb7: 21710529	14.215	5.16	12	−0.212	1.617
BC	SFvb4_29399865	Fvb4: 29399865	53.184	8.8	20.5	−0.053	−0.552
BC	SFvb5_25913832	Fvb5: 25913832	65.718	9.78	22.5	0.151	−0.582
RU	SFvb4_29399865	Fvb4: 29399865	53.184	6.56	15	0.536	−1.382
RU	SFvb5_25814268	Fvb5: 25814268	64.908	6.13	14.1	0.899	−1.271

Abbreviations: BC, branch crowns; FT, flowering time; RU, runners The most significant markers for each QTL, their positions and different genetic parameters are shown. Negative or positive value for dominance indicate that the allele of 'H4' or 'FV' parent has larger effect on the phenotype, respectively.

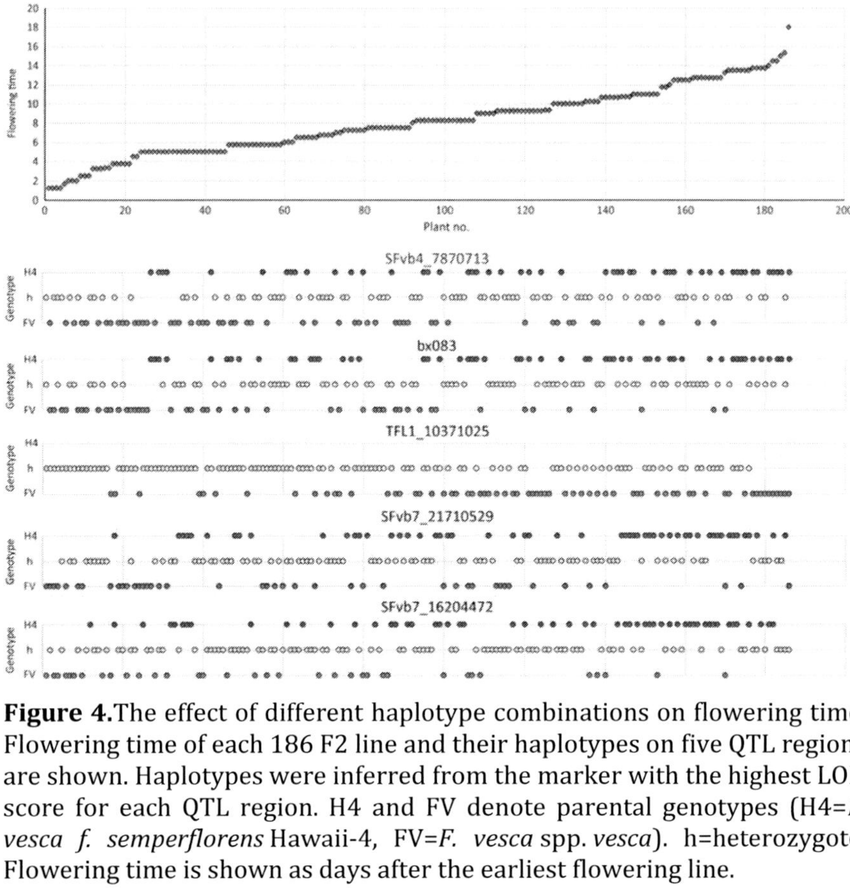

Figure 4. The effect of different haplotype combinations on flowering time. Flowering time of each 186 F2 line and their haplotypes on five QTL regions are shown. Haplotypes were inferred from the marker with the highest LOD score for each QTL region. H4 and FV denote parental genotypes (H4=*F. vesca f. semperflorens* Hawaii-4, FV=*F. vesca* spp. *vesca*). h=heterozygote. Flowering time is shown as days after the earliest flowering line.

The analysis of field phenotypes for flowering time revealed highly significant QTL only on LG4 (Figure 5). The peak of the LOD of this QTL was associated with marker bx083 as in the greenhouse dataset, but there was also another equally high peak ~1 Mb upstream of bx083. This QTL explained almost 25% of the phenotypic variance. Another minor QTL just above the genome-wide significance threshold of 3 was detected on LG2 with a peak of the LOD associated with the marker CFVCT020 at 11.94 cM. However, no QTLs were found on LG6 or LG7 in the field using phenotypic data of 186 F2 lines.

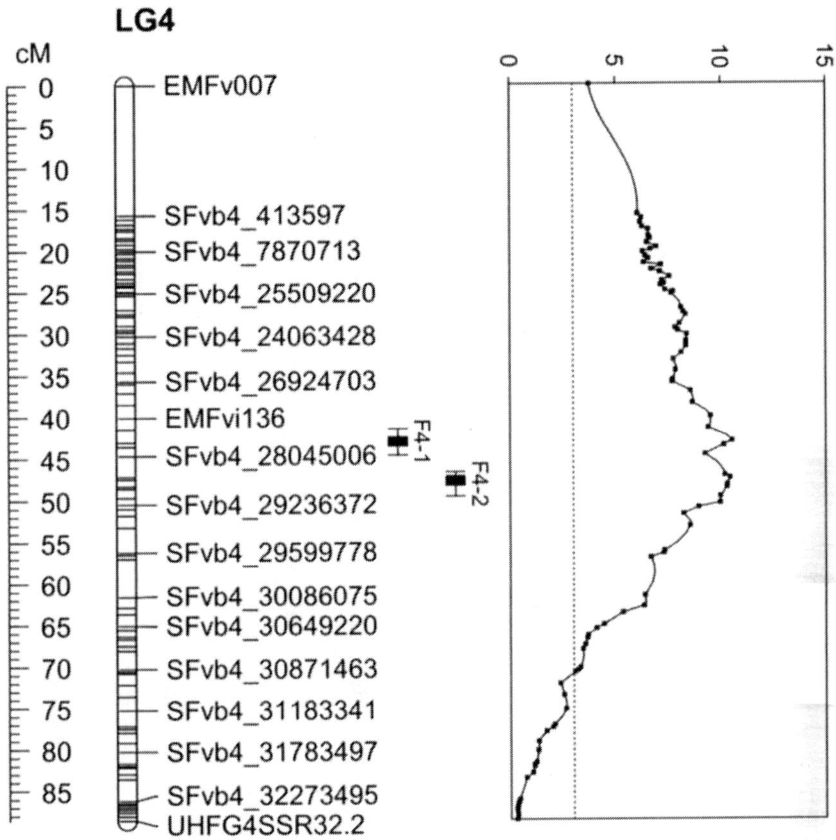

Figure 5. Flowering time QTL on the chromosome 4 in the field experiment. The genetic map (left) and LOD scores (right) are shown. Bars denote areas with the highest LOD scores (QTL regions). H4×FV F2 lines (n=186) were grown in the field and flowering time was observed in summer 2011.

Runnering and branch crown evaluation

QTL were revealed on LG4 and LG5 associated with branch crown formation and runnering. The peak of the LOD of the significant QTL for the number of branch crowns was associated with the marker SFvb4_29399865 at 53.184 cM on LG4 and with the marker SFvb_25913832 at 65.718 cM on LG5 (Figures 6a and b), and QTLs for the number of runners were located in the same regions (Supplementary Figure 5). Both QTLs explained over 20% of the variance in the number of branch crowns, but their effect on runnering was less pronounced (Table 2). Increased numbers of runners and branch crowns were associated with 'FV' alleles on both LG4 and LG5.

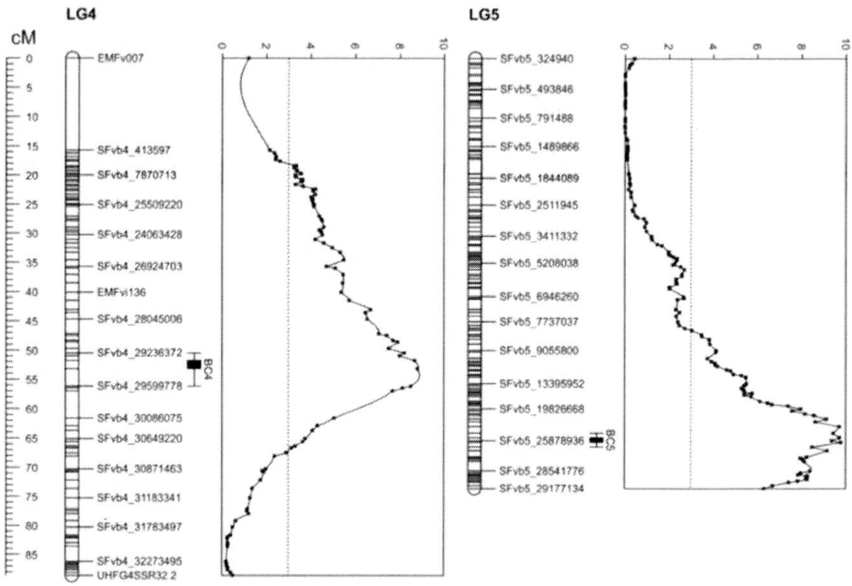

Figure 6. QTLs for the number of branch crowns in the greenhouse experiment. Two QTLs for the number of branch crowns (BC4 and BC5) were detected in chromosomes 4 and 5. The genetic map (left) and LOD scores (right) are shown for each chromosome. Bars denote areas with the highest LOD scores (QTL regions). Field-grown plants were taken into the greenhouse in the middle of December 2011 for growth observations. Four clones of each of 186 F2 plants were phenotyped and mean values were used for the QTL analysis.

Candidate genes for QTLs

To identify candidate genes around the QTLs, BLAST searches were carried out on the genomic regions between the flanking markers. In most cases, no obvious candidate genes were detected inside the narrow QTL peaks, but promising candidates were found very close to the highest LOD values. *FvFT2* (mrna04680.1-v1.0-hybrid) that was also identified in another QTL mapping study,69 was found within the ~400 kb QTL region in the end of the LG4. However, no candidate genes were searched for the other QTL on the LG4 because the physical position of the QTL was uncertain. On LG7, a homolog of *EARLY FLOWERING 6* (*ELF6*, gene23255-v1.0-hybrid)70 was located about 600 kb upstream of the first QTL peak, and a gene encoding a homolog of floral repressor TFL1 (*FvCENTRORADIALIS1, FvCEN1*; gene13304-v1.0-hybrid)32 was found ~650 kb upstream of the second QTL peak.

Candidate genes were also identified within the QTL intervals controlling branch crown and runner formation. On LG4, a TCP transcription factor (FvTCP7, gene04759-v1.0-hybrid)71 was found 150 kb upstream of the QTL peak. Furthermore, genes encoding two closely related MADS transcription factors that are homologous to DORMANCY ASSOCIATED MADS BOX (DAM, mrna12119.1-v1.0-hybrid, mrna12120.1-v1.0-hybrid),2,72,73 were found inside the QTL region on the LG5. No non-synonymous variation was found in the coding sequence of any candidate gene between the parental genome sequences (data not shown).

DISCUSSION

Extension of the strawberry production season in the open field is of significant economic importance, and two different breeding strategies have been implemented to maximize the length of the season. Several breeding programs are focusing on EB cultivars that can produce berries for several months during a single season,6,8,42 whereas the breeding of early and late ripening cultivars is an alternative approach to extend the ripening season.7 Previous QTL mapping studies in strawberry have focused on the perpetual flowering habit,5,6,8,42 and no QTLs have been reported for earliness. Here, we report the identification of additive QTLs on three LGs that explain over 50% of the observed 18-day variance in flowering time in woodland strawberry mapping population previously reported by Koskela *et al.*32 Moreover, we have found two QTLs that affect branch crown formation, the trait that has a direct effect on strawberry yield potential.45,47,48

Additive QTLs on three linkage groups control flowering time in woodland strawberry

QTL mapping using replicated greenhouse flowering time data in our H4×FV F2 population revealed one new QTL on the LG4, two on the LG7, and two additional QTLs that co-localized with previously mapped QTLs on the LG4 and LG6. Four of these QTLs were also detected in a replicated growth chamber experiment using 32 F2 selected lines that exhibited extreme flowering time phenotypes in a greenhouse experiment, indicating that these QTLs are robust at least in controlled climate. In the non-replicated field experiment, the analysis using GBS data on 186 lines revealed major QTL only on LG4. However, QTL mapping on a larger population of 335 lines using SSR markers showed LOD values close to 4.0 in both LG6 and LG7 (data not shown), but a replicated field experiment is needed to confirm these QTLs in the field.

On the LG6, the QTL mapped on the same region than a previously characterized *FvTFL1* gene that encodes a repressor of flowering.[32] In cultivated strawberry, a major QTL for perpetual flowering and runnering (*PFRU*) was previously identified on the LG4.[6,8,42] A recent fine-mapping study placed this QTL close to marker bx083,[69] and the same marker was located in the QTL peak also in our greenhouse and field experiments. Therefore, our data suggest a new role for *PFRU* in the control of earliness, in addition to its role in the flowering period, at least in woodland strawberry. The effect of *PFRU* on flowering time seems to be robust in different environments. However, the other QTL, that is located upstream of *PFRU* on LG4, was only detected in controlled climate experiments indicating that it may function only in specific environments.

We found other QTLs on LG7 where no flowering related QTLs have been previously detected in strawberries. These QTLs showed high LOD values in the greenhouse experiment with the F2 population of 186 lines (GBS data), but not in the field indicating that they were also influenced by the environment. One of these QTLs matched also with the QTL region detected in our growth chamber experiment (Table 1). This QTL was associated with the marker BFaCT044 at 22 166 857 bp on LG7 in a set of 32 early and late flowering lines in both growth chamber and greenhouse, and about 400 kb upstream (the most significant marker SFvb7_21710529) in the GBS dataset of 186 lines.

The analysis of allele combinations on QTL positions demonstrated their additive effects on flowering time (Figure 4; Table 2). Clearly, 'H4' specific markers on LG4 and LG7 QTL regions, in both heterozygous and homozygous condition delayed flowering. Furthermore, homozygous *FvTFL1* alleles from 'FV' parent caused additional delay compared to heterozygous plants. Plants homozygous for 'H4' alleles of *FvTFL1* were excluded from this study because they exhibit perpetual flowering habit.[32]

Candidate genes for flowering time QTLs

We searched for candidate flowering time genes in the genomic regions containing the markers with the highest LOD if the physical and genetic maps on the region were mostly co-linear (Figure 2). This was not the case for the newly identified QTL on the LG4, and therefore, a better genome assembly is needed to enable reliable candidate gene search on this region.

Candidate genes for *PFRU*, that co-localize with the QTL identified in our greenhouse and field studies on the LG4, has been recently reported.69 These genes include *FvFT2* and *FvCDF2* that are homologs of known regulators of photoperiodic flowering. In *Arabidopsis*, CDFs are floral repressors that contribute to photoperiodic flowering by down-regulating both *CO* and *FT* mRNA expression especially in the morning.74 Furthermore, specific alleles of potato *CDF* gene were selected during domestication allowing the cultivation of this SD crop in LD conditions in northern latitudes.75 Tissue specific gene expression analyses in woodland strawberry showed that *FvFT2* is mainly expressed in flowers and fruits.32,76 However, both FvFT2 and FvCDF2 may also have important roles in the control of photoperiodic development in strawberries and are thus good candidates for future functional studies.

FvELF6 was identified as a candidate gene for the QTL on LG7. ELF6 was originally identified as a repressor of photoperiodic flowering in *Arabidopsis*.70 It is a histone demethylase that control gene expression through chromatin regulation.77,78 As histone methylation is a common mechanism in the control of flowering time,79 and a homologous gene to *ELF6* co-localizes with a QTL for spring heat requirement for flowering on almond LG2,2 thus, *ELF6* is an interesting candidate for future studies.

The homolog of *TFL1/CEN* is located in the QTL region for flowering time in peach 72,80 and gene functional studies showed that *TFL1* homologs encode major floral repressors in several rosaceous species.32,33,34,35 Interestingly, *TFL1/CEN* homologs were identified as candidate genes for two QTL regions in this study. A QTL on the LG6 co-localize with *FvTFL1* that encodes a major floral repressor,32 and closely related *FvCEN1* is located near the second QTL peak on LG7 and awaits for functional validation. Our finding that plants containing only one functional *FvTFL1* copy flowered earlier than plants homozygous for functional 'FV' alleles provides interesting opportunities to control flowering time, particularly in polyploid crops such as the octoploid cultivated strawberry. In fact, a recent report showed that FaTFL1 is a major floral repressor also in cultivated strawberry, and that altered regulation of *FaTFL1* mRNA expression is associated with different flowering responses in this species.43 To enable more efficient breeding of flowering time, further studies are needed to uncover whether the variation in *FaTFL1* expression between cultivars is caused by allelic variation of *FaTFL1* itself or by upstream regulators. Among these regulators, the expression of *FT1* and *SOC1* homologs correlate with *TFL1* mRNA levels only in specific conditions in woodland and cultivated strawberries suggesting that *TFL1* alleles or its other regulators should be major targets of future investigations.[39,43] Whether some of the candidate genes identified in this study control flowering time through *FvTFL1* is an interesting open question.

Identification of QTLs controlling crown branching and runner formation

Although we mainly focused on flowering time, we also identified two QTLs on the LG4 and LG5 controlling the number of branch crowns and runners in woodland strawberry. These traits showed a strong positive correlation because the production of branch crowns increases the number of axillary buds that are able to form runners under LD conditions. Therefore, these QTL primarily control the branching of the leaf rosette in woodland strawberry. On the LG4, the QTL co-localized with previously mapped *PFRU* locus.[6,69] However, the peak of the LOD was located about 700 kb downstream of the flowering time QTL mapped in this study. This indicates that two closely located QTLs may control flowering and axillary bud differentiation to runners and branch crowns in our population, but additional studies are needed to confirm this hypothesis.

We identified a gene encoding TCP transcription factor close to the most significant marker on the LG4. This TCP, recently named as FvTCP7,[71] shows the highest similarity with *Arabidopsis* TCP14 and TCP15, which control cell proliferation and internode elongation.[81] As runners are long shoots with elongated internodes this is a promising candidate gene for the strawberry axillary bud differentiation. Initial characterization of *FvTCP7* indicated that it is highly expressed in vegetative tissues including runners, but also in flower buds, and the protein is localized to nucleus.[71] However, the function of FvTCP7 in strawberry is unknown. Further studies are needed to elucidate what is the relationship between the branching QTL detected here and *PFRU* that controls runner formation, in addition to perpetual flowering, in cultivated strawberry.[6,8]

Also the newly identified QTL region on the LG5 contained interesting candidate genes for future studies including two genes homologous to *DAM* and *SHORT VEGETATIVE PHASE* (*SVP*) that encode major regulators of dormancy in Rosaceae and floral repressor in *Arabidopsis*, respectively.[2,72,73,82] Although strawberries do not have true dormancy, their growth vigor is strongly reduced under SDs in autumn including the cessation of runner formation, and a long period of chilling is needed to resume normal vegetative development.[83] Functional studies would reveal whether identified FvDAM transcription factors control dormancy and what is their role in the control of axillary bud differentiation. Furthermore, the possible interaction of the branching/runnering QTLs identified in this study with the FvSOC1 and gibberellin pathway, that were previously shown to control runnering in woodland strawberry,[37,49] should be elucidated.

CONCLUSIONS

Here, we have reported the mapping of three new flowering time QTLs in woodland strawberry, which co-localize with homologs of known flowering time regulators that have not been functionally characterized in this species. In addition, we have shown that *PFRU*, another QTL that was previously reported to control the duration of flowering and the number of runners in cultivated strawberry,[6,69] affects flowering time, crown branching and runner formation in woodland strawberry. Although we were able to detect four out of five flowering time QTLs in two independent experiments in controlled climate, only one of these showed a LOD value above 3 in the non-replicated field experiment. Therefore, further replicated experiments are needed to reveal the role of these QTLs in different environments. On the basis of the evidence shown here, identified QTLs control flowering time additively along with a previously identified floral repressor FvTFL1 that has quantitative effect on flowering time. As *TFL1* homologs likely integrate both light and temperature signals to control flowering time in woodland and cultivated strawberries,[39,43] we suggest that candidate genes on the LG4 and LG7 QTL regions may control flowering time through *TFL1*. However, further studies are needed to clarify causal genes and to reveal their roles in the control of flowering time and vegetative traits. As cultivated strawberry is an octoploid species with at least two subgenome donors,[27,84] we hypothesize that it may contain significant genetic variation in these regulators that may enable fine tuning of flowering time and shoot architecture in different environments through marker assisted selection or genomic selection strategies that have recently been developed in this species.[85]

REFERENCES

1. Andrés F, Coupland G. The genetic basis of flowering responses to seasonal cues. Nat Rev Genet 2012; 13: 627–639.
2. Sánchez-Pérez R, Dicenta F, Martinez-Gomez P. Inheritance of chilling and heat requirements for flowering in almond and QTL analysis. Tree Genet Genomes 2012; 8: 379–389.

3. Hibrand-Saint Oyant L, Crespel L, Rajapakse S, Zhang L, Foucher F. Genetic linkage maps of rose constructed with new microsatellite markers and locating QTL controlling flowering traits. Tree Genet Genomes 2007; 4: 11–23.
4. Zhebentyayeva TN, Fan S, Chandra A et al. Dissection of chilling requirement and bloom date QTLs in peach using a whole genome sequencing of sibling trees from an F2 mapping population. Tree Genet Genomes 2014; 10: 35–51.
5. Weebadde CK, Wang D, Finn C et al. Using a linkage mapping approach to identify QTL for day-neutrality in the octoploid strawberry. Plant Breed 2008; 127: 94–101.
6. Gaston A, Perrotte J, Lerceteau-Köhler E et al. PFRU, a single dominant locus regulates the balance between sexual and asexual plant reproduction in cultivated strawberry. J Exp Bot 2013; 64: 1837–1848.
7. Bestfleisch M, Möhring J, Hanke MV, Peil A, Flachowsky H. A diallel crossing approach aimed on selection for ripening time and yield in breeding of new strawberry (Fragaria×ananassa Duch.) cultivars. Plant Breed 2014; 133: 115–120.
8. Castro P, Bushakra JM, Stewart P et al. Genetic mapping of day-neutrality in cultivated strawberry. Mol Breed 2015; 35: 1–16.
9. Opstad N, Sønsteby A, Myrheim U, Heide OM. Seasonal timing of floral initiation in strawberry: effects of cultivar and geographic location. Sci Hortic 2011; 129: 127–134.
10. Heide OM. Photoperiod and temperature interactions in growth and flowering of strawberry. Physiol Plant 1977; 40: 21–26.
11. Durner EF. Photoperiod and temperature effects on flower and runner development in day-neutral, Junebearing, and everbearing strawberries. J Amer Soc Hort Sci 1984; 109: 396–400.
12. Manakasem Y, Goodwin PB. Responses of dayneutral and Junebearing strawberries to temperature and daylength. J Hortic Sci Biotechnol 2001; 76: 629–635.
13. Heide OM, Sønsteby A. Interactions of temperature and photoperiod in the control of flowering of latitudinal and altitudinal populations of wild strawberry (Fragaria vesca). Physiol Plant 2007; 130: 280–289.
14. Bradford E, Hancock JF, Warner RM. Interactions of temperature and photoperiod determine expression of repeat flowering in strawberry. J Am Soc Hortic Sci 2010; 135: 102–107.
15. Sønsteby A, Heide OM. Dormancy relations and flowering of the strawberry cultivars Korona and Elsanta as influenced by photoperiod and temperature. Sci Hortic 2006; 110: 57–67.
16. Nishiyama M, Kanahama K. Effects of temperature and photoperiod on flower bud initiation of day-neutral and everbearing strawberries. Acta Hortic 2002; 567: 253–255.
17. Sønsteby A, Heide O. Long-day control of flowering in everbearing strawberries. J Hortic Sci Biotechnol 2007; 82: 875–884.
18. Brown T, Wareing PF. the genetical control of the everbearing habit and three other characters in varieties of Fragaria vesca. Euphytica 1965; 14: 97–112.
19. Sargent DJ, Kuchta P, Girona EL et al. Simple sequence repeat marker development and mapping targeted to previously unmapped regions of the strawberry genome sequence. Plant Genome J 2011; 4: 165–177.
20. Sargent DJ, Davis TM, Tobutt KR, Wilkinson MJ, Battey NH, Simpson DW. A genetic linkage map of microsatellite, gene-specific and morphological markers in diploid Fragaria. Theor Appl Genet 2004; 109: 1385–1391.
21. Sargent DJ, Clarke J, Simpson DW et al. An enhanced microsatellite map of diploid Fragaria. Theor Appl Genet 2006; 112: 1349–1359.
22. Ashley MV, Wilk JA, Styan SMN et al. High variability and disomic segregation of microsatellites in the octoploid Fragaria virginiana Mill. (Rosaceae). Theor Appl Genet 2003; 107: 1201–1207.

23. Tabone T, Mather DE, Hayden MJ. Temperature switch PCR (TSP): robust assay design for reliable amplification and genotyping of SNPs. BMC Genomics 2009; 10: 580.
24. Schuelke M. An economic method for the fluorescent labeling of PCR fragments. Nat Biotechnol 2000; 18: 233–234.
25. Urrutia M, Bonet J, Arús P, Monfort A. A near-isogenic line (NIL) collection in diploid strawberry and its use in the genetic analysis of morphologic, phenotypic and nutritional characters. Theor Appl Genet 2015; 128: 1261–1275.
26. Shulaev V, Sargent DJ, Crowhurst RN et al. The genome of woodland strawberry (Fragaria vesca). Nat Genet 2011; 43: 109–116.
27. Tennessen Ja, Govindarajulu R, Ashman T-L, Liston A. Evolutionary origins and dynamics of octoploid strawberry subgenomes revealed by dense targeted capture linkage maps. Genome Biol Evol 2014; 6: 3295–3313.
28. Bassil NV, Davis TM, Zhang H et al. Development and preliminary evaluation of a 90 K Axiom SNP array for the allo-octoploid cultivated strawberry Fragaria×ananassa. BMC Genomics 2015; 16: 155.
29. Mahoney LL, Sargent DJ, Abebe-Akele F et al. A high-density linkage map of the ancestral diploid strawberry, constructed with single nucleotide polymorphism markers from the IStraw90 Array and genotyping by sequencing. Plant Genome 2016; 9: 1–14.
30. Deng C, Davis TM. Molecular identification of the yellow fruit color (c) locus in diploid strawberry: a candidate gene approach. Theor Appl Genet 2001; 103: 316–322.
31. Albani MC, Battey NH, Wilkinson MJ. The development of ISSR-derived SCAR markers around the SEASONAL FLOWERING LOCUS (SFL) in Fragaria vesca. Theor Appl Genet 2004; 109: 571–579.
32. Koskela EA, Mouhu K, Albani MC et al. Mutation in TERMINAL FLOWER1 reverses the photoperiodic requirement for flowering in the wild strawberry Fragaria vesca. Plant Physiol 2012; 159: 1043–1054.
33. Iwata H, Gaston A, Remay A et al. The TFL1 homologue KSN is a regulator of continuous flowering in rose and strawberry. Plant J 2012; 69: 116–125.
34. Flachowsky H, Szankowski I, Waidmann S, Peil A, Tränkner C, Hanke MV. The MdTFL1 gene of apple (Malus×domestica Borkh.) reduces vegetative growth and generation time. Tree Physiol 2012; 32: 1288–1301.
35. Freiman A, Shlizerman L, Golobovitch S et al. Development of a transgenic early flowering pear (Pyrus communis L.) genotype by RNAi silencing of PcTFL1-1 and PcTFL1-2. Planta 2012; 235: 1239–1251.
36. Kurokura T, Mimida N, Battey NH, Hytönen T. The regulation of seasonal flowering in the Rosaceae. J Exp Bot 2013; 64: 4131–4141.
37. Mouhu K, Kurokura T, Koskela EA et al. The Fragaria vesca Homolog of SUPPRESSOR OF OVEREXPRESSION OF CONSTANS1 represses flowering and promotes vegetative growth. Plant Cell 2013; 25: 3296–3310.
38. Rantanen M, Kurokura T, Mouhu K et al. Light quality regulates flowering in FvFT1/FvTFL1 dependent manner in the woodland strawberry Fragaria vesca. Front Plant Sci 2014; 5: 271.
39. Rantanen M, Kurokura T, Jiang P, Mouhu K, Hytönen T. Strawberry homologue of TERMINAL FLOWER1 integrates photoperiod and temperature signals to inhibit flowering. Plant J 2015; 82: 163–173.
40. Ahmadi H, Bringhurst RS, Voth V. Modes of inheritance of photoperiodism in Fragaria. J Amer Soc Hort Sci 1990; 115: 146–152.
41. Sugimoto T, Tamaki K, Matsumoto J, Yamamoto Y, Shiwaku K, Watanabe K. Detection of RAPD markers linked to the everbearing gene in Japanese cultivated strawberry. Plant Breed 2005; 124: 498–501.
42. Honjo M, Nunome T, Kataoka S et al. Simple sequence repeat markers linked to the everbearing flowering gene in long-day and day-neutral cultivars of the octoploid cultivated strawberry Fragaria×ananassa. Euphytica 2016; 209: 291–303.

43. Koskela E, Sønsteby A, Flachowsky H et al. TERMINAL FLOWER1 is a breeding target for a novel everbearing trait and tailored flowering responses in cultivated strawberry (Fragaria×ananassa Duch.). Plant Biotechnol J 2016; 14: 1852–1861.
44. Nakano Y, Higuchi Y, Yoshida Y, Hisamatsu T. Environmental responses of the FT/TFL1 gene family and their involvement in flower induction in Fragaria×ananassa. J Plant Physiol 2015; 177: 60–66.
45. Konsin M, Voipio I, Palonen P. Influence of photoperiod and duration of short-day treatment on vegetative growth and flowering of strawberry (Fragaria×ananassa duch). J Hortic Sci Biotechnol 2001; 76: 77–82.
46. de Camacaro MEP, Camacaro GJ, Hadley P, Battey NH, Carew JG. Pattern of growth and development of the strawberry cultivars Elsanta, Bolero, and Everest. J Am Soc Hortic Sci 2002; 127: 901–907.
47. Hytönen T, Palonen P, Mouhu K, Junttila O. Crown branching and cropping potential in strawberry (Fragaria×ananassa Duch.) can be enhanced by daylength treatments. J Hortic Sci Biotechnol 2004; 79: 466–471.
48. Hytönen T, Mouhu K, Koivu I, Junttila O. Prohexadione-calcium enhances the cropping potential and yield of strawberry. Eur J Hortic Sci 2008; 73: 210–215.
49. Hytönen T, Elomaa P, Moritz T, Junttila O. Gibberellin mediates daylength-controlled differentiation of vegetative meristems in strawberry (Fragaria×ananassa Duch). BMC Plant Biol 2009; 9: 18.
50. Cipriani G, Testolin R. Isolation and characterization of microsatellite loci in Fragaria. Mol Ecol Notes 2004; 4: 366–368.
51. Monfort A, Vilanova S, Davis TM, Arus P. A new set of polymorphic simple sequence repeat (SSR) markers from a wild strawberry (Fragaria vesca) are transferable to other diploid Fragaria species and to Fragaria x ananassa. Mol Ecol Notes 2006; 6: 197–200.
52. Sargent DJ, Hadonou AM, Simpson DW. Development and characterization of polymorphic microsatellite markers from Fragaria viridis, a wild diploid strawberry. Mol Ecol Notes 2003; 3: 550–552.
53. Sargent DJ, Passey T, Surbanovski N et al. A microsatellite linkage map for the cultivated strawberry (Fragaria×ananassa) suggests extensive regions of homozygosity in the genome that may have resulted from breeding and selection. Theor Appl Genet 2012; 124: 1229–1240.
54. Rozen S, Skaletsky H. Primer3 on the WWW for general users and for biologist programmers. Methods Mol Biol 1999; 132: 365–386.
55. Elshire RJ, Glaubitz JC, Sun Q et al. A robust, simple genotyping-by-sequencing (GBS) approach for high diversity species. PLoS ONE 2011; 6: e19379.
56. Glaubitz JC, Casstevens TM, Lu F et al. TASSEL-GBS: a high capacity genotyping by sequencing analysis pipeline. PLoS ONE 2014; 9: e90346.
57. Bradbury PJ, Zhang Z, Kroon DE, Casstevens TM, Ramdoss Y, Buckler ES. TASSEL: software for association mapping of complex traits in diverse samples. Bioinformatics 2007; 23: 2633–2635.
58. Danecek P, Auton A, Abecasis G et al. The variant call format and VCFtools. Bioinformatics 2011; 27: 2156–2158.
59. Browning BL, Browning SR. Improving the accuracy and efficiency of identity-by-descent detection in population data. Genetics 2013; 194: 459–471.
60. Ward JA, Bhangoo J, Fernández-Fernández F et al. Saturated linkage map construction in Rubus idaeus using genotyping by sequencing and genome-independent imputation. BMC Genomics 2013; 14: 2.
61. Van Ooijen JW. JoinMap 4, Software for the Calculation of Genetic Linkage Maps in Experimental Populations. Kyazma B.V.: Wageningen, Netherlands, 2006.
62. Voorrips RE. MapChart: software for the graphical presentation of linkage maps and QTLs. J Hered 2002; 93: 77–78.
63. Van Ooijen JW. MapQTL 5, Software for the Mapping of Quantitative Trait Loci in Experimental Populations. Kyazma B.V.: Wageningen, Netherlands, 2004.

64. Martin M. Cutadapt removes adapter sequences from high-throughput sequencing reads. EMBnet.journal 2011; 17: 10.
65. Li H. Aligning sequence reads, clone sequences and assembly contigs with BWA-MEM. arXiv Prepr arXiv13033997; 2013.
66. Li H, Handsaker B, Wysoker A et al. The sequence alignment/Map format and SAMtools. Bioinformatics 2009; 25: 2078–2079.
67. Li H. A statistical framework for SNP calling, mutation discovery, association mapping and population genetical parameter estimation from sequencing data. Bioinformatics 2011; 27: 2987–2993.
68. Conesa A, Götz S, García-Gómez JM, Terol J, Talón M, Robles M. Blast2GO: a universal tool for annotation, visualization and analysis in functional genomics research. Bioinformatics 2005; 21: 3674–3676.
69. Perrotte J, Gaston A, Potier A, Petit A, Rothan C, Denoyes B. Narrowing down the single homoeologous FaPFRU locus controlling flowering in cultivated octoploid strawberry using a selective mapping strategy. Plant Biotechnol J 2016; 14: 2176–2189.
70. Noh B, Lee S-H, Kim H-J et al. Divergent roles of a pair of homologous jumonji/zinc-finger-class transcription factor proteins in the regulation of Arabidopsis flowering time. Plant Cell 2004; 16: 2601–2613.
71. Wei W, Hu Y, Cui M-Y, Han Y-T, Gao K, Feng J-Y. Identification and transcript analysis of the TCP transcription factors in the diploid woodland strawberry Fragaria vesca. Front Plant Sci 2016; 7: 1–18.
72. Fan S, Bielenberg DG, Zhebentyayeva TN et al. Mapping quantitative trait loci associated with chilling requirement, heat requirement and bloom date in peach (Prunus persica). N Phytol 2010; 185: 917–930.
73. Bielenberg DG, Wang Y, Li Z et al. Sequencing and annotation of the evergrowing locus in peach [Prunus persica (L.) Batsch] reveals a cluster of six MADS-box transcription factors as candidate genes for regulation of terminal bud formation. Tree Genet Genomes 2008; 4: 495–507.
74. Song YH, Smith RW, To BJ, Miller AJ, Imaizumi T, Millar AJ. FKF1 conveys timing information for CONSTANS stabilisation in photoperiodic flowering. Science 2012; 336: 1045–1049.
75. Kloosterman B, Abelenda JA, del Mar Carretero Gomez M, Oortwijn M, de Boer JM, Kowitwanich K. Naturally occurring allele diversity allows potato cultivation in northern latitudes. Nature 2013; 495: 246–250.
76. Kang C, Darwish O, Geretz A, Shahan R, Alkharouf N, Liu Z. Genome-scale transcriptomic insights into early-stage fruit development in woodland strawberry Fragaria vesca. Plant Cell 2013; 25: 1960–1978.
77. Jiang D, Yang W, He Y, Amasino RM. Arabidopsis relatives of the human lysine-specific demethylase1 repress the expression of FWA and FLOWERING LOCUS C and thus promote the floral transition. Plant Cell 2007; 19: 2975–2987.
78. Jeong JH, Song HR, Ko JH et al. Repression of FLOWERING LOCUS T chromatin by functionally redundant histone H3 lysine 4 demethylases in Arabidopsis. PLoS ONE 2009; 4: e8033.
79. Jeong HJ, Yang J, Yi J, An G. Controlling flowering time by histone methylation and acetylation in Arabidopsis and rice. J Plant Biol 2015; 58: 203–210.
80. Romeu JF, Monforte AJ, Sánchez G, Granell A, García-Brunton J, Badenes ML et al. Quantitative trait loci affecting reproductive phenology in peach. BMC Plant Biol 2014; 14: 52.
81. Kieffer M, Master V, Waites R, Davies B. TCP14 and TCP15 affect internode length and leaf shape in Arabidopsis. Plant J 2011; 68: 147–158.
82. Hartmann U, Höhmann S, Nettesheim K, Wisman E, Saedler H, Huijser P. Molecular cloning of SVP: a negative regulator of the floral transition in Arabidopsis. Plant J 2000; 21: 351–360.
83. Heide OM, Stavang JA, Sønsteby A. Physiology and genetics of flowering in cultivated and wild strawberries—a review. J Hortic Sci Biotechnol 2013; 88: 1–18.

84. Sargent DJ, Yang Y, Šurbanovski N et al. HaploSNP affinities and linkage map positions illuminate subgenome composition in the octoploid, cultivated strawberry (Fragaria×ananassa). Plant Sci 2016; 242: 140–150.
85. Gezan SA, Osorio LF, Verma S, Whitaker VM. An experimental validation of genomic selection in octoploid strawberry. Hortic Res 2017; 4: 16070.

CHAPTER 12

Agricultural Waste as Growing Media Component for The Growth and Flowering of *Gerbera Jamesonii* Cv. Hybrid Mix

Atif Riaz[1] • Adnan Younis[1] • Imran Ghani[1] • Usman Tariq[1] • Muhammad Ahsan[1]

Institute of Horticultural SciencesUniversity of Agriculture Faisalabad Pakistan

ABSTRACT

Purpose

This study was performed to investigate the impact of various agricultural wastes as potting media on growth, yield, and quality flower production of *Gerbera jamesonii* L. cv. hybrid mix, and to estimate hazardous impact of chemical fertilizers.

Method

The effect of four agricultural substrates viz. farm yard manure, coconut coir dust, Lahore compost (local produce), and leaf compost combined with conventional media i.e., garden soil, sand, and silt in equal proportion was studied for commercial production of gerbera. All the treatment combinations were arranged into completely randomized design with three replicates.

Result

Visible variations in morphological characters were evidenced among conventional medium and agricultural waste as substrates with significant superiority of flower quality and plant growth. The presented results showed that treatment combination of silt, coconut coir dust, and top soil (SCT) produced maximum plant height while mixture of silt, Lahore compost and top soil (SLT) gave maximum number of leaves per plant and maximum flower stalk thickness. The highest values for leaf area, number of roots, flower fresh to dry weight ratio and number of flowers per plant were achieved in combination of silt, farm yard manure, and top soil (SFT) whereas silt, sand, and top soil (SST) presented maximum flower diameter, maximum fresh, and dry weight of flowers. Chemical characteristics of growing media showed superiority of silt, farm yard manure, and top soil (SFT) and silt, sand, and top soil (STL) for available phosphorus and potassium with values of 27.0 and 500 ppm, respectively.

Conclusion

It was concluded that the application of waste materials in combination with silt provided positive results for vegetative and reproductive growth of *Gerbera jamesonii* plants.

Keywords: Agricultural waste Farm yard manure Flower quality *Gerbera jamesonii* Leaf compost

INTRODUCTION

Gerbera (*Gerbera jamesonii*), is a herbaceous perennial plant with flowering stalks produced in between the leaves, ranging from 12 to 37 cm in height depending upon the cultivar (Weeraratne et al. 2012). Single daisy like flowers of gerbera are used for garden and home decoration in early stages but rarely used as cut flower because of their weak pedicle, bent neck, and poor vase life. The introduction of new varieties having double flowers (Torrey 1951) is considered better for cut purposes in all respects, encouraging the cut flower growers to enhance their production. It's potential as a cut flower is now widely accepted because of its ability to propagate through seed, rapid growth, and the demand for cheap labor supplies. For a better production of gerbera cut flowers, it is necessary to estimate all the growth factors like nutrition, growth environment, growth media, etc.

Growth media are considered among one of the most important factors affecting the production of Gerbera (Nowak and Strojny 2003; Sindhu et al. 2010; Khalaj et al. 2011; Ahmad et al. 2012). Natural soil along with peat are the most common growing substrates used for the container production of annual and perennial ornamental plants since many years (Guerrero et al. 2002; Hernandez-Apaolaza et al. 2005). In order to enhance quality and production of flowering plants chemical fertilizer are frequently applied to soils but the excessive applications of chemical fertilizers have harmful effects on soil flora, fauna, and enzymes. This ultimately leads to decrease their activity for maintaining the natural fertility of soil (Gupta et al. 2014) and increase the dependence on chemical fertilizers. As a result, there is a growing concern to reduce the application of chemical fertilizer to soils, and in order to use soil nutrients more efficiently and to overcome pollution hazards the application of organic matter is gaining acceptance among farming community.

In the past few years, usage of peat has been reduced due to environmental concerns for conservation (Abad et al. 2002). Therefore, researchers showed their interest in other organic materials like agricultural wastes which could be viable source of organic nutrients for the production of container grown plants. Agricultural waste such as rice hulls, coir, sawdust, and compost as growing media has been proved as an important substitute of simple soil for ornamental plant production (Younis et al. 2007; Wilson et al. 2009; Tariq et al. 2012) including gerbera. Furthermore, production of large sized gerbera flowers is entirely based on the structure, texture, pH, organic matter contents, as well as nitrogen, phosphorus, and potassium levels of the growth media (Rader 1998; Lang and Pannuk 1998). Organic media that are light in weight, rich in nutrients and well-drained considered ideal choices for growth and development of flowering plants like gerbera (Awang et al. 2009; Dubey et al. 2013), as these plants are notoriously ranked as heavy feeders and require abundant quantities of plant food for maximum development. It thrives best in a media with a pH slightly below or above the neutral point which helps the plants to face the seasonal hazards (Abad et al. 2002).

Currently, in Pakistan less work has been carried out on this flowering plant using different growth media sources. Most of the farmers are using chemical fertilizers and peat in order to attain maximum production of flowering plants. Both ways are eco destructive in several ways and growers in Pakistan are less familiar about the usage of other organic sources for better crop production. A less expensive and more adoptable growing media from agriculture waste not only benefits the environment but also attains meaningful production outcomes for Gerbera growers. Keeping in view, environmental concerns, economic constraints, and growers need, this study was performed to investigate the impact of various agricultural wastes as potting media on morphological characters as well as yield and quality flower production of Gerbera.

MATERIALS AND METHODS

The comparative study for evaluation of different potting media for *Gerbera jamsonii* cv. Hybrid Mix plants production was conducted at Floriculture Research Area, Institute of Horticultural Sciences, University of Agriculture, Faisalabad (latitude 31°30′N, longitude 73°10′E and altitude 213 m). Eight-week-old seedlings were purchased from local nursery in the month of November and 108 plants of uniform size (3 cm) having two sets of true leaves were selected to grow in pots having different combinations of FYM, coconut coir dust, Lahore compost (local produce), sand, leaf compost, silt, and garden soil. Six treatment combinations including control were made with the similar ratio by volume (Table 1).

Table 1. Different treatment combinations of agricultural wastes used as growing media

Treatments	Combinations	Ratio by volume (*v:v:v*)
T_0	Control (garden soil = GS)	
T_1	Silt + farm yard manure + top soil (SFT)	1:1:1
T_2	Silt + coconut coir dust + top soil (SCT)	1:1:1
T_3	Silt + Lahore compost + top soil (SLT)	1:1:1
T_4	Silt + sand + top soil (SST)	1:1:1
T_5	Silt + top soil + leaf compost (STL)	1:1:1

All treatment combinations were arranged in completely randomized design (CRD) with three replicates and six plants were planted in each replication. Management practices like irrigation, weeding, and pesticide application remained same for the all treatments.

Plant growth characteristics

Following observations on plant growth characteristics were evaluated during the course of study.

Plant height The height of plants was measured in centimeters, from base to the top leaf of the plant.

Number of leaves per plant The total number of leaves per plant was counted for each media after every 15 days.

Leaf area The leaf area (cm^2) was measured with the help of LI-3000 Portable Leaf Area Meter, USA.

Days to flower emergence Number of days were counted from the date of sowing to emergence of first flower on the plant.

Number of flowers per plant Flowers per plant were counted regularly during blooming period of the crop and then average was calculated.

Flower diameter Diameter of all the flowers was measured in centimeters (cm) with the help of vernier caliper and average was estimated.

Flower stalk thickness Flower stalk thickness was measured in centimeters (cm) with the help of vernier caliper and average was calculated.

Fresh weight of a flower One flower was collected at random from each plant and weighed on an electric balance (in grams) and average was computed.

Dry weight of a flower Flowers were dried in oven at 65 °C for 48 h, and then weighed again on electric balance. In this way, dry weight of flowers was recorded in grams and average was computed.

Fresh and dry weight ratio Fresh and dry weight ratio of a flower was calculated by dividing the fresh weight to dry weight of the flower.

Number of roots per plant The number of roots of plants was counted individually by uprooting the plant from the pots at mature stage.

Chemical characteristics of growing medium

The pH and Electrical conductivity (EC) in dS m^{-1} was determined by HI 9811-5 portable pH/EC/TDS/°C meter (Mclean 1982) and digital electrical conductivity (EC) meter (Lovibond Senso Direct 150). Organic matter was determined using the method of Walkley (1947), while, available phosphorus was estimated using Spectrophotometer, model spectronic 21 USA (Olsen et al. 1954). The flame photo metric method was used for estimation of Potassium (U.S. Salinity Lab. Staff 1954).

Statistical analysis

The collected data were subjected to Fisher's analysis of variance technique (Steel et al. 1997). The Duncan's multiple range (DMR) test at 0.05 % level of probability was employed to compare the differences among treatment means.

RESULTS

Plant growth characteristics

Results regarding plant height showed significant superiority of SCT (T_2) treatment, which produced maximum plant height with 40.22 % increase as compared to control (T_0) followed by 30.50 and 20.66 % increase with SLT (T_3) and STL (T_5), respectively. Mean comparison of treatments in Table 2 regarding number of leaves per plant showed significant superiority of SLT (T_3) again with 34.85 % increase in number of leaves followed by STL (T_5) with 30.30 % more leaves, compared to minimum number of leaves (3.30) in garden soil (GS) (T_0). Therefore, maximum leaf area (39.24 cm^2) was produced in mixture of SFT (T_1) with 164.96 % increase followed by STL (T_5), SCT (T_2), and SST (T_4) with 160.50, 125.79, and 107.63 % rise, respectively as compared to conventional media i.e., GS (T_0). Growing media combination in the form of STL (T_5) presented the highest value for number of roots (6.70) followed by SST (T_4) and SCT (T_2) treatments and finally up to 148.15 % improvement over GS (T_0) was noted (Table 2). Size of flower also improved in SST (T_4) significantly with maximum increases in flower diameter (20.75 %), in comparison to all growing media used in this study. Whereas, plants grown in SCT (T_2), SLT (T_3), and SFT (T_1) behaved similar in case of flower diameter (Figure 1). The maximum flower stalk thickness was produced in SLT (T_3) growth media, while the least flower stalk thickness was attained in plants grown in SCT (T_2) after GS (Figure 2). The utmost but statistically similar stalk thickness obtained in SST (T_4) and STL (T_5). There was 50 % improvement in flower stalk thickness in plants grown in SLT as compared to control treatment. On average, maximum fresh weight of flower with 31.87 % increase in weight was attained in SST (T_4) growing media compared to control (Figure 3). The smallest count of fresh weight of flower was observed in SFT (T_1) as this growing media caused 5.49 % decrease in weight when compared with control. Similarly, results regarding dry weight of flowers showed that the SST (T_4) and SLT (T_3) gave the highest results without any significant difference. Whereas, the lowest dry weight of flowers was observed in STL (T_5) with 43.77 % decrease from control (GS) (Figure 4). The STL (T_5) mixture possessed the highest fresh to dry weight ratio, which was 104.56 % greater than the control treatment, whereas minimum fresh/dry weight ratio was produced again by GS (T_0) (see Figure 5).

Table 2. Effect of growing substrates on plant height, number of leaves per plant, leaf area and number of roots of *Gerbera jamesonii*

Treatments	Plant height (cm)	Number of leaves per plant	Leaf area (cm^2)	Number of roots
T_0	7.31 ± 0.81f	3.30 ± 0.88d	14.81 ± 2.06f	2.70 ± 0.45f
T_1	7.86 ± 1.36e	4.09 ± 0.34c	39.24 ± 1.79a	3.66 ± 0.41e
T_2	10.25 ± 0.75a	4.15 ± 0.40c	33.44 ± 1.69c	4.45 ± 0.20c
T_3	9.54 ± 1.04b	4.45 ± 0.70a	30.75 ± 2.00e	4.07 ± 0.04d
T_4	8.44 ± 0.94d	4.11 ± 0.36c	32.24 ± 1.49d	5.79 ± 0.03b
T_5	8.82 ± 1.32c	4.30 ± 0.75b	38.58 ± 2.08b	6.70 ± 0.57a
F value	1436**	104**	26,562**	512**
p value	3.16×10^{-16}	1.23×10^{-11}	7.97×10^{-24}	2.13×10^{-13}

NS non significant, ±SE

* Significant ($P < 0.05$)

** Highly significant ($P < 0.01$)

Letters (a–f) exhibit significant difference among means at $P < 0.05$

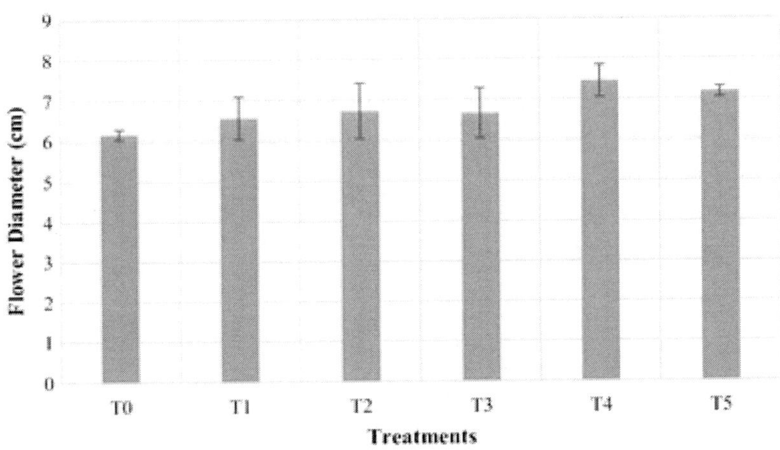

Figure 1. Effect of different growing media on flower diameter of *Gerbera jamesonii*

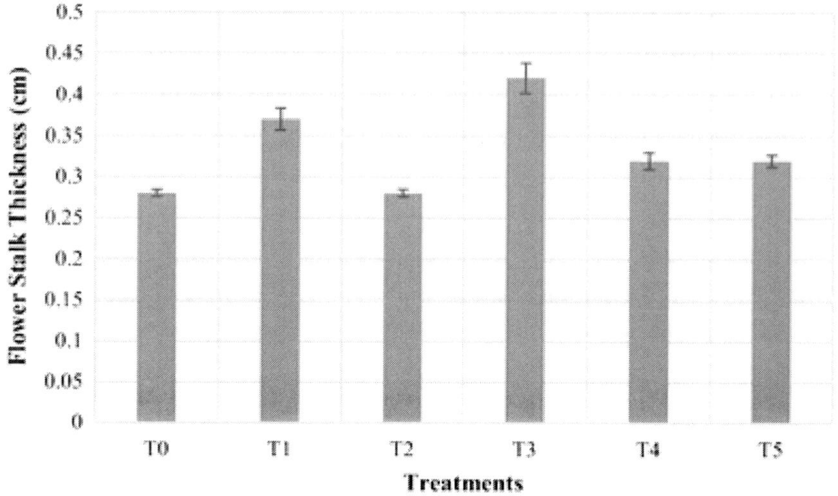

Figure 2. Effect of different growing media on flower stalk thickness of *Gerbera jamesonii*

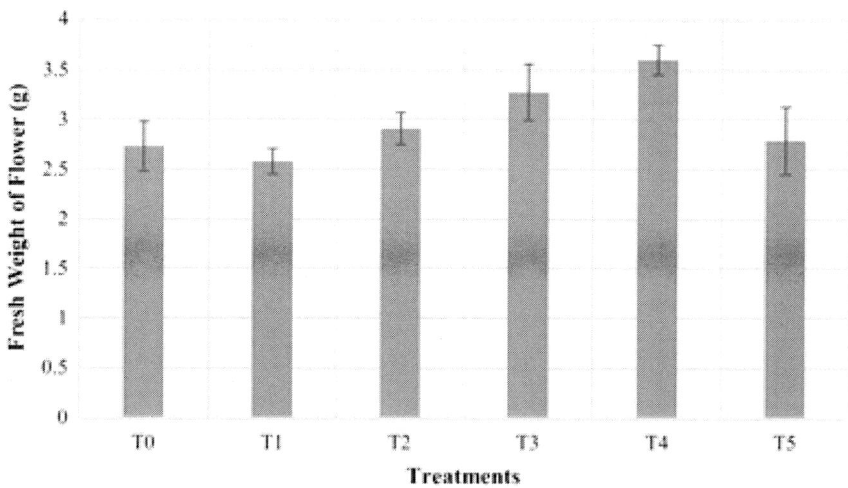

Figure 3. Effect of different growing media on fresh weight of flower of *Gerbera jamesonii*

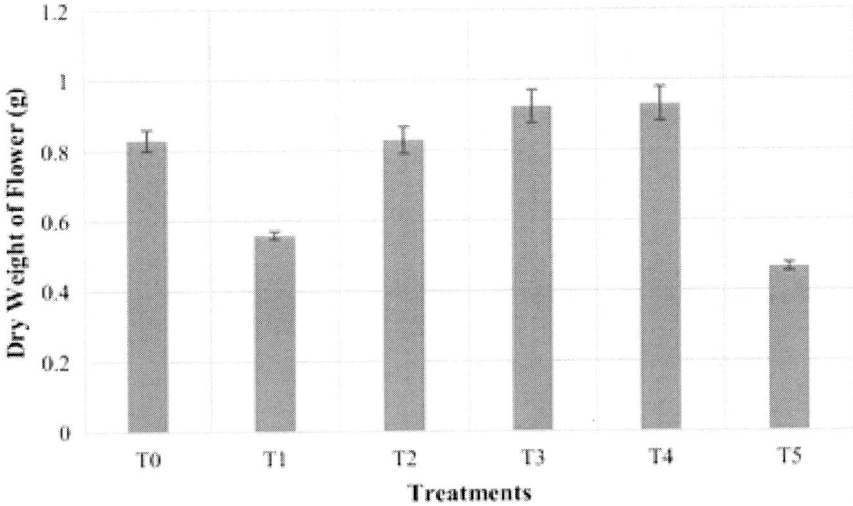

Figure 4. Effect of different growing media on dry weight of flower of *Gerbera jamesonii*

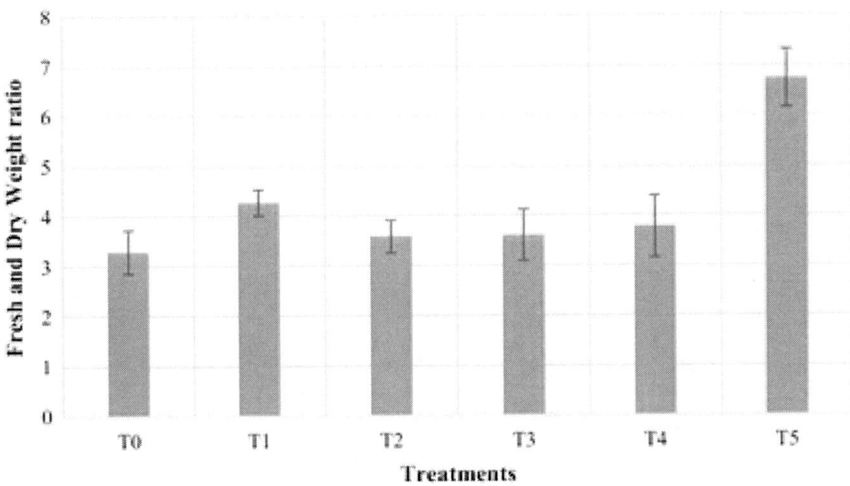

Figure 5. Effect of different growing media on fresh and dry weight ratio of flower of *Gerbera jamesonii*

However, difference for days to flower emergence was found to be non-significant but in general, STL (T_5) mixture attained least days to first flower emergence and showed 8.30 % earliness in this regard than control (Figure 6). Likewise, the mixture STL (T_5) again proved the superiority (34.85 %) regarding flowers per plant over control (GS) which produced the least flowers per plant (Figure 7). Other mixtures including SCT (T_2), SST (T_4), and SFT (T_1) were statistically not different from each other with 23 to 30 % increase in flowers per plant compared to control (GS) (T_0).

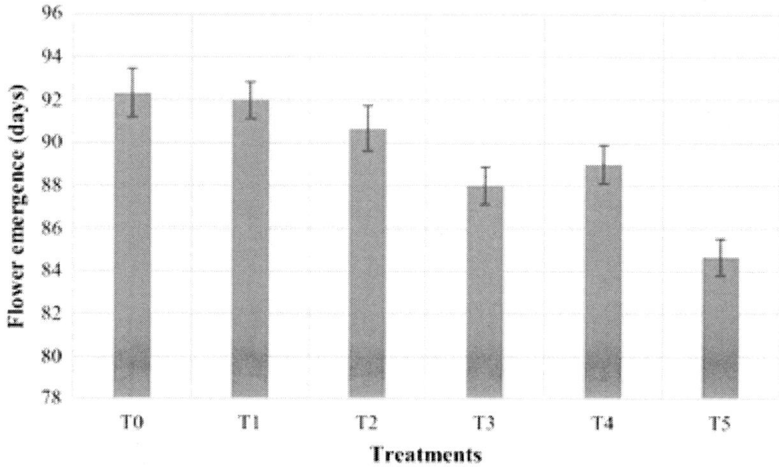

Figure 6. Effect of different growing media on flower emergence of *Gerbera jamesonii*

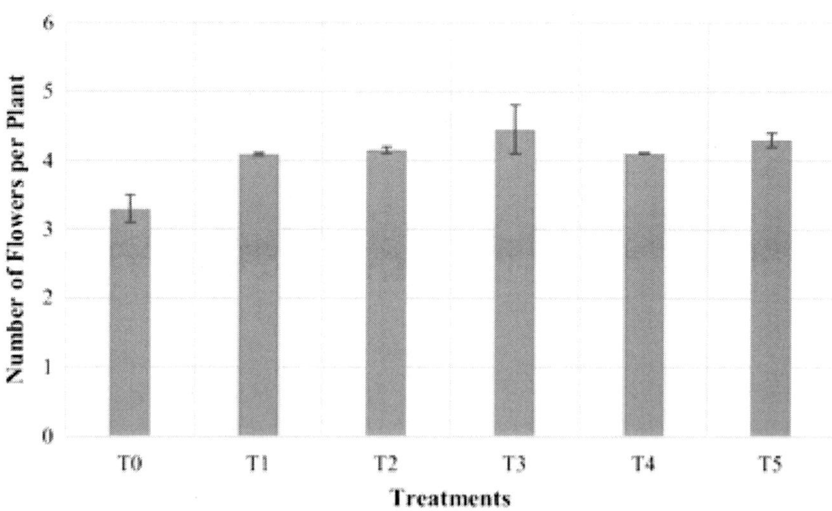

Figure 7. Effect of different growing media on flowers per plant of *Gerbera jamesonii*

Chemical characteristics of growing medium

The highly significant differences among organic matter contents, available potash (K), available phosphorus (P), pH and EC were observed in all the growth media used in this investigation. Growing media with SFT (T_1) combination provided better P (27 %) and K (72 %) contents compared to other growing media. While, maximum increase (up to 86.54 % over control) in organic matter contents was found again in SFT (T_1) and STL (T_5) mixtures without any significant difference. The SCT potting media gave the least percentage of available organic matter compared to other growing media treatments. Results regarding pH level of growing media exhibited that GS (T_0) gave the maximum value (9.54) while SST (T_4) and SFT (T_1) occupied 2nd position. The minimum level regarding pH (29 % less than control) was recorded in treatment containing SCT (T_2). The highest EC value (0.73) with 58.70 % increase over control was noticed with SFT (T_1) potting media whereas, GS (T_0) and SST (T_4) gave the lowest EC values (0.36) (see Table 3).

Table 3. Effect of growing substrates on organic matter, available potash, available phosphorus, pH and EC of *Gerbera jamesonii*

Treatments	Organic matter %	Available potash (ppm)	Available phosphorus (ppm)	pH	EC of media (dSm^{-1})
T_0	52 ± 1.07d	290 ± 11.99b	21.3 ± 0.76b	9.54 ± 0.39e	0.46 ± 0.048c
T_1	97 ± 1.93a	500 ± 8.91a	27 ± 1.36a	7.36 ± 0.49c	0.73 ± 0.02a
T_2	41 ± 8.02e	230 ± 8.49c	16.8 ± 1.42c	6.75 ± 0.67a	0.59 ± 0.05b
T_3	82 ± 3.21b	500 ± 13.89a	15.2 ± 1.19c	6.81 ± 0.22b	0.49 ± 0.004c
T_4	62 ± 2.13c	220 ± 6.94c	16.5 ± 2.72c	7.94 ± 0.71d	0.36 ± 0.01d
T_5	93 ± 7.01a	500 ± 10.02a	25 ± 2.02a	6.82 ± 0.1b	0.59 ± 0.08b
F value	1597**	952**	18.1**	154**	123**
p value	1.67×10^{-16}	3.70×10^{-15}	3.22×10^{-05}	1.88×10^{-10}	6.86×10^{-10}

* Significant ($P < 0.05$)

** Highly significant ($P < 0.01$). ±SE

Letters (a–f) exhibit significant difference among means at $P < 0.05$

DISCUSSION

Improved quality characters and growth traits of gerbera plants observed in this study with different growing substrates having garden soil, farm yard manure, coconut coir dust, a local made organic compost, and leaf compost are related to the previous findings (Saijeen et al. 2009; Ahmad et al. 2012). Appropriate concentration of nutrients in conjunction with growing medium has advancing effects on best flower growth and yield.

Results exhibited in previous section showed that balanced rooting media using agricultural waste can greatly affect the plant height and are essential for attaining maximum of it. These results are in accordance with Treder (2008) and Dubsky and Sramek (2008) who reported excellent growth of perennials in leaf mold and peat substrate with the addition of rock wool and observed maximum plant height. Though, number of leaves is influenced mainly by environmental conditions, nutrients present in growth media are also one of the factors which have prime importance in this regard. More number of leaves in plants reflect good vigor and their suitability to environment and growth media. Since nitrogen in growing media significantly affects the plant growth, increase in number of leaves of plants can also be due to adequate availability of nitrogen content in growing substrate (Benito et al. 2005; Khayyat et al. 2007; Riaz et al. 2008). More number of leaves could not be considered as factor contributing toward leaf area but it is mainly due to adequate supply of nutrients. Therefore, the nutrient rich substrate can be used to achieve significant results as maximum increase in size of leaves shows adaptability of plants to soil or growth media (Cardens et al. 2006; Riaz et al. 2014). Increase in number of shoots per plant highlighted the vigorous vegetative plant growth and only nutrient-rich growing media promote such growth in plants. Above mentioned results help to conclude that growing media in a combination of STL presented the best results for the number of shoots (Bashir et al. 2007).

Results regarding number of roots reflect that treatments containing mainly leaf compost represented superiority in promoting roots number. Similar trend has also been noticed by Lopez (2006) who found that root growth of geranium, petunia, and coleus was increased as the proportion of sewage sludge plus straw compost increased up to 50 % in peat. Maximum flowering as well as increase in flower bud diameter and subsequently increase in flower diameter indicates suitability of that growing medium, where organic matter in growing substrate and optimum amount of NPK manipulate plant growth and flower size (Kiran et al. 2007). Growing media containing silt and Lahore compost exhibited their superiority for stalk thickness, stem diameter, flower fresh and dry weights, and its ratio contrast to conventional media which proves that this kind of media can be used for gerbera production effectively. These results are in consistent with Grassotti et al. (2003) and Caballero et al. (2009) who observed that growing media having combination of silt + compost and manure can produce the maximum flower stalk thickness, fresh, and dry weights of gladiolus and lily. Similarly, the combination of sand, mushroom compost, coconut coir, and peat produced the best quality of plants with considerable increase in most of the growth parameters (Younis et al. 2013). Khalaj et al. (2011) also observed that maximum stem diameter of Gerbera plants attained with media in combination of perlite and peat. Rich supply of nutrients like K, N, and P in growing substrate is proved to be sufficient for production of good quality flowers (Younis et al. 2014) and adequate availability of these nutrients in growing media gave significant results regarding growth and flowering indices in this experiment. Organic matter and P content in growing substrate affected the quality of flowers as well. Growing media with excellent nutrient quality influences plant growth as availability of P and K contents in growth media has positive relationship with flowering indices. In current research, excellent nutrient quality with maximum phosphorus and potassium contents (27.00 and 500 ppm, respectively) was achieved in SFT growing media. Good quality flowers obtained in this combination is a confirmatory with Grassotti et al. (2003). Treder (2008) reported that plants had long stem, and maximum flowering in growing media with optimum P and K contents, where potting media comprising farm yard manure can give maximum amount of P and K contents (Strojny and Nowak 2004; Younis et al. 2007). Organic matter in growing media can individually promotes the plant growth due to high availability of nutrients (Fascella and Zizzo 2005; Younis et al. 2010), where maximum organic matter was found in farm yard manure containing substrates while other treatments showed no significant results. Physio-chemical attributes of high organic matter encourage the plant growth and improves substrate quality (Abad et al. 2001) which has been seen in this research work.

Various substrates having farm yard manure and compost are responsible to reduce pH and EC due to high nutrient supply (Bi et al. 2009; Diacono and Montemurro 2010). Actually, accessibility of nutrients to plants in growing media is coupled with the changes in media pH as increase or decrease in pH alters the solubility of nutrients and has direct effect on plant growth and development. Changes in pH above or lower than the optimum range adversely affected plants by damaging roots and decreasing nutrients availability (Altland 2006; Awang et al. 2009). The best range of dissolved salt contents i.e., EC in soil supports the better plant growth (Awang et al. 2009). Higher EC levels were not found to inhibit plant growth however; it decreased the biological activity of soil than the threshold value which negatively impacts plant growth (Miller 2001; Ribeiro et al. 2002). Suitable use of substrates for nutrient accessibility by plants not only depends upon nutrient composition but also on pH, nutrient form, organic compounds, and adsorption capacity of the growing media due to the presence of high organic matter content, P and K.

CONCLUSION

Results of this study indicated that application of coconut coir dust in combination with silt produced maximum plant height while addition of Lahore compost to substrate gave maximum number of leaves per plant and maximum flower stalk thickness. The addition of farm yard manure to silt and top soil produced maximum leaf area, highest value for number of roots, maximum fresh to dry weight ratio, and maximum flowers per plant, whereas maximum flower diameter, maximum fresh and dry weights of flower, resulted with mixture of sand with silt and top soil. All these organic waste substrates showed positive attribute to growth and flowering of gerbera with readily availability of nutrients for uptake. Therefore, it is concluded that suggested agricultural waste material can be used as growing media for better *Gerbera jamesonii* growth and can be explored further for other combinations.

ACKNOWLEDGMENTS

The Higher Education Commission (HEC) is highly acknowledged for providing funds for this research project.

REFERENCES

1. Abad M, Noguera P, Bures S (2001) National inventory of organic wastes for use as growing media for ornamental potted plant production: case study in Spain. Bioresour Technol 77:197–200
2. Abad M, Noguera P, Puchades R, Maquieira A, Noguera V (2002) Physico-chemical and chemical properties of some coconut dusts for use as a peat substitute for containerized ornamental plants. Bioresour Technol 82:241–245
3. Ahmad I, Ahmad T, Gulfam A, Saleem M (2012) Growth and flowering of gerbera as influenced by various horticultural substrates. Pak J Bot 44:291–299
4. Altland JE (2006) Substrate pH, a tricky topic. Digger 50:42–47
5. Awang Y, Shaharom AS, Mohamad RB, Selamat A (2009) Chemical and physical characteristics of cocopeat based media mixtures and their effects on the growth and development of Celosia cristata. Am J Agric Biol Sci 4:63–71
6. Bashir MA, Ahmad M, Anjum MA (2007) Effect of various potting media on growth of rooted jojoba (Simmondsia chinensis) cuttings. Int J Agri Biol 9:147–151
7. Benito M, Masaguer A, Antonio RD, Moliner A (2005) Use of pruning waste compost as a component in soil less growing media. Bioresour Technol 96:597–603
8. Bi L, Zhang B, Liu G, Li Z, Liu Y, Ye C, Yu X, Lai T, Zhang J, Yin J, Liang Y (2009) Long-term effects of organic amendments on the rice yields for double rice cropping systems in subtropical China. Agri Ecosyst Environ 129(4):534–541

9. Caballero R, Pajuelo P, Ordovas J, Carmona E, Delgado A (2009) Evaluation and correction of nutrient availability to Gerbera jamesonii H. Bolus in various compost-based growing media. Sci Horti 122:244–250
10. Cardens MCA, Gomez IFR, Ronacancio VJF, Cordoba BC, Canola PW (2006) Growth analysis of standard carnation cv. 'Nelson' in different substrate. Acta Hort 718:623–630
11. Diacono M, Montemurro F (2010) Long-term effects of organic amendments on soil fertility. A review. Agron Sustain Dev 30:401–422
12. Dubey RK, Simrat-Singh Kukal SS, Kalsi HS (2013) Evaluation of different organic growing media for growth and flowering of petunia. Commun Soil Sci Plant Anal 44:1777–1785
13. Dubsky M, Sramek F (2008) Crushed rockwool as a component of growing substrates. Acta Hort 779:491–495
14. Fascella G, Zizzo GV (2005) Effect of growing media on yield and quality of soilless cultivated rose. Acta Hort 697:133–138
15. Grassotti A, Nesi B, Malrtta M, Magnani G (2003) Effects of growing media and planting time on lily hybrids in soilless culture. Acta Hort 603:395–399
16. Guerrero F, Gasco JM, Hernandez-Apaolaza L (2002) Use of pinebark and sewage sludge compost as components of substrates for Pinus pinea and Cupressus arizonica production. J Plant Nutr 25(1):129–141
17. Gupta R, Yadav A, Garg VK (2014) Influence of vermicompost application in potting media on growth and flowering of marigold crop. Int J Recycl Organ Waste Agri 3(1):1–7
18. Hernandez-Apaolaza L, Gascó AM, Gascó JM, Guerrero F (2005) Reuse of waste materials as growing media for ornamental plants. Bioresour Technol 96(1):125–131
19. Khalaj MA, Amiri M, Sindhu SS (2011) Study on the effect of different growing media on the growth and yield of Gerbera (Gerbera jamesonii L.). J Ornam Hortic Plants 1(3):185–189
20. Khayyat M, Nazari F, Salehi H (2007) Effects of different pot mixtures on pothos (Epipremnum aureum Lindl. and Andre 'Golden Pothos') growth and development. Am-Eurasian J Agric Environ Sci 2:341–348
21. Kiran M, Din J, Waseem K, Jilani MS, Khan MQ (2007) Effect of different growing media on the growth and development of Dahlia (Dahlia pinnata) under the agro-climatic condition of Dera Ismail Khan. Pak J Biol Sci 10:4140–4143
22. Lang HJ, Pannuk TR (1998) Effects of fertilizer concentration and minimum-leach drip irrigation on the growth of New Guinea Impatiens. Hort Sci 33:83–688
23. Lopez A (2006) Media influence on post harvest container plant quality in a retail nursery setting. Dissertation, University of Agriculture and Mechanical College, USA
24. McLean EO (1982) Soil pH and lime requirement. In: Page AL, Miller RH, Keeney DR (eds) Methods of soil analyses, part 2, chemical and microbiological properties, 2nd edn. American Society Agronomy, Madison, pp 199–224
25. Miller M (2001) Fulfilling special needs of nurseries. Biocycle 42:65–67
26. Nowak JS, Strojny Z (2003) Effect of different container media on the growth of Gerbera. Acta Hort 608:59–63
27. Olsen SAR, Sant MD, Gislerod HR, Solbraa K (1954) Nitrogen balance in bark composts used as growing media. Acta Hort 150:202–203
28. Rader JS (1998) Petunia. In: Ball V (ed) Ball Redbook. Ball Publishing, Batavia, IL, USA, pp 682–686
29. Riaz A, Arshad M, Younis A, Raza A, Hameed H (2008) Effects of different growing media on growth and flowering of Zinnia elegans cv. Blue point. Pak J Bot 40:1579–1585
30. Riaz A, Farooq U, Younis A, Karim A, Taj AR (2014) Growth responses of Zinnia to different organic media. Acta Hort 1018:565–572
31. Ribeiro HM, Vasconcelos E, Santos JQ (2002) Fertilization of potted geranium with municipal solid waste compost. Bioresour Technol 73:247–249

32. Saijeen S, Kaewman O, Suksawat M (2009) Evaluation of media, organic and chemical fertilizer applications on growth of pot gerbera (Gerbera jamesonii). Asian J Food Agro-Ind 2:S51–S56
33. Sindhu SS, Gholap DB, Singh MC, Dhiman MR (2010) Effect of medium amendments on growth and flowering in gerbera. Indian J Hort 67:391–394
34. Steel RGD, Torrie JH, Dicky DA (1997) Principles and procedures of statistics: a biometric approach. Mc Graw Hill Inc, New York
35. Strojny Z, Nowak JS (2004) Effect of different growing media on the growth of some bedding plants. Acta Hort 644:157–162
36. Tariq U, Rehman S, Khan MA, Younis A, Yaseen M, Ahsan M (2012) Agricultural and municipal waste as potting media components for the growth and flowering of Dahlia hortensis 'Figaro'. Turk J Bot 36:378–385
37. Torrey JG (1951) Cambial formation in isolated pea roots following decapitation. Am J Bot 38:596–604
38. Treder J (2008) The effects of cocopeat and fertilization on the growth and flowering of oriental lily 'star gazer'. J Fruit Ornam Plant Res 16:361–370
39. U.S. Salinity Lab Staff (1954) Diagnosis and improvement of saline and alkali soils. USDA Handbook 60, Washington DC, USA
40. Walkley A (1947) A critical examination of a rapid method for determining organic carbon in soils. Effect of variations in digestion conditions and of organic soil constituents. Soil Sci 63:251–263
41. Weeraratne TP, Daundasekera WAM, Wijesundara DSA (2012) Field survey of postharvest handling of cut flowers produced in the Up country of Sri Lanka for the local market. Ceylon J Sci 41:67–70
42. Wilson SB, Muller KL, Wilson PC, Incer MR, Stoffella PJ, Graetz DA (2009) Evaluation of new container media for Aglaonema production. Commun Soil Sci Plant Anal 40:2673–2687
43. Younis A, Ahmad M, Riaz A, Khan MA (2007) Effect of different potting media on the growth and flowering of Dahlia coccinia cv. Mignon. Acta Horti 804:191–196
44. Younis A, Riaz A, Wasim M, Khan MA, Nadeem M (2010) Production of quality croton (Codiaeum variegatum) plants by using different growing media. Am-Eurasian J Agric Environ Sci 7:232–237
45. Younis A, Riaz A, Siddique MI, Lim KB, Hwang YJ, Khan MA (2013) Anatomical and morphological variation in Dracaena reflexa 'Variegata' grown in different organic potting substrates. Flower Res J 21:162–171
46. Younis A, Anjum S, Riaz A, Hameed M, Tariq U, Ahsan M (2014) Production of quality dahlia (Dahlia variabilis cv. Redskin) flowers by efficient nutrients management. Am-Eurasian J Agric Environ Sci 14:137–142

CHAPTER 13

Transforming the Snapdragon Aurone Biosynthetic Genes into Petunia Alters Coloration Patterns in Transgenic Flowers

Chen-Kuen Wang, Yi-Chin Chin, Chih-Yu Lin, Po-Yen Chen, Kin-Ying To*

Agricultural Biotechnology Research Center, Academia Sinica, Taiwan

ABSTRACT

Aurones belong to a small class of flavonoids that provide yellow color in some floricultural plants including snapdragon. To explore novel flower coloration, two full-length cDNAs encoding chalcone 4'-O-glucosyltransferase (designated as SRY4'CGT) and aureusidin synthase (designated as SRYAS1) in the aurone biosynthetic pathway were cloned from yellow flowers of snapdragon (Antirrhinum majus cv. Ribbon Yellow). Binary vectors were constructed and transformed separately into Petunia hybrida harboring blue flowers. Only a few flowers in 4 out of 9 transgenic SRY4'CGT plants showed variegated blue-white sectors; as time passed, amounts of variegated flowers and proportion of white sectors in the background blue color of the new-born flowers gradually increased, until finally, the petal color was completely white in all late-born flowers. In contrast, a few flowers in 3 out of 13 transgenic SRYAS1 plants showed variegated blue-white sectors; but, the amounts of variegated flowers did not increase over the whole flowering stage, and no complete white flowers were observed. RNA samples isolated from blue and white sectors of T_1 transgenic SRY4'CGT plants were analyzed by reverse transcription-PCR, transgenic SRY4'CGT transcripts were detected in both sectors; however, transcripts of an upstream gene, chalcone synthase (CHS), were abundantly detected in the blue sectors but largely reduced in the white sectors, suggesting that the expression of CHS gene was suppressed in white sectors of transgenic plants. Furthermore, HPLC coupled with mass spectrometry demonstrated cyandin, malvidin and their derivatives were absent in white sectors, causing the white phenotype. Our findings may be attractive to molecular breeders.

Keywords: Agrobacterium-Mediated Transformation, Aurone, Compound Identification, Flower Color, Genetic Engineering, HPLC Profiling, Suppression, Transgene-Induced Gene Silencing

1. INTRODUCTION

Flower color is an important feature in floricultural plants, contributing to the major value in the flower market. In general, flower color is determined by three classes of pigments: flavonoids, carotenoids and betalains. Flavonoids are the major contributors in the determination of a range of colors such as yellow, orange, red and purple; and the carotenoids are the red, orange and yellow lipid-soluble pigments found embedded in the membranes of choroplasts and chromoplasts [1] [2] . The third class of pigments, betalains, can be found only in certain plant species of the Carophyllales and in some higher fungi such as fly agaric [2] [3] . Over the past few decades, almost all the key genes in the main biosynthesis pathways of flavonoids and carotenoids have been cloned from a range of plant species, and genetic engineering has been carried out to breed new cultivars with novel flower colors in many floricultural plants [2] -[8] . Some branch/modifying enzymes are limited to certain plant species; however, most of these genes are not yet known.

Flavonoids, starting from the general phenylpropanoid pathway, can be subdivided into several major groups according to their structures: chalcones, aurones, isoflavonoids, flavones, flavonols, flavan-3-ols, anthocyanins, condensed tannins (proanthocyanins) and phlobaphenes [3] [9] . Among these groups, anthocyanins, flavonols, chalcones and aurones contribute color to plants. Aurones are the bright yellow pigments found in some ornamental flowers such as snapdragon (Antirrhinum majus), cosmos and coreopsis [10] . Biochemical and flavonoid mutant studies from A. majus clearly revealed that aurones were produced from chalcones [4] [10] . The first common flavonoid compound is 2',4',6',4-tetrahydroxychalcone (THC) or also known as naringenin chalcone, which is the catalytic product of the key enzyme chalcone synthase (CHS). In many floricultural plants, THC is converted into naringenin by chalcone isomerase (CHI), and naringenin is further converted into different kinds of anthocyanins, flavonols and flavones by several enzymes in the flavonoid biosynthesis pathway. In yellow- flowering plants such as snapdragon, THC is converted into THC 4'-O-glucoside by UDP-glucose: chalcone 4'-O-glucosyltransferase (4'CGT), and THC 4'-O-glucoside is further converted into aureusidin 6-O-glucoside (yellow pigment) by aureusidin synthase (AS) [11] . In addition, in vitro activity of AS, catalyzing THC into aureusidin (yellow pigment), has also been reported [10] . Both genes, designated as AmAS1 and Am4'CGT, have been cloned from yellow snapdragon (A. majus cv. Yellow Butterfly) and it has been demonstrated that both genes are essential for aurone biosynthesis and yellow coloration [10] [11] . Subsequently, coexpression of the AmAS1 and Am4'CGT transgenes together with suppression of endogenous anthocyanin biosynthetic gene [flavanone 3-hydroxylase (F3H) or dihydroflavonol 4-reductase (DFR)] produced yellow flowers in transgenic lines of Torenia hybrid cv. Summer Wave Blue, a cultivar has blue flowers [11] .

We are interested in molecular breeding of flower color since flower color is one of the most attractive characteristics in garden ornamentals and contributes to the major value in the floricultural markets [2] [12] . Petunia is one of the most popular ornamentals worldwide, and has been used as one of the model plants for various studies including flavonoid biosynthesis, floral development, male sterility, self-incompatibility, retroelement activity, transposons, epigenetics, transformation, senescence, and so on [13] [14] . In this study, we try to generate bright yellow flowers in petunia. As a first step to achieve our goal, here we report the cloning and construction into expression vectors of AS1 and 4'CGT cDNAs from yellow flowers in another snapdragon cultivar, Antirrhinum majus cv. Ribbon Yellow. These constructs were individually introduced into Petunia hybrida harboring blue flowers by using Agrobacterium-mediated method. Although yellow flowers were not observed; however, a few flowers in several transgenic plants showed variegated blue-white sectors and attractive coloration patterns. Here we just summarize what we have done and observed in single gene transformation. Breeding transgenic plants coexpressing both transgenes are in progress. The findings reported herein will be useful for breeding new cultivars with attractive coloration patterns.

2. MATERIALS AND METHODS

2.1. Plant Materials

Seeds of Antirrhinum majus cv. Ribbon Yellow and Petunia hybridacv. Extra Blue were purchased from Known- You Seed Company, Taiwan. For gene cloning, seeds of A. majus were germinated, and young seedlings were transferred into pots and grown in a greenhouse until maturity. For plant transformation, seeds of P. hybrida were sterilized with 1% sodium hypochlorite for 20 min, washed thoroughly with sterile water, and germinated on MS basal medium containing MS salts [15] , 2% sucrose, 0.8% bacto-agar, pH 5.7. The cultures were then incubated in a 26°C growth chamber under a 16-h illumination (100 mmol/m^2/sec) and 8-h dark cycle. In vitro explants grown for 2 to 3 weeks were used to establish the plant transformation system.

2.2. Gene Cloning and Construction of Plant Expression Vector

To clone the aurone biosynthetic genes, RNA from different developmental stages of yellow flowers of A. majus was isolated by TRIzol protocol as previously described [16] . Primers SRY4'CGT-F1 (5'-ATGGGAGAAGAA-TACAAGAAA-3') and SRY4'CGT-R1 (5'-TTAACGAGTGACCGAGTTGAT-3') specific for A. majus chalcone 4'-O-glucosyltransferase (4'CGT), and primers SRYAS1-F1 (5'-ATGTTCAAAAATCCTAATATC-3') and SRYAS1-R1 (5'-TTAGCCATCAAGCTCAATCTT-3') specific for A. majus aureusidin synthase (AS1) were synthesized to amplify full-length reading frames of SRY4'CGT (1374 bp) and SRYAS1 (1689 bp), respectively. RT-PCR was carried out as previously described [16] . PCR products were analyzed by 1% agarose gel electrophoresis. After purification, PCR products were cloned into pGEM-T Easy vector (Promega), and then sequenced by an automatic DNA sequencer (ABI PRISM 3730 DNA sequencer, Perkin-Elmer). To construct the expression vectors, PCR products of snapdragon 4'CGT and AS1 were cloned into Gateway binary vectors pH2GW7 (carrying the hygromycin selection marker) and pK2WG7 (carrying the kanamycin selection marker), respectively, by using the pCR8/GW/TOPO TA cloning kit (Invitrogen). The resulting vectors pSRY4'CGT and pSRYAS1 were isolated, screened by PCR analysis, and confirmed by DNA sequencing analysis.

2.3. Agrobacterium-Mediated Transformation

To introduce the vector DNA into Agrobacterium, 50 ml competent cells of Agrobacterium tumefaciens strain GV3101 were mixed with 100 ng plasmid DNA (pSRY4'CGT or pSRYAS1) and incubated on ice for 1 to 2 min. The mixture was then transferred into an ice-cold 0.1-cm electroporation cuvette. Electroporation was carried out in a Bio-Rad Gene Pulser II (USA) using the following parameters: 25 mF, 200 W, and 1.25 kV. After electroporation, 1 ml of YEP medium (10 g/l peptone; 10 g/l yeast extract; 5 g/l NaCl; pH 7.0) was added immediately into the mixture, incubated at 28°C for 2 h, and 200 ml of bacterial mixture was transferred onto selective YEP plates containing 100 mg/l spectinomycin. The plates were incubated at 28°C for 2 days. A single colony of plasmid-carrying Agrobacterium was picked up and incubated in 4 ml YEP medium containing 100 mg/l spectinomycin at 28°C overnight. The value of OD_{600} value should be 0.1 - 0.3. One ml of overnight culture was added to 50 ml AB medium (5 g/l glucose; 1 g/l NH_4Cl; 0.3 g/l $MgSO_4 \cdot 7H_2O$; 0.15 g/l KCl; 10 mg/l $CaCl_2$; 2.5 mg/l $FeSO_4 \cdot 7H_2O$; 3 g/l K_2HPO_4; 1.15 g/l $NaH_2PO_4 \cdot H_2O$) supplemented with 100 mg/l spectinomycin, and incubated for 1 day at 28°C. Acetosyringone (Fluka Chemika, Switerland) and glucose were added to the AB/ spectinomycin medium to yield a final concentration of 100 mM and 5%, respectively, and the culture was further incubated 4 h at 28°C. The bacterial culture was then centrifuged at 5000 rpm for 5 min, and the pellet was dissolved in KCMS medium (MS salts; 0.9 mg/l thiamine; 0.2 mg/l 2,4-dichlorophenylacetic acid (2,4-D); 200 mg/l KH_2PO_4; pH 5.6) supplemented with 100 mM acetosyringone and 5% glucose. The suspension was then diluted such that the final OD_{600} reading was 0.8 - 1.0.

Agrobacterium-mediated transformation was carried out as previously described [17] with modifications. For pre-culture prior to Agrobacterium infection, cotyledons or leaves were excised from 2- to 3-week-old in vitro plantlets of P. hybrida one day before infection, and incubated in KCMS solid medium supplemented with 100 mM acetosyringone for one day. Infection was carried out by adding 15 ml of diluted Agrobacterium suspension to the pre-cultured explants for 1 h. Excess Agrobacterium was blotted from the explants on sterile filter papers, then the infected explants were transferred onto KCMS solid medium containing 100 mM acetosyringone and 5% glucose, and left to grow for 2 days in a dark growth chamber at 26°C. Explants were then washed 2-3 times with shaking (the first shaking was at 220 rpm, and subsequent shakings were at 100 rpm, 30 min each time) in KCMS liquid medium supplemented with 200 mg/l timentin. Explants were dried with sterile filter papers, and transferred to MSZ medium (MS salts; Nitsch vitamins; 1% sucrose; 2% mg/l zeatin; 0.7% bacto-agar; pH 6.0) supplemented with 200 mg/l timentin to inhibit Agrobacterium growth and 10 mg/l hygromycin (for selection in transforming vector pSRY4'CGT) or 100 mg/l kanamycin (for selection in transforming vector pSRYAS1), and incubated at 26°C in a growth chamber with a 16-h light (100 - 200 mmol/m^2/sec) and 8-h dark cycle. Explants were transferred to a new MSZ medium supplemented with 200 mg/l timentin and suitable antibiotic (10 mg/l hygromycin or 100 mg/l kanamycin) every two weeks. Formation of the initial calli could be seen on the wound site of explants after 3 weeks of transformation, and shoots were regenerated after 6 weeks. Single shoot was excised from the callus and transferred onto the MS basal medium supplemented with 100 mg/l timentin. The rooted plantlets were transferred into pots and grown in a greenhouse.

2.4. Reverse Transcription-PCR Analysis

Total RNA was extracted from transgenic flowers, treated with DNase I, and 100 ng of treated RNA was used to perform reverse transcription-PCR (RT-PCR) analysis as previously described [12] . For amplification of gene-specific region, primers SRYAS1-F1 (5'-ATGTTCAAAAATCCTAATATC-3') and SRYAS1-R2 (5'-ACCAG- GTGAATCATAGTTCCA-3') were employed to amplify the 624-bp-long DNA fragment of snapdragon AS1 (GenBank accession no. EF650014), primers SRY4'CGT-F2 (5'-ATGGGAGAAGA ATACAAGAAAAC-3') and SRY4'CGT-R3 (5'-CATGCTGGTGGATACTTC-3') were employed to amplify the 390-bp-long DNA fragment of snapdragon 4'CGT (GenBank accession no. EF650015), and primers CHS-C1 (5'-ACAACAAGGGCGCTC- GAG-3') and CHS-C2 (5'-CAAGCCCTTCACCAGTAG-3') were employed to amplify the 558-bp-long DNA fragment of Petunia chalcone synthase (GenBank accession No. AF233638).

2.5. Seedling Assay for Kanamycin or Hygromycin Resistance

T_1 seeds from self-fertilized transgenic plants were sterilized with 1% sodium hypochlorite as mentioned above. They were germinated in MS basal medium supplemented with 10 mg/l hygromycin (for transgenic SRY4'CGT lines) or 100 mg/l kanamycin (for transgenic SRYAS1 lines) for 3 weeks. Seedlings with white or pale green cotyledons, no true leaf development, inhibition of root extension and plant development were considered to be hygromycin or kanamycin sensitive, while seedlings with green cotyledons and healthy development of leaves and roots were considered to be hygromycin or kanamycin resistant.

2.6. Sample Preparation

Freeze-dried petals from transgenic Petunia flowers carrying SRY4'CGT or SRYAS1 transgene were homogenized by using liquid nitrogen with a pestle and mortar, transferred into a tube and dissolved in 50% acetonitrile containing 0.1% (vol/vol) trifluoroacetic acid (TFA) to a final concentration of 0.1 g/ml (fresh weight/vol). The mixture was rotated for 3 h at room temperature, and then briefly centrifuged. The supernatant was transferred into a clean tube, and then vacuum-dried. The extracts were resuspended in 50% acetonitrile containing 0.1% (vol/vol) formic acid (FA), the final concentration was adjusted to 0.1 g/ml (fresh weight), and used for HPLC-DAD-MS/MS analysis.

2.7. HPLC-DAD-MS/MS Analysis

The Agilent 1100 series liquid chromatograph system (Agilent, Waldbronn, Germany) coupled with a diode- array detector (DAD) and a Thermo Finnigan LCQ Advantage ion-trap mass spectrometer (San Jose, CA, USA) was employed for analysis at the Metabolomics Core Facility of the Agricultural Biotechnology Research Center, Academia Sinica, Taiwan. The extracts were separated by HPLC with a RP-C18 column (Phenomenex Luna 3 μm C18, 150 mm × 2.0 mm) at a flow rate of 0.2 ml/min. The mobile phase consisted of water containing 0.1% (v/v) FA (mobile phase A) and acetonitrile containing 0.1% (v/v) FA (mobile phase B). The following conditions were used for the gradient elution: 10% - 23% B for 3 min, next with 23% B for 17 min, 23% - 80% B for 2 min and then 80% B for 3 min. The sample injection volume was 10 ml. Eluting peaks were monitored by DAD under the following conditions: Chalcones, aurones, anthocyanidins, and flavones were monitored respectively at 360 nm, at 400 nm, at 520 nm, and at 360 nm [11] . Both positive and negative ion polarity modes were used for compound ionization. The MS optimized parameters in positive ion mode were as follows: source voltage, 4.5 kV; sheath gas (N_2), 50 arbitrary units; capillary temperature, 250°C; capillary voltage, 3 V; tube lens offset voltage, 30 V; in the negative ion mode, source voltage, −4 kV; capillary voltage, −47 V; tube lens offset voltage, −45 V. For full scan MS analysis, spectra were recorded in the range of m/z 100 - 2000.The data-de- pendent program was set so that the three most intense ions in each survey scan were selected and subjected to tandem mass spectrometry. The HPLC-DAD?MS/MS system was controlled by Xcalibur 2.0 SR2 software (Thermo Finnigan, San Jose, CA, USA).

3. RESULTS

3.1. Cloning and Sequence Analysis of Full-Length cDNAs in the Aurone Biosynthetic Pathway from Yellow Snapdragon Flowers

Yellow flowers of snapdragon (Antirrhinum majus cv. Ribbon Yellow) at different stages from tiny flower bud to mature flower (Figure 1) were harvested, total RNA was isolated, and RT-PCR was carried out to clone the full-length cDNAs encoding chalcone 4'-O-glucosyltransferase (designated as SRY4'CGT) and aureusidin synthase (designated as SRYAS1) in aurone biosynthetic pathway. The SRY4'CGT clone is 1374 bp in length and contains a reading frame of 457 amino acids. Our clone is the same length as the published A. majus Am4'CGT nucleotide sequence ([11] ; GenBank accession No. AB198665); however, it differed from the published clone by 16 nucleotides and 8 amino acids (Figure 2). To examine the accuracy of the sequence of our SRY4'CGT cDNA clone, the corresponding genomic DNA fragment was amplified by PCR analysis using genomic DNA isolated from A. majus cultivar Ribbon Yellow and primer set (SRY4'CGT-F1 and SRY4'CGT-R1) and then sequenced. As shown in Figure 2, nucleotide sequences of our cDNA and the genomic clones were identical, indicating that the genomic SRY4'CGT gene has no intron. This result is consistent with the finding that there is no intron in the Am4'CGT gene of A. majus [11] . The differences between the published clone and our clone reflect the two different cultivars (Ribbon Yellow versus Yellow Butterfly) examined. The full-length cDNA and genomic DNA encoding SRY4'CGT in cultivar Ribbon Yellow were submitted to GenBank database with accession numbers EF650015 and JQ234673, respectively. It was reported that chalcone 4'-O-glucosyltransferase catalyzes 2',4',6', 4-tetrahydroxychalcone (THC) to THC 4'-O-glucoside (colorless) or 2',4',6',3,4-pentahydroxychalcone (PHC) to PHC 4'-O-glucoside (colorless) in aurone biosynthetic pathway [11] .

The other clone, SRYAS1, is 1689 bp in length and contains a reading frame of 562 amino acids. Our SRY- AS1 differed from the published snapdragon AmAS1 nucleotide sequence ([10] ; GenBank accession no. AB- 044884) by two bases in the coding region at nucleotide positions 285 and 892; however, the amino acids were identical to the corresponding published AmAS1 protein sequence (Figure 3). The full-length cDNA encoding SRYAS1 in snapdragon cultivar Ribbon Yellow was submitted to GenBank database with accession number EF650014. It was reported that aureusidin synthase catalyzes THC 4'-O-glucoside (colorless) or PHC 4'-O- glucoside (colorless) to produce aureusidin 6-O-glucoside (yellow pigment) or bracteatin 6-O-glucoside (yellow pigment), respectively, in the aurone biosynthetic pathway [11] . The deduced amino acid sequence of SRYAS1 protein showed high sequence similarity to the plant polyphenol oxidase family [10] , which has been demonstrated to be responsible for flower coloration [10] [11] [18] .

The relevant cDNA fragments were cloned separately into Gateway vectors, obtaining the expression vectors pSRY4'CGT and pSRYAS1 (Figure 4), which were confirmed by restriction enzyme digestion and DNA sequencing.

3.2. Isolation and Phenotypic Observation of Transgenic Petunia Plants Carrying SRY4'CGT or SRYAS1 Transgene

Agrobacterium-mediated transformation was carried out to transform pSRY4'CGT and pSRYAS1 separately into P. hybrida. Finally, 9 putative transgenic plants were obtained from a total of 186 cotyledon explants by transforming the vector pSRY4'CGT and using 10 mg/l hygromycin as selection agent (Table 1); and 13 putative transgenic plants were obtained from a total of 205 cotyledon explants by transforming the vector pSR-YAS1 and using 100 mg/l kanamycin as selection agent (Table 2). Transformation efficiency for SRY'4CGT and SRYAS1 transgenes was 4.8% and 6.3%, respectively. All transgenic plants were grown in a greenhouse until maturity.

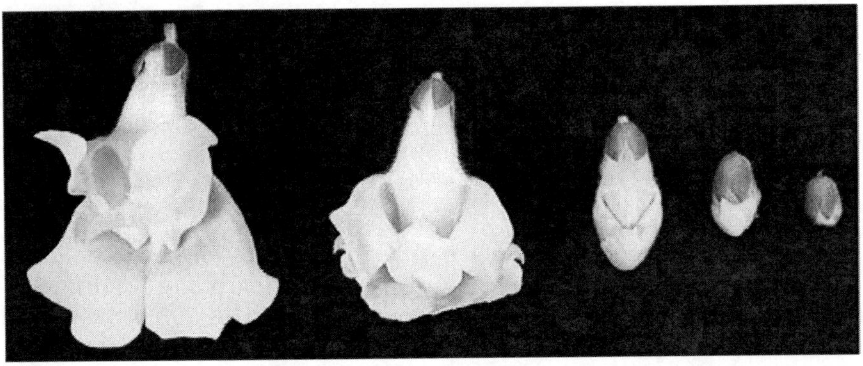

Figure 1. Flower phenotype of snapdragon. Seeds of snapdragon (Antirrhinum majus cv. Ribbon Yellow) were germinated, and young seedlings were transferred into pots and grown in greenhouse until maturity. RNA was isolated from different stages of tiny flower bud to mature flower.

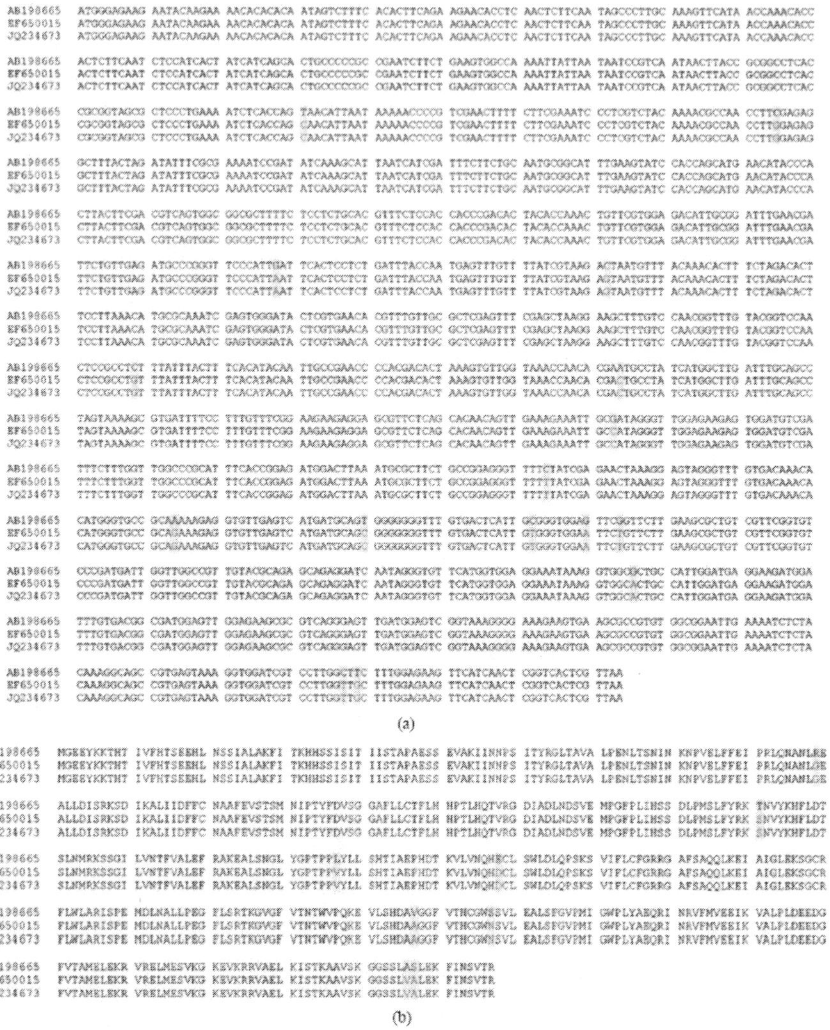

Figure 2. Nucleotide sequence (panel (a)) and deduced amino acid sequence (panel (b)) of snapdragon cDNA encoding chalcone 4'-O-glucosyltransferase (4'CGT). Accession number AB198665 is the published Am4'CGT sequence from Antirrhinum majus cv. Yellow Butterfly [11] , EF650015 is our cDNA sequence from A.majus cv. Ribbon Yellow (SRY4'CGT), and JQ234673 is our genomic sequence. Nucleotides or amino acids that differed from the published sequence [11] are indicated by highlights.

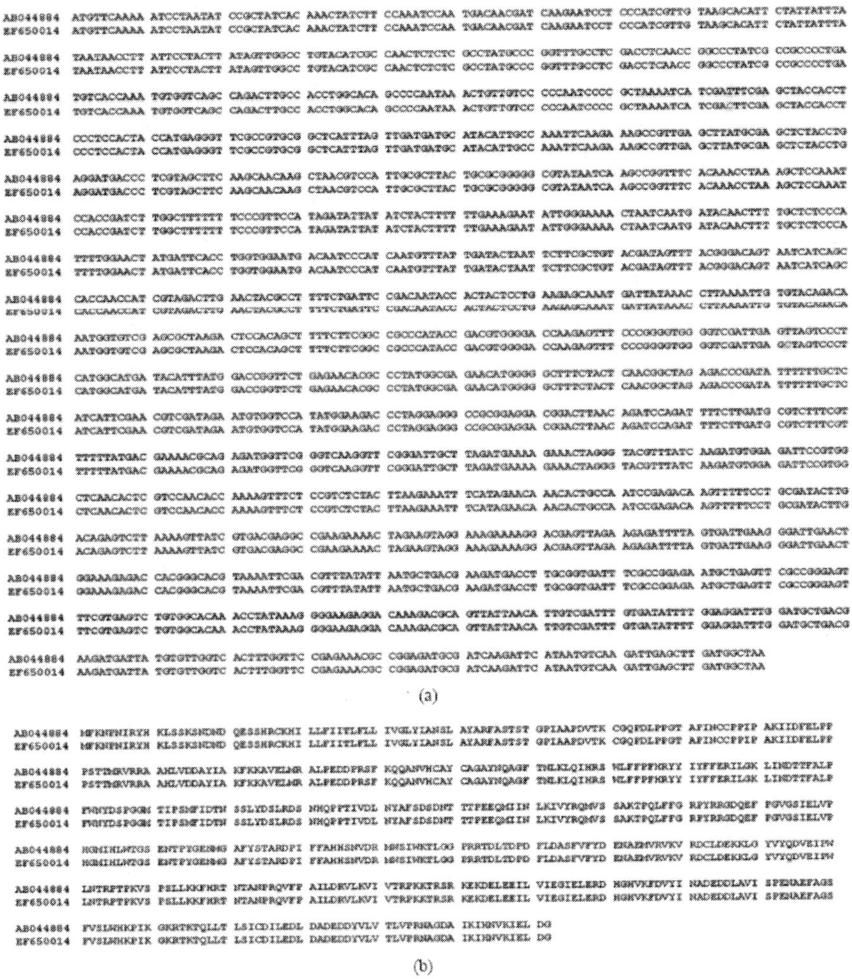

Figure 3. Nucleotide sequence (panel (a)) and deduced amino acid sequence (panel (b)) of snapdragon cDNA encoding aureusidin synthase (AS1). Accession number AB044884 is the published AmAS1 sequence from Antirrhinum majus cv. Yellow Butterfly [10] , EF650014 is our cDNA sequence from A.majus cv. Ribbon Yellow (SRYAS1). Nucleotides that differed from the published sequence [10] are indicated by highlights.

Of the 9 transgenic Petunia plants carrying the SRY4'CGT transgene, only a few flowers (from transformants SRY4'CGT-1, -6, -7, -10) showed variegated blue-white sectors (Table 1). Those transformants that produced variegated blue-white flowers, at the late stage of flowering, more and more new-born flowers produced variegated blue-white flowers, and the areas of white sectors increased in the blue background of the petals. As time passed, most of the old flowers without alteration in pigmentation faded, and the number of new-born flowers carrying more white sectors/regions in the background blue color increased. Finally, petal color was changed to completely white in all the late-born flowers (Figure 5). In contrast, of the 13 transgenic Petunia plants carrying SRYAS1 transgene, three (SRYAS1-4, -6, -12) showed variegated blue-white flowers at the late stage of flowering (Table 2): a few variegated flowers carrying small portions of white sectors in the background blue petals were observed in these 3 transgenic lines. However, complete white flowers were not

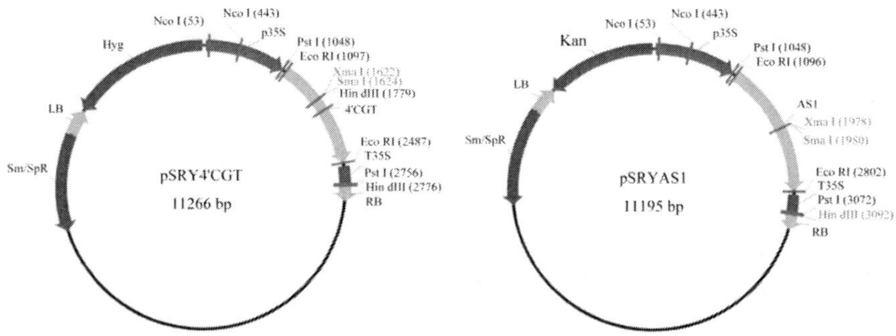

Figure 4. Restriction maps of transformation vectors pSRY4'CGT (a) and pSRYAS1 (b).

Table 1. Alteration in flower color and inheritance in T_1 progeny in transgenic Petunia plants carrying SRY4'CGT transgene.

Transgenic plant	Flower color	No. of resistant (R) and sensitive (S) seedlings from T_0 selfed seeds		Test ratio (R:S)	Remarks
		R	S		
SRY4'CGT-1	Blue-white				No seeds
SRY4'CGT-2	Blue	178	172	1:1	
SRY4'CGT-3	Blue				No seeds
SRY4'CGT-4	Blue	282	170	2:1	
SRY4'CGT-5	Blue				No seeds
SRY4'CGT-6	Blue-white	265	308	1:1	
SRY4'CGT-7	Blue-white	577	191	3:1	
SRY4'CGT-10	Blue-white	138	43	3:1	
SRY4'CGT-12	Blue	189	98	2:1	

Seeds were germinated on MS medium containing 10 mg/l hygromycin for 3 weeks.

observed. The growth pattern of a typical transgenic SRYAS1 plant carrying variegated flowers is shown in Figure 6. In addition, the amount and appearance of variegated blue-white flowers during the late stage of flowering (or plant senescence) in transgenic SRYAS1 plants were less than and later than, respectively, those of transgenic SRY4'CGT plants.

3.3. Gene Expression in T_0 Transgenic SRYAS1 and SRY4'CGT Plants Carrying Variegated Blue-White Flowers

To examine the presence and expression of transgenes in transgenic plants, total RNA was isolated from variegated blue-white flowers in three transgenic SRYAS1 plants (SRYAS1-4, -6, -12) and four transgenic SRY4'CGT plants (SRY4'CGT-1, -6, -7, -10). In addition, total RNA was also isolated from blue flowers in wild-type P. hybrida. RT-PCR analysis was carried out to investigate the expression of transgene in transgenic plants (Figure 7). As expected, no transcripts were detected in wild-type Petunia flowers when gene-specific SRYAS1 or SRY4'CGT primers were employed. Transgenic SRYAS1 transcripts were detected in all three transgenic SRYAS1 plants, and transgenic SRY4'CGT transcripts were detected in all four transgenic SRY4'CGT plants (Figure 7). Thus, these results demonstrate clearly the presence and expression of transgenes in transgenic plants we examined.

3.4. Inheritance in T_1 Progeny of Transgenic Plants

The stable integration of the T-DNA region in two transformation vectors (Figure 4) into the petunia chromosomal genome was investigated by germinating T_1 seeds from self-fertilized T_0 transgenic plants on MS medium containing 10 mg/l hygromycin (Table 1) or 100 mg/l kanamycin sulfate (Table 2). In transgenic plants carrying SRY4'CGT transgene (Table 1), T_1 seeds were not available from three transformants (SRY4'CGT-1, -3, -5). Hygromycin-resistant and hygromycin-sensitive T_1seedlings segregated in a 3:1 ratio in 2 transformants (SRY4'CGT-7 and SRY4'CGT-10), indicating the presence of single copy of the hygromycin-resistant transgene in the nuclear genomes of the 2 lines (Table 1). Since the transformation vector pSRY4'CGT contains expression cassettes for the selection marker gene (hygromycin resistant gene) and snapdragon SRY4'CGT transgene within the T-DNA region (Figure 4(a)), it is reasonable to predict that the same copy number of the hygromycin resistant transgene and SRY4'CGT transgene will be found in each transformant.

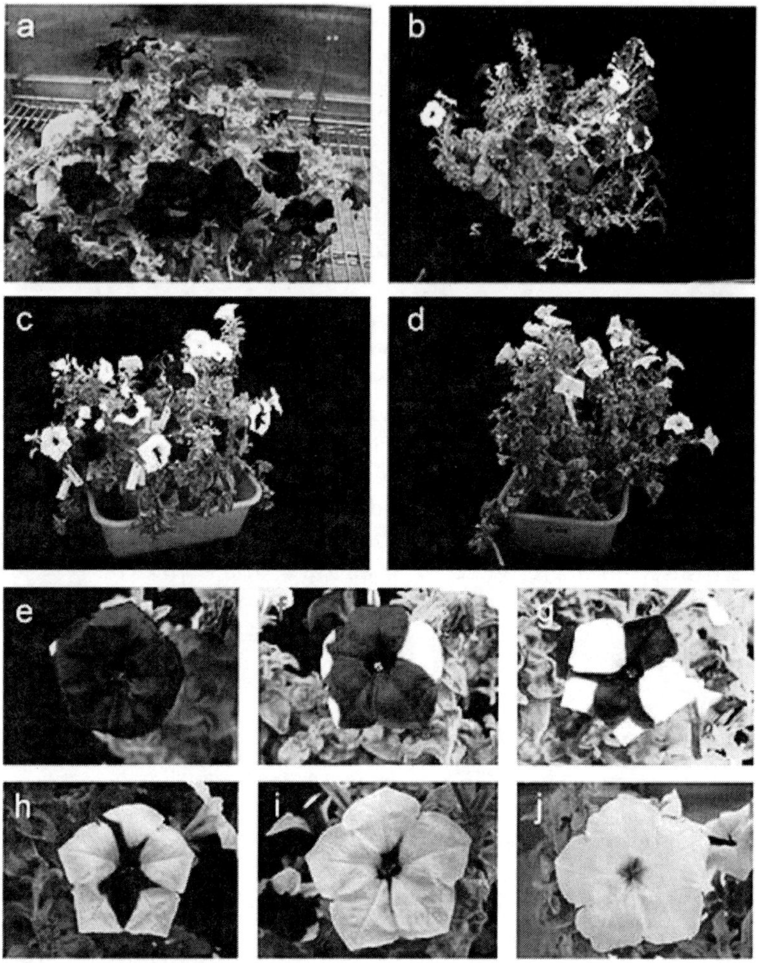

Figure 5. Alternation of flower coloration in transgenic Petunia plants carrying SRY4'CGT transgene. (a) At the beginning of flowering, only a few flowers (indicated by arrow) showed white sectors at the edge of the blue petals. (b) and (c) Over time, flowers with regular blue color faded, and at the middle and late stages of flowering variegated flowers carrying white sectors in a blue background, or complete white flowers, appeared. (d) During the late stage of flowering, variegated blue-white flowers gradually faded and new flowers emerging at that stage were almost or all white in color. As a result, all the flowers at the late-flowering stage in transgenic plants looked white in color. (e) to (j) Flower phenotypes from a spot of white color in the blue background to complete white color observed in the same transgenic plant.

Table 2. Alteration in flower color and inheritance in T_1 progeny in transgenic Petunia plants carrying SRYAS1 transgene.

Figure 6. Flower phenotype in transgenic Petunia plants carrying SRYAS1 transgene. (a) Most transgenic flowers were blue in color, similar to wild-type flowers. (b) A small number of transgenic flowers had white sectors on the edge of a blue petal.

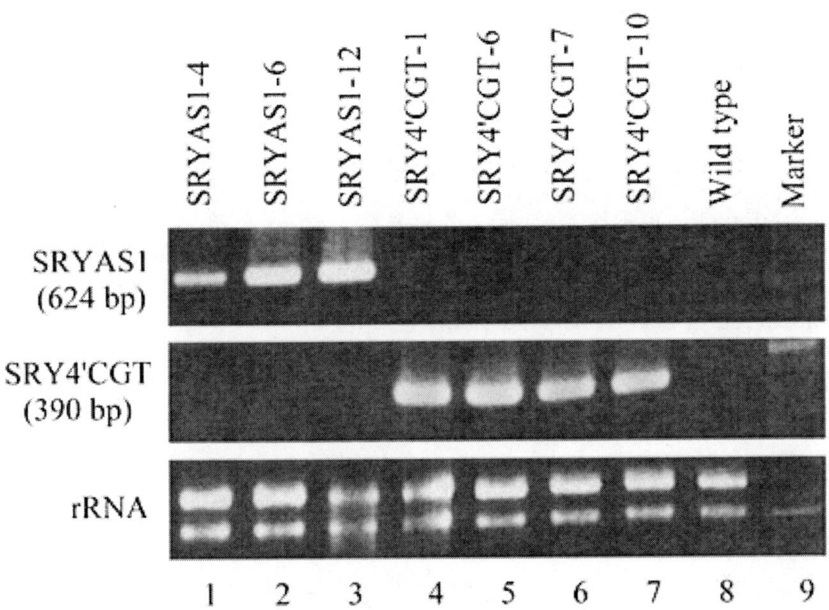

Figure 7. Gene expression in T_0 transgenic SRYAS1 and SRY4'CGT plants carrying variegated blue-white flowers. Total RNA was isolated from variegated blue-white flowers in three transgenic SRYAS1 plants (SRYAS1-4, -6, -12) and four transgenic SRY4'CGT plants (SRY4'CGT-1, -6, -7, -10). "Wild type" represented RNA sample isolated from wild-type Petunia flowers (blue color). RT-PCR was conducted for 25 cycles using F1 and R2 for amplification of SRYAS1 gene (expected size of this fragment was 624 bp), and F2 and R3 primers for amplification of SRY4'CGT gene (expected size of this fragment was 390 bp). The rRNA visualized by ethidium bromide staining is indicated as a loading control.

All transgenic plants carrying SRYAS1 transgene are shown in Table 2. No seeds were obtained from two transformants, SRYAS1-8 and SRYAS1-13. All T_1 seedlings from the two transformants SRYAS1-7 and SRYAS1-11 were sensitive to the selection agent kanamycin, suggesting that these two lines may not contain the foreign T-DNA fragment and can be considered to be false-positive clones. Thus, transformation efficiency in transgenic SRYAS1 plants was recalculated as 5.4%. Single insertion of the nptII transgene was determined in three transgenic plants SRYAS1-2, SRYAS1-4 and SRYAS1-12, as the segregation ratio of resistant and sensitive seedlings in these 3 transgenic plants was 3:1 (Table 2). Similarly, as transformation vector pSRYAS1 contains expression cassettes for the selection marker gene (nptII) and snapdragon SRYAS1 transgene within the T-DNA region (Figure 4(b)), it is reasonable to predict the same copy number of the nptII transgene and SRYAS1 transgene will be found in each transformant.

3.5. Gene Expression in T_1 Transgenic SRY4'CGT Plants

Five T_1 hygromycin-resistant seedlings were randomly selected in hygromycin-containing medium from T_0 selfed seeds of transgenic plant SRY4'CGT-7 that showed significant alteration in flower coloration, and grown in a greenhouse until maturity. Instead of no phenotypic change in 3 T_1transgenic plants, obvious variegated blue-white flowers were still observed in 2 T_1 transgenic plants (Figure 8). However, no complete white flowers

Figure 8. Flower phenotype and gene expression patterns in T_1 transgenic SRY4'CGT-7 plants. (a) Mature plant of T_1 transgenic SRY4'CGT-7-T_1-1. Five hygromycin-resistant T_1seedlings from transgenic SRY4'CGT-7 T_0 plant

were randomly-selected and grown until maturation. Among them, 2 T_1 progenies, designed as SRY4'CGT-7-T_1-1 and SRY4'- CGT-7-T_1-4, showed variegated blue-white flowers as parental SRY4'CGT-7 T_0 plant, whereas no alteration in flower color was observed in the rest of the three T_1 progenies. (b) An enlarged flower in transgenic SRY4'CGT-7-T_1-1 plant. (c) An enlarged flower in transgenic SRY4'CGT-7-T_1-4 plant. (d) Gene expression patterns in T_1 transgenic SRY4'CGT-7 flowers. RNA was isolated from blue sectors of T_1 transgenic flowers of SRY4'CGT-7-T_1-1 (abbreviated as 7-1-b) or white sectors of the same plant (abbreviated as 7-1-w), or from blue sectors of T_1transgenic flowers of SRY4'CGT-7-T_1-4 (abbreviated as 7-4-b) or white sectors of the same plant (abbreviated as 7-4-w). RT-PCR was conducted for 25 cycles using F2 and R3 primers for amplification of SRY4'CGT gene (expected size of this fragment was 390 bp), and using C1 and C2 primers for amplification of CHS gene (expected size of this fragment was 558 bp).

were found even at the late stage of flowering in any lines. To examine the gene expression in these variegated flowers, white sectors and blue sectors were dissected from variegated flowers in 2 T_1 transgenic lines, RNA was isolated, and RT-PCR was performed using gene-specific primers (Figure 8(d)). As predicted, no SRY4'- CGT transcripts were detected in wild-type Petunia flowers. Transgenic SRY4'CGT transcripts were detected in both blue and white sectors from these two transgenic lines. When specific primers were employed for an upstream gene chalcone synthase (CHS) in the flavonoid/aurone biosynthetic pathway, CHS transcripts were only detected in blue sectors but not in white sectors of the transgenic plants examined (Figure 8(d)), suggesting that the expression of CHS gene was suppressed in the white sectors of transgenic plants.

3.6. High-Performance Liquid Chromatography (HPLC) Profile in T_1 Transgenic Flowers

To examine the pigment profile in transgenic flowers, and more importantly, to investigate whether the appearance of white sectors in blue background petals is due to the loss of anthocyanins, white sectors and blue sectors were dissected from variegated flowers in T_1 transgenic SRY4'CGT and SRYAS1 lines, and ground into a powder by liquid nitrogen. After extraction, crude extracts were analyzed by HPLC. For each sample, similar HPLC patterns were obtained when monitoring was performed at 360 nm (for measurement of chalcones, flavones), 400 nm (for measurement of aurones) or 520 nm (for measurement of anthocyanidins). As shown in Figure 9, an absorption peak was detected in all samples when retention time (RT) approached 2.51 to 2.59 min. However, two peaks at RT = 11.57

min and RT = 11.78 min were detected in blue sectors from transgenic SRY4'CGT- 7-T_1-1 flowers (designed as B1) but not in white sectors from transgenic SRY4'CGT-7-T_1-1 flowers (designed as W1). Again, two peaks at RT = 11.71 min and RT = 11.82 min were detected in blue sectors from transgenic SRY4'CGT-7-T_1-4 flowers (designed as B2) but not in white sectors from transgenic SRY4'CGT-7-T_1-4 flowers (designed as W2). Only one peak at RT = 11.85 min was detected in blue sectors from transgenic SRYAS1-3- T_1-5 flowers (designed as B3) but not in white sectors from transgenic SRYAS1-3-T_1-5 flowers (designed as W3); however, the HPLC pattern in B3 was distinguishable from the HPLC patterns in B1 and B2, which represented two individual T_1 plants from the same T_0 parent SRY4'CGT-7 plant. Besides the constitutive peak at RT = 2.51 - 2.59 min in all blue and white samples, almost no significant peaks were detected from white sectors (W1, W2, W3), suggesting that colored pigments may not exist or may accumulate at very low level in these white sectors. HPLC patterns monitored at 360 nm, 400 nm and 520 nm are shown in Figures 9(a)-(c), respectively.

3.7. Compound Identification by Tandem Mass Spectrometry (MS/MS)

To further identify the major compounds in these peaks, tandem mass spectrometry (MS/MS) was employed to identify fragmentation patterns of different anthocyanins. As shown in Figure 10 and Figure 11, several major peaks obtained from different retention times were chosen and then analyzed. The molecular weight of each fragment was measured as mass-to-charge ratio (m/z) in the ionization process, and the most abundant fragment in each RT was considered as 100%. Compound identification, mainly based on molecular ions and product ions, was performed with the aid of existing literatures spectral analysis regarding flavonoids /anthocyanins [19] - [22] . Cyanidin triglucoside was found as major compound in the RT = 2.62 min fraction (Figure 10; Figure 11). Fragmentation pattern of cyanidin 3,5,5'-triglucoside is shown in Figure 12(a).

Next, major compounds were identified in peaks around RT = 11.51 - 12.96 min (Figure 10; Figure 11). It is worth noting that these peaks can be detected only in the blue sectors of transgenic flowers but not in white sectors. The red color of cyanidin and its glucoside derivatives (cyanidin glucoside, cyanidin diglucoside, cyanidin malonyldiglucoside), and mauve (pale purple) color of malvidin and its derivative malvidin-3-(p-coumaroyl)- rutinoside-5-glucoside were identified in these tissues. Fragmentation patterns of these anothcyanins are shown in Figure 12(b), Figure 13(a) and Figure 13(b). These data demonstrated clearly that the major compounds of anthocyanins extracted from the blue sectors of transgenic SRY4'CGT and SRYAS1 flowers are cyanidin, malvidin and their derivatives. In contrast, cyanidin or malvidin was not detected or was accumulated at a very low level in the white sectors of transgenic flowers, as a result, white color was observed.

Finally, to evaluate whether aurones (the target yellow pigment) can be detected in our analysis system, yellow petals from snapdragon were excised, and sample preparation and HPLC-DAD-MS/MS were performed as previously described. As shown in Figure 14(a), several peaks including RT = 2.51 min were detected in HPLC analysis. Compound identification in the fraction at RT = 11.25 min revealed that aureusidin glucoside and its dimer are the major compounds in that fraction (Figure 14(b)). It is worth pointing out that aureusidin is a yellow pigment belonging to the aurone, and aureusidin 6-O-glucoside (yellow pigment) occurs in the yellow petals of snapdragon. In addition, aureusidin 6-O-glucoside is the product of aureusidin synthase, one of our target genes (SRYAS1) in this study. The fragmentation pattern aureusidin glucoside is shown in Figure 14(c). In conclusion, yellow pigment of aureusidin glucoside can be detected in our analysis system. However, aureusidin or aureusidin glucoside was not found in our transgenic Petunia flowers. In addition, genistin 7-O-glucuronide (an isoflavone derivative) was identified as a major compound in fraction at RT = 17.89 min (data not shown).

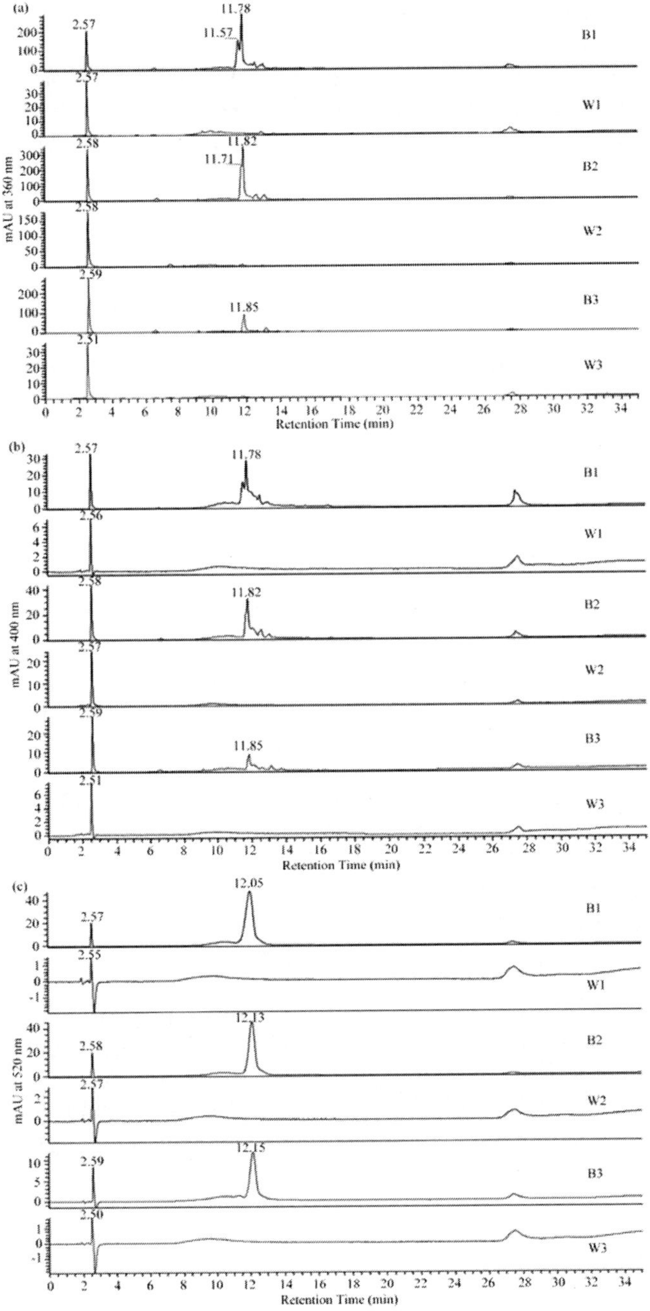

Figure 9. HPLC chromatograms of transgenic Petunia flowers detected at different absorption spectra. (a) 360 nm; (b) 400 nm; (c) 520 nm.

Abbreviations: B1, Blue sectors from transgenic SRY4'CGT-7-T_1-1 flowers; W1, White sectors from transgenic SRY4'CGT-7-T_1-1 flowers; B2, Blue sectors from transgenic SRY4'CGT-7-T_1-4 flowers; W2, white sectors from transgenic SRY4'CGT-7-T_1-4 flowers; B3, Blue sectors from transgenic SRYAS1-3-T_1-5 flowers; W3, white sectors transgenic SRYAS1-3-T_1-5 flowers.

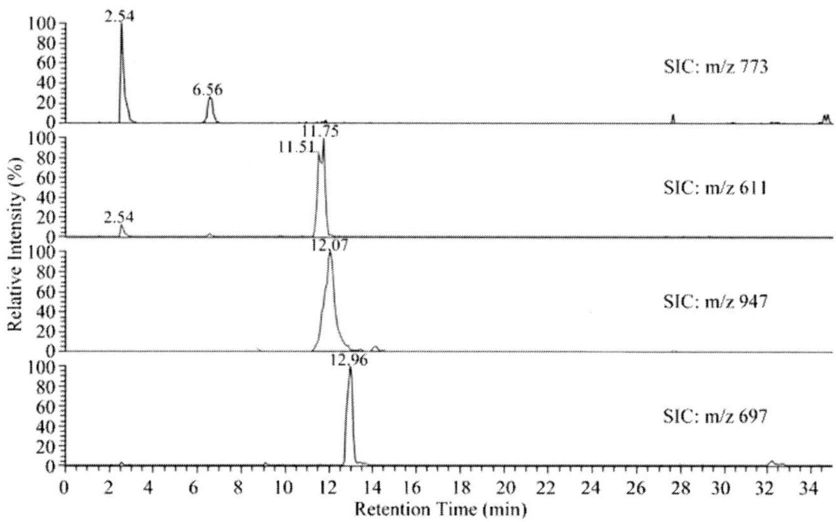

Figure 10. Selected ion chromatograms (SIC) of Electrospray Ionization (ESI) mass spectra detecting precursor ions from cyanidin 3,5,3'-triglucoside (m/z 773), cyanidin 3,5-diglucoside (m/z 611), malvidin-3-(p-coumaroyl)-rutinoside-5-glucoside (m/z 947), and cyanidin malonyldiglucoside (m/z 697).

HPLC coupled with tandem mass spectrometry was employed to analyze the blue and white sectors from variegated flowers of transgenic SRY4'CGT and SRYAS1 plants, and the results clearly demonstrated that cyanidin, malvidin, and their derivatives are present in the blue sectors, but none of anthocyanins are present in white sectors, causing the white phenotype.

4. DISCUSSION

Flavonoids/anthocyanins are major contributors to flower coloration. As a result, most studies on genetic engineering of flower color have been carried out via the flavonoid biosynthetic pathway [3] [6] [7] [23] [24] .

Introduction of an additional CHS gene into different varieties of Petunia hybrida resulted in variegated flowers with white sectors in some transgenic lines and complete white flowers in other transgenic lines [25] [26] . Molecular dissection of these transformants showed that both the introduced and endogenous CHS genes were "turned down" or silenced to different degrees, and as a result, loss of pigmentation in transgenic petals was observed [25] [26] . CHS is a key enzyme catalyzing one molecule of 4-coumaroyl-CoA and three molecules of malonyl- CoA into one molecule of naringenin chalcone, which is the central intermediate in the biosynthesis of different classes of flavonoids including aurones. In this study, two key genes of snapdragon (SRY4'CGT and SRYAS1) in the aurone biosynthetic pathway were cloned, and transformed separately into P. hybrida by an Agrobacterium- mediated method. Some transformants carrying the SRY4'CGT or SRYAS1 transgene showed variegated blue- white flowers (Table 1; Table 2), which were similar to the previous transformants carrying the CHS transgene [25] [26] . A dramatic change in flower color from almost complete blue to variegated blue-white sectors and then complete white in certain transformants carrying SRY4'CGT transgene during plant senescence was also observed (Figure 5). This pattern of alteration in flower coloration has not been previously reported.

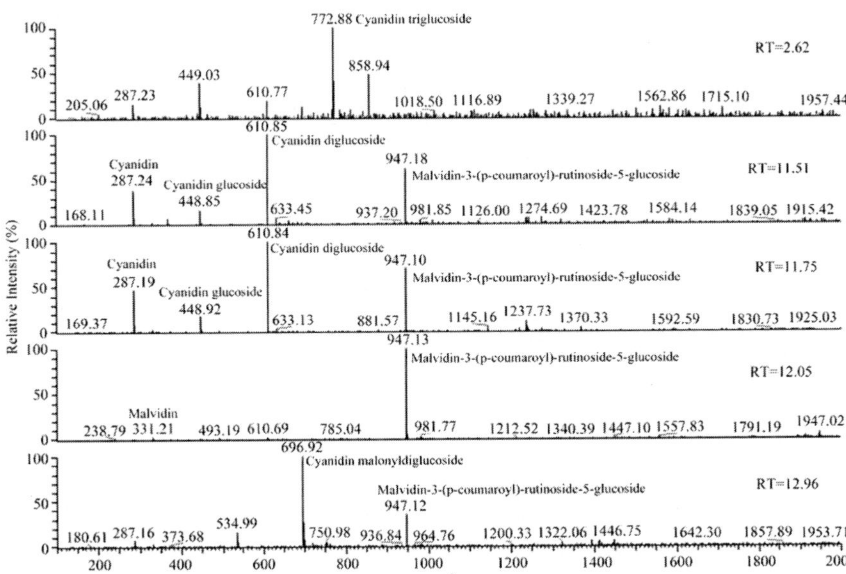

Figure 11. Mass spectra at different retention times.

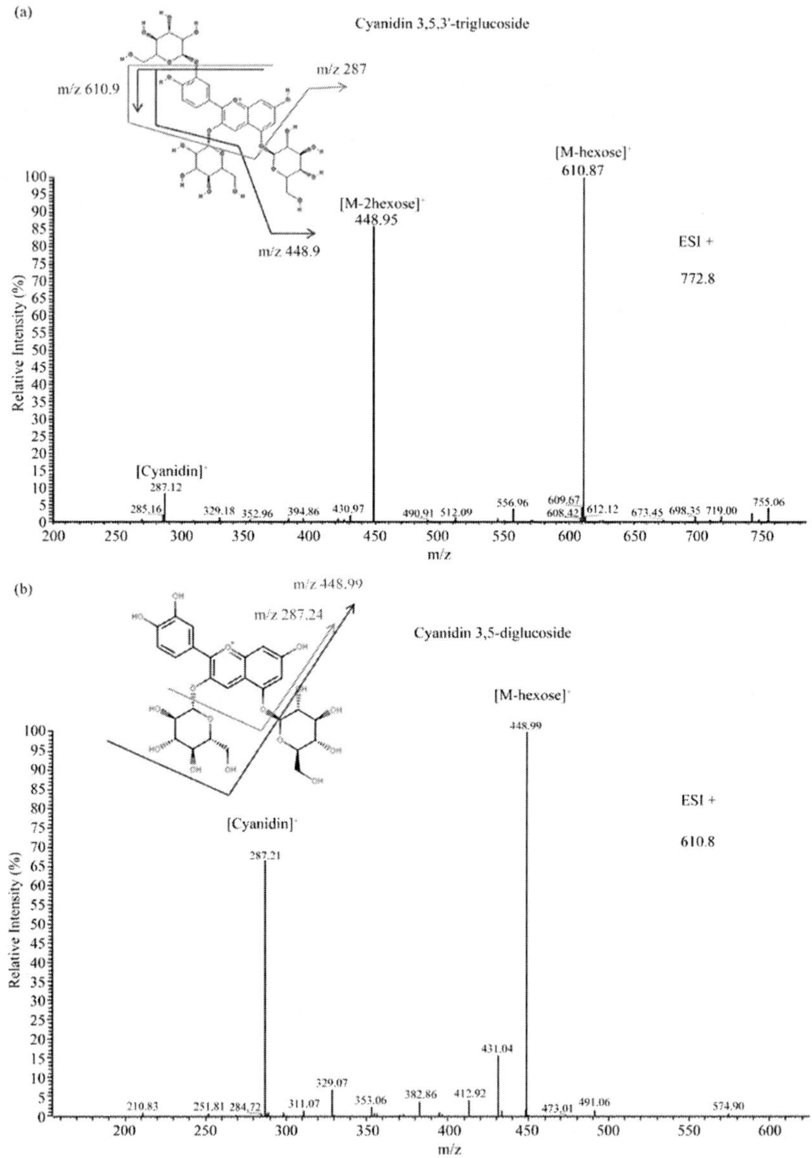

Figure 12. (a) MS/MS production ion spectrum of m/z 772.8 ion observed in the MS spectrum for cyanidin 3,5',3'- triglucoside. (b) MS/MS production ion spectrum of m/z 610.8 ion observed in the MS spectrum for cyanidin 3,5- diglucoside.

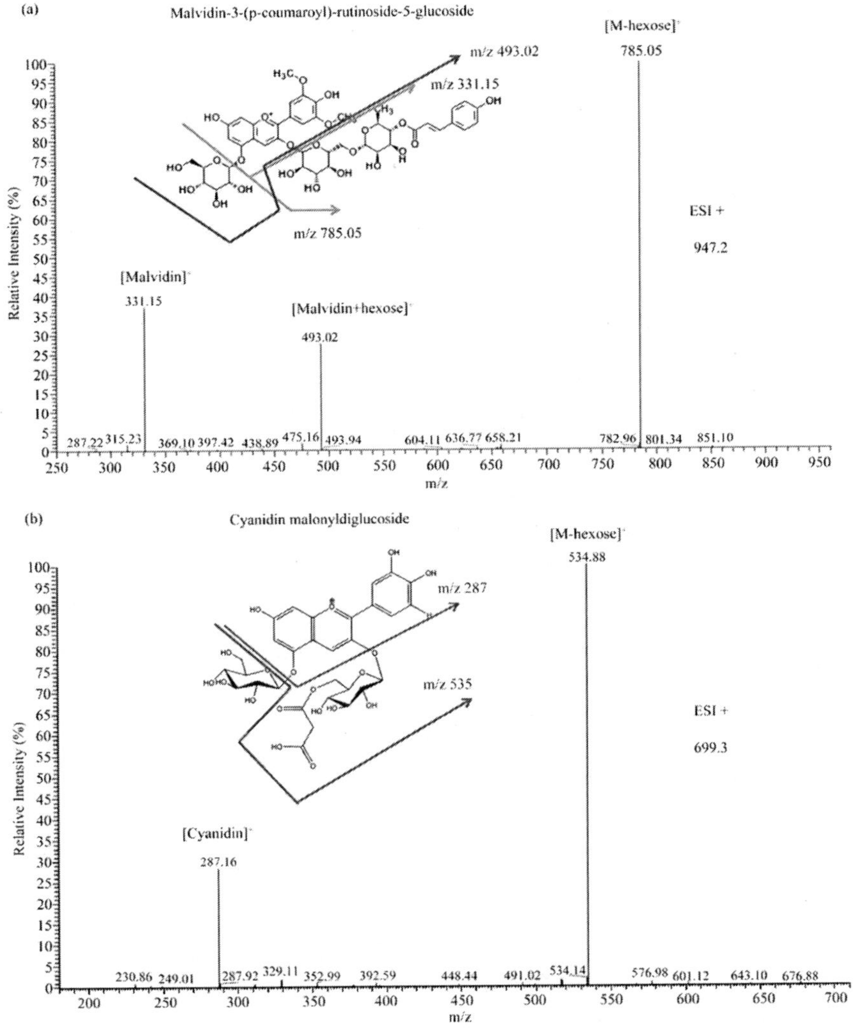

Figure 13. (a) MS/MS production ion spectrum of m/z 947.2 ion observed in the MS spectrum for malvidin-3-(p-couma- royl)-rutinoside-5-glucoside. (b) MS/MS production ion spectrum of m/z 699.3 ion observed in the MS spectrum for cyanidin malonyldiglucoside.

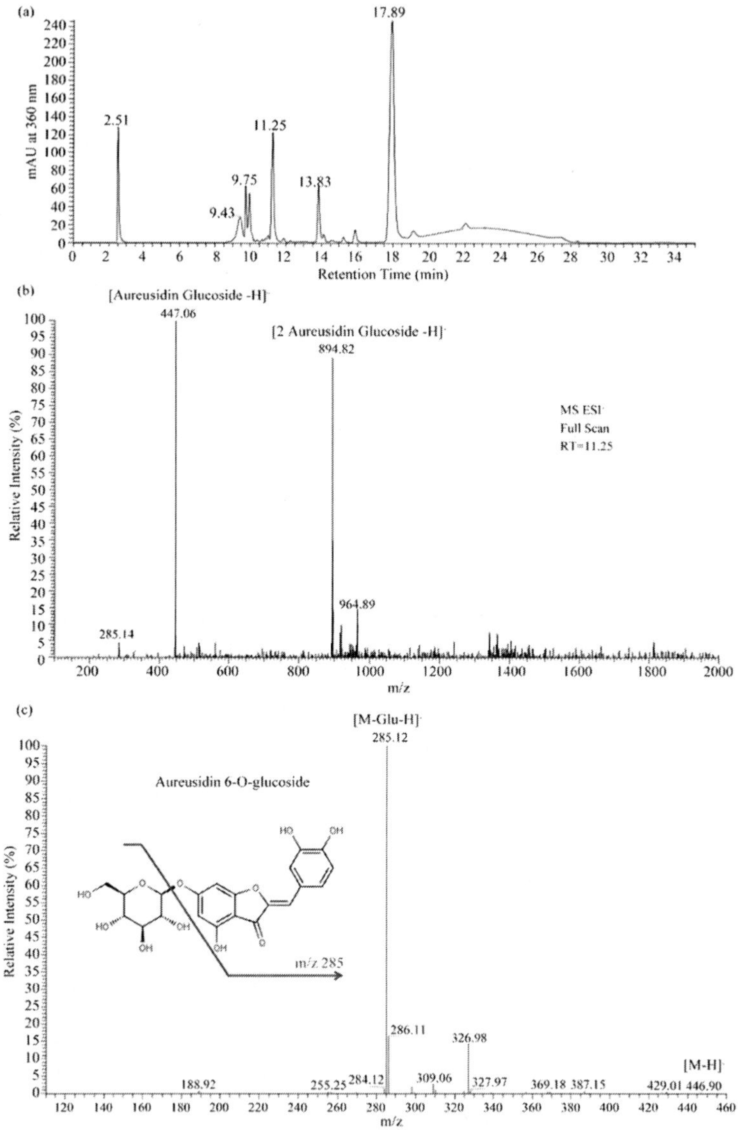

Figure 14. (a) HPLC chromatogram of yellow flowers from snapdragon detected at 360 nm; (b) Aureusidin glycoside and its dimer are the major compounds in the fraction at RT = 11.25 min. (c) MS/MS product ion spectrum of m/z 446.9 ion observed in the MS spectrum for aureusidin 6-O-glucoside.

Transgene-induced gene silencing is a quite common phenomenon in plants, for example, host genes can be silenced as a consequence of the presence of a homologous transgene [27] [28] . However, in this study, no sequence similarity was found between the host CHS gene and the SRY4'CGT transgene (or between the host CHS gene and the SRYAS1 transgene). It is interesting to find that expression of SRY4'CGT transgene can also suppress the expression of the non-homologous host CHS gene in the white sectors (Figure 8(d)). On the other hand, suppression of endogenous CHS expression is related to the transgene-induced gene silencing, which is considered to be sequence homology-dependent event [27] [28] . For example, an expression vector pCHS carrying Petunia CHS transgene was transformed into tobacco, nucleotide sequence analysis revealed that both Petunia and tobacco CHS cDNAs contained 1170 bp in length and sequence identity were 91%; as a result, four out of seven transgenic plants produced completely white flowers as compared to the pink flowers in wild-type tobacco, and it was further demonstrated that cosuppression of CHS expression resulted the loss of cyanidin pigments in those transgenic white flowers [12] . It has been reported that overexpression of apple anthocyanidin reductase (ANR) genes in tobacco caused inhibition of expression of both chalcone isomerase (CHI) and DFR genes in transgenic flowers, leading to loss of anthocyanin and resulting in white, pale pink and white/red variegated flowers as compared to wild-type red flowers [29] . Since low sequence identity between apple ANS gene and tobacco DFR gene, the authors concluded that cosuppression of tobacco DFR gene might not be related to RNA silencing, and might be due to competition or interaction among different enzymes involved in the flavonoid biosynthetic pathway [29] . Therefore, our transgenic plants should be good material for studying gene expression and regulation in the aurone/flavonoid biosynthetic pathway.

In T_1 progeny assay, a number of abnormal segregation ratios (i.e., R:S = 1:1 or R:S = 2:1) or no seed production was observed (Table 1; Table 2). We noticed that pollen was extremely rare in transgenic SRY4'CGT-1 and SRYAS1-13 plants; this may explain why no seeds were produced in these plants (Table 1; Table 2) (Figure 15). In brief, some of the transgenic plants we obtained in this study seem abnormal. The genus Petunia, belonging to the family Solanaceae, comprises more than 30 species. Most petunias are diploid (2n = 14) and are interfertile with different petunia species. The origin of P. hybrida is unknown, but there is general agreement that it was derived from interspecific crossing between a white-flo- wered species (probablyP. axillaris) and a pigmented species (probablyP. integrifolia) near 190 years ago [30] . The complicated genetic background of P. hybrid commercial varieties may cause those abnormal phenotypes we observed in this study.

From the point of view of biotechnology application, as the loss of the newly acquired traits in subsequent generations or under certain environmental conditions has been reported elsewhere [6] [31] [32] , the next important task is to ensure the reliability and stability of phenotypes reported here after several generations of propagation. Furthermore, it is very important to breed transgenic petunia plant carrying both SRYAS1 and SRY4'- CGT by crossing experiment, which is in progress. In addition, it will be very interesting to examine whether the same phenotypes can be achieved by transforming pSRYAS1 or pSRY4'CGT vector into other floricultural plants, such as Cleome spinosa. Agrobacterium-mediated transformation has been established in this value plant [33] .

Figure 15. Abnormal flower phenotype showing an extra petal-like structure in two anthers out of five anthers in some transgenic Petunia flowers. (a) Dissection of an abnormal flower. Arrows indicated an extra petal-like structure in two out of five anthers. The fifth anther harboring normal phenotype is not shown in this panel. (b) An enlarged photo of one abnormal anther. The extra petal-like structure appear at the top of the anther. (c) and (d) The abnormal petal-like structure is still presented and even the pollen sac is broken at the late stage of flower development.

ACKNOWLEDGEMENTS

The authors are grateful to the Metabolomics Core Facility, Agricultural Biotechnology Research Center, Academia Sinica, for HPLC-DAD-MS/MS assistance. We also thank Miranda Loney for critical reading and editing of the manuscript. This work was financially supported by Academia Sinica, Taiwan, to Dr. Kin-Ying To.

CITE THIS PAPER

Chen-Kuen Wang, Yi-Chin Chin, Chih-Yu Lin, Po-Yen Chen, Kin-Ying To (2015) Transforming the Snapdragon Aurone Biosynthetic Genes into Petunia Alters Coloration Patterns in Transgenic Flowers. *Advances in Bioscience and Biotechnology*,**06**,702-722.

REFERENCES

1. 1. Bartley, G.E. and Scolnik, P.A. (1995) Plant Carotenoids: Pigments for Photoprotection, Visual Attraction, and Human Health. Plant Cell, 7, 1027-1038.
2. 2. To, K.Y. and Wang, C.K. (2006) Molecular Breeding of Flower Color. In: Teixeira da Silva, J.A., Ed., Floriculture, Ornamental and Plant Biotechnology, Volume 1, Global Science Books, London, 300-310.
3. Nishihara, M. and Nakatsuka, T. (2011) Genetic Engineering of Flavonoid Pigments to Modify Flower Color in Floricultural Plants. Biotechnology Letters, 33, 433-441.
4. Schwinn, K.E. and Davies, K.M. (2004) Flavonoids. In: Davies, K.M., Ed., Plant Pigments and Their Manipulations, Blackwell Publishing Ltd., Oxford, 92-149.
5. Tanaka, Y., Katsumoto, Y., Brugliera, F. and Mason, J. (2005) Genetic Engineering in Floriculture. Plant Cell, Tissue and Organ Culture, 80, 1-24.
6. Tanaka, Y., Brugliera, F., Kalc, G., Senior, M., Dyson, B., Nakamura, N., Katsumoto, Y. and Chandler, S. (2010) Flower Color Modification by Engineering of the Flavonoid Biosynthetic Pathway: Practical Perspectives. Bioscience, Biotechnology and Biochemistry, 74, 1760-1769.
7. Vogt, T. (2010) Phenylpropanoid Biosynthesis. Molecular Plant, 3, 2-20.
8. Chandler, S.F. and Brugliera, F. (2011) Genetic Modification in Floriculture. Biotechnology Letter, 33, 207-214.
9. Winkel-Shirley, B. (2001) Flavonoid Biosynthesis: A Colorful Model for Genetics, Biochemistry, Cell Biology, and Biotechnology. Plant Physiology, 126, 485-493.
10. Nakayama, T., Yonekura-Sakakibara, K., Sato, T., Kikuchi, S., Fukui, Y., Fukuchi-Mizutani, M., Ueda, T., Nakao, M., Tanaka, Y., Kusumi, T. and Nishino, T. (2000) Aureusidin Synthase: A Polyphenol Oxidase Homolog Responsible for Flower Coloration. Science, 290, 1163-1166.
11. Ono, E., Fukuchi-Mizutani, M., Nakamura, N., Fukui, Y., Yonekura-Sakakibara, K., Yamaguchi, M., Nakayama, T., Tanaka, T., Kusumi, T. and Tanaka, Y. (2006) Yellow Flowers Generated by Expression of the Aurone Biosynthetic Pathway. Proceedings

of the National Academy of Sciences of the United States of America, 103, 11075-11080.

12. Wang, C.K., Chen, P.Y., Wang, H.M. and To, K.Y. (2006) Cosuppression of Tobacco Chalcone Synthase Using Petunia Chalcone Synthase Construct Results in White Flowers. Botanical Studies, 47, 71-82.
13. Gerats, T. and Vandenbussche, M. (2005) A Model System for Comparative Research: Petunia. Trends in Plant Science, 10, 251-256.
14. Lutke, W.K. (2006) Petunia (Petunia hybrid). Methods in Molecular Biology, 344, 339-349.
15. Murashige, T. and Skoog, F. (1962) A Revised Medium for Rapid Growth and Bioassays with Tobacco Tissue Cultures. Physiologia Plantarum, 15, 473-497.
16. Wang, H.M., Yin, W.C., Wang, C.K. and To, K.Y. (2009) Isolation of Functional RNA from Different Tissues of Tomato Suitable for Developmental Profiling by Microarray Analysis. Botanical Studies, 50, 115-125.
17. Wang, C.K., Hsu, S.Y., Chen, P.Y. and To, K.Y. (2012) Transformation and Characterization of Transgenic Bidens pilosa L. Plant Cell, Tissue and Organ Culture, 109, 457-464.
18. Nakayama, T., Sato, T., Fukui, Y., Yonekura-Sakakibari, K., Hayashi, H., Tanaka, Y., Kusumi, T. and Nishino, T. (2001) Specificity Analysis and Mechanism of Aurone Synthesis Catalyzed by Aureusidin Synthase, a Polyphenol Oxidase Homolog Responsible for Flower Coloration. FEBS Letters, 499, 107-111.
19. Giusti, M.M., Rodriguez-Saona, L.E., Griffin, D. and Wrolstad, R.E. (1999) Electrospray and Tandem Mass Spectroscopy as Tools for Anthocyanin Characterization. Journal of Agricultural and Food Chemistry, 47, 4657-4664
20. Schütz, K., Persike, M., Carle, R. and Schieber, A. (2006) Characterization and Quantification of Anthocyanins in Selected Artichoke (Cynara scolymus L.) Cultivars by HPLC-DAD-ESI-MSn. Analytical and Bioanalytical Chemistry, 384, 1511-1517.
21. Hillebrand, S., Naumann, H., Kitzinski, N., Köhler, N. and Winterhalter, P. (2009) Isolation and Characterization of Anthocyanins from Blue-Fleshed Potatoes (Solanum tuberosum L.). Food, 3, 96-101.
22. Zhao, H., Hu, X., Chen, X., Shi, S., Jiang, X., Chen, W. and Zhang, S. (2015) Analysis and Improved Characterization of Minor Antioxidants from Leaves of Malus doumeri Using a Combination of Major Constituents' Knockout with High-Performance Liquid Chromatography-Diode Array Detector-Quadrupole Time-of-Flight Tandem Mass Spectrometry. Journal of Chromatography A, 1398, 57-65.
23. Nakatsuka, T., Mishiba, K.I., Kubota, A., Abe, Y., Yamamura, S., Nakamura, N., Tanaka, Y. and Nishihara, M. (2010) Genetic Engineering of Novel Flower Colour by Suppression of Anthocyanin Modification Genes in Gentian. Journal of Plant Physiology, 167, 231-237.
24. Chandler, S.F. and Sanchez, C. (2012) Genetic Modification; the Development of Transgenic Ornamental Plant Varieties. Plant Biotechnology Journal, 10, 891-903.
25. Napoli, C., Lemieux, C. and Jorgensen, R. (1990) Induction of a Chimer Chalcone Synthase Gene into Petunia Results in Reversible Co-Suppression of Homologous Genes in trans. Plant Cell, 2, 279-289.
26. van der Krol, A., Mur, L.A., Beld, M., Mol, J.N.M. and Stuitje, A.R. (1990) Flavonoid Genes in Petunia: Addition of a Limited Number of Gene Copies May Lead to a Suppression of Gene Expression. Plant Cell, 2, 291-299.
27. Vaucheret, H., Béclin, C., Elmayan, T., Feuerbach, F., Godon, C., Morel, J.-B., Mourrain, P., Palauqui, J.-C. and Vernhettes, S. (1998) Transgene-Induced Gene Silencing in Plants. Plant Journal, 16, 651-659.
28. Eamens, A., Wang, M.B., Smith, N.A. and Waterhouse, P.M. (2008) RNA Silencing in Plants: Yesterday, Today, and Tomorrow. Plant Physiology, 147, 456-468.
29. Han, Y., Vimolmangkang, S., Soria-Guerra, R.E. and Korban, S.S. (2012) Induction of Apple ANR Genes into Tobacco Inhibits Expression of both CHI and DFR Genes in Flowers, Leading to Loss of Anthocyanin. Journal of Experimental Botany, 63, 2437-2447.

30. Peltier, D., Farcy, E., Dulieu, H. and Bervillé, A. (1994) Origin, Distribution and Mapping of RAPD Markers from Wild Petunia Species in Petunia hybrida Hort Lines. Theoretical and Applied Genetics, 88, 637-645.
31. Dietz-Pfeilstetter, A. (2010) Stability of Transgene Expression as a Challenge for Genetic Engineering. Plant Science, 179, 164-167.
32. auser, M.T., Aufsatz, W., Jonak, C. and Luschnig, C. (2011) Transgenerational Epigenetic Inheritance in Plants. Biochimica et Biophysica Acta, 1809, 459-468.
33. Tsai, Y.T., Chen, P.Y. and To, K.Y. (2012) Plant Regeneration and Stable Transformation in the Floricultural Plant Cleome spinosa, a C3 Plant Closely Related to the C4 Plant C. gynandra. Plant Cell Reports, 31, 1189-1198.

CHAPTER 14

Functional Analysis of *GmCPDs* and Investigation of Their Roles in Flowering

Miao Wang[1,2], Xin Xu[1], Xinxin Zhang[1], Shi Sun[1], Cunxiang Wu[1], Wensheng Hou[1], Qingyu Wang[2] *, Tianfu Han[1] *

[1] *Ministry of Agriculture Key Laboratory of Soybean Biology (Beijing), Institute of Crop Sciences, the Chinese Academy of Agricultural Sciences, Beijing, 100081, China,*

[2] *College of Plant Science, Jilin University, Changchun, 130062, Jilin, China*

ABSTRACT

The onset of floral development is a pivotal switch in the life of soybean. Brassinosteroids (BRs), a group of steroidal phytohormones with essential roles in plant growth and development, are associated with flowering induction. Genes involved in BR biosynthesis have been studied to a great extent in Arabidopsis, but the study of these genes has been limited in soybean. In this study, four CPD homologs (GmCPDs) catalyzing BR synthesis were isolated from soybean. Transcripts were mainly confined to cotyledons and leaves and were down-regulated in response to exogenous BR. Bioinformatic analysis showed strong sequence and structure similarity between GmCPDs and AtCPD as well as CPDs of other species. Overexpression of *GmCPDs* in an Arabidopsis BR-deficient mutant rescued the phenotype by restoring the biosynthesis pathway, revealing the functional roles of each *GmCPDs* in. Except for the rescue of root development, leaf expansion and plant type architecture, *GmCPDs* in expression also complemented the late flowering phenotype of Arabidopsis mutants deficient in CPD. Further evidence in soybean plants is that the expression levels of *GmCPDs* in are under photoperiod control in *Zigongdongdou*, a photoperiod-sensitive variety, and show a sudden peak upon floral meristem initiation. Together with increased *GmCPDs* in expression in the leaves and cotyledons of photoperiod-insensitive early-maturity soybean, it is clear that *GmCPDs* in contribute to flowering development and are essential in the early stages of flowering regulation.

INTRODUCTION

Flowering is one of the most important events in the life cycle of plants, with optimal timing being especially crucial. Therefore, flowering is controlled by numerous interacting endogenous and environmental cues to ensure appropriate conditions for seed production. At least four signaling pathways have been demonstrated to regulate flowering in concert, involving length of day (photoperiodism), winter cold (vernalization), regulation by gibberellins (GAs), and autonomous floral initiation occurring in the absence of any effective environmental signals [1]. In addition, other factors such as ascorbic acid, ethylene, ambient temperature and light quality have been found to be critical in mediating flowering [2–5]. Brassinosteriod is one of these factors currently awaiting study.

Brassinosteroids (BRs) are a class of plant-specific steroid hormones. The effects of BRs span multiple physiological processes and responses, including the induction of cell elongation, xylem differentiation, root development, leaf bending/expansion, senescence and male fertility [6–8]. In Arabidopsis, most BR-deficient mutants, such as the BR-biosynthetic mutants *det2* [9], *dwf4* [10] and *cpd* [11,12], exhibit a delayed flowering time. These mutants over-accumulate different BR precursors as a result of the blocked BR-biosynthetic pathway [10,13], indicating that changes to endogenous BR and BR precursor levels in Arabidopsis affects flowering time. In addition, BRs have been found to modify the flowering induction networks of Arabidopsis by regulating critical flowering-time genes. The BR-insensitive mutant *bri1* was reported to significantly delay flowering of the autonomous-pathway mutant *ld* and *fca* by elevating *FLC* expression [12]. Moreover, *bdbrd1–1*, a BR-biosynthetic mutant from *Brachypodium distachyon*, was found to suppress the expressions of autonomous pathway genes such as *FCA, FY, FLD, FVE* and *LD* [14,15]. Recently, *BZR1*, a positive regulator in the BR signaling, was confirmed to regulate *FLD* expression by directly binding to *FLD* promoter to mediate flowering [16]. All above raise the possibility that BRs promote flowering through autonomous pathway. Another BR related gene, *CONSTITUTIVE PHOTOMORPHOGENESIS AND*

DWARFISM (*CPD*), can be regulated by light and may modulate flowering through the photoperiod pathway.

The *CPD* gene was identified from an Arabidopsis T-DNA insertional mutation that causes the constitutive photomorphogenesis and dwarfism (*cpd*) [17]. This gene encodes CYP90A1/ CPD, which belongs to the cytochrome P450 family [17]. CYP90A1/CPD was thought to be an enzyme that catalyzes C-23 hydroxylation because C23-hydroxylated BR precursors but not 22-hydroxylated cathasterone (CT) rescued the *cpd* mutant [17]. However, recent evidence has suggested that the C-23 hydroxylation reaction is catalyzed by CYP90C1 and CYP90D1 [18]. Actually, (22S)-22-hydroxycampesterol (22-OHCR) is a favored substrate of CYP90A1/CPD [19]. It has been demonstrated that CYP90A1/CPD encoded by *CPD* is a C-3 oxidase that is required for the conversion of 22-OHCR to (22S)-22-hydroxycampest-4-en-3-one (22-OH-4-en-3-one). The BR biosynthesis network in Arabidopsis proceeds from campesterol (CR) to brassinolide (BL), which is biologically the most active form of BR [20–22]. The originally proposed BR synthesis route is considered to start at the conversion of CR to campestanol (CN)[22]. However, the latest study has indicated that these steps are non-essential and that a CN-independent BR pathway is the main route based on enzyme substrate preferences [19]. In the CN-independent BR synthesis pathway, CR is first hydroxylated to 22-OHCR, followed by the oxidation to 22-OH-4-en-3-one. Obviously, CYP90A1/CPD encoded by *CPD* catalyzes the early step of the BR biosynthesis pathway, suggesting its pivotal role in BR biosynthesis.

Interestingly, expression of the BR-biosynthetic gene *CPD* was reported to possess diurnal rhythmicity with light regulation superimposed upon circadian control [23]. These transcriptional changes are independent of BR feedback regulation but are accompanied by the diurnal variation of endogenous BR content [23]. Similar to most of the light-responsive genes of GA synthesis, *CPD* is under photoreceptor-specific control mainly through phytochrome signaling, suggesting a mechanism in which light controls physiological functions via BRs [24–27]. The CIRCADIAN CLOCK ASSOCIATED 1 (CCA1) transcription factor has been shown to mediate circadian control and phytochrome-regulated gene expression [28,29]. As a potential binding sequence of CCA1, the AAAATCT motif was therefore speculated to be present in *CPD* promoters [23]. *CCA1* is a major gene involved in the circadian clock. In Arabidopsis, the circadian clock has dramatic effects on flowering time through the CO–FT photoperiodic flowering pathway modulated by its core CCA1–LHY/TOC1 [30–33]. The interaction between *CPD* and *CCA1* may hold clues to the causes of the late-flowering phenotype of *cpd* mutants [12,34]. Furthermore, it has been reported that BR can modulate circadian rhythms and promote the periodicity of the circadian clock genes *CHLOROPHYLL A/B BINDING*

PROTEIN (*CAB2*), *COLD AND CIRCADIAN-REGULATED 2* (*CCR2*) and *CCA1* [35]. This interaction is consistent with the observation that the period of *CCR2* is prolonged in *cpd* mutants [35]. Thus, the above findings suggest that BR regulates flowering time through the circadian clock system, a crucial mechanism in photoperiod pathway.

Soybean is a short-day crop of agricultural and economic importance. Soybean flowering is largely regulated by photoperiod, with many varieties highly photoperiod-sensitive. Typically, *Zigongdongdou* will not initiate flowering until short-day induction; this variety even undergoes flowering reversion in which the floral meristem developing the floral organs reverts to produce leaves when the photoperiod is altered from a short day to a long day [36–40]. This high sensitivity restricts the adaptability of soybean to diverse environmental conditions, limiting the season and region available to many high yield varieties, negatively impacting soybean production [41]. In addition, the photoperiod sensitivity is diverse among soybean varieties, leading to multiple maturity periods. Consequently, many varieties with good behavior cannot be hybridized with each other as a result of asynchronous florescence. Therefore, it has been long recognized by breeders that controlling flowering time is crucial to ensuring soybean yield [42].

In the current study, four soybean *CPD* homologous genes belonging to the BR biosynthesis pathway are found to be associated with soybean flowering. These GmCPDs are extremely similar with AtCPD in sequence and structure and can complement the AtCPD function in Arabidopsis mutants deficient in AtCPD. The expression levels of these *GmCPDs* all exhibit a sudden peak upon floral meristem initiation in soybean and are increased in a photoperiod-insensitive soybean variety, suggesting a relationship between BR biosynthesis genes and floral transition.

MATERIALS AND METHODS

Plant Growth Conditions

Soybean varieties *Williams 82* and *Zigongdongdou* were grown in a chamber at day/night temperatures of 26/24°C. *Zigongdongdou* and *Heihe27* used for the analysis of *GmCPDs* expression in soybean varieties with different photoperiod sensitivities were grown at the temperature of constant 25°C. *Williams 82* plants were cultivated under a short-day condition (12/12 h day/night cycle). *Zigongdongdou* and *Heihe27* plants were cultivated under either short-day or long-day (16/8 h day/night cycle) conditions depending on the experiment.

Arabidopsis accessions Col-0 and *cpd-91* were grown at 22°C under a long-day condition (16/8 h day/night cycle) in potting soil or in half strength MS agar plates with 1% (w/v) sucrose. All plates were axenically cultured and packed with silver papers in the dark treatment.

Brassinosteroid Treatment

In the BR response assay, 5-day-old *Williams 82* seedlings were cultivated in Hoagland solution after germination in soil. BR treatment was undertaken 10 days later by adding 1 μM 2,4-epibrassinolide (C28H48O6; TCR, Toronto, ON, Canada) to the solution. The treatment lasted for 2 hours, and the samples were collected every half hour.

In the root inhibition assay, Arabidopsis seeds were planted on vertically oriented plates containing half-strength MS medium supplemented with 1% sucrose in the absence or presence of 100 nM 2,4-epibrassinolide (24-epiBL). Root lengths were measured after seedlings were grown for 10 days.

Sampling and RNA Isolation

The entire *Williams 82* plant was sampled for *GmCPDs* gene cloning. For tissue-specific expression analysis, hypocotyls, cotyledons and roots of *Williams 82* were collected from 7-day-old seedlings, and the leaves, stems and shoot apices were collected from 20-day-old adult plants. The flowers were tagged on the day of anthesis, and the pods were harvested when 0.5–2 cm long. After BR treatment, the leaves of *Williams 82* were collected every half hour and labeled 0.5 h (0.5 hour after treatment), 1 h, 1.5 h and 2 h. When the cotyledons of *Zigongdongdou* and *Heihe27* opened, SD (short-day), LD (long-day) photoperiod treatments were carried out. The SD13d-LD (transfer to an LD condition after a 13-day SD treatment) condition was applied only to *Zigongdongdou*. Plant leaves were collected every other day until the 25th day following photoperiod treatment. Cotyledons were obtained at 3 d (3 days after photoperiod treatment), 6 d, and 9 d with leaves removed after cotyledon opening.

All samples were a mixture of more than five individual plants and were ground into powder in liquid nitrogen. Total RNA was extracted using TRIzol Reagent (Invitrogen, Carlsbad, CA, USA). The RNA from a whole Arabidopsis plant sample was isolated using the same method.

Analysis of mRNA Expression Level by Real-Time PCR

cDNA for PCR was prepared using 1 μg of total RNA with a mixture of random primers. RT-qPCR analysis was performed on an ABI7900 instrument (Applied Biosystems, Foster City, CA, USA) using Takara SYBR Premix ExTaq (Takara, Shiga, Japan) for 40 cycles (95°C for 5 s; 60°C for 30 s; 72°C for 30 s). All reactions were carried out at least three times. Quantification of mRNA level was based on Ct (threshold cycle) values using a comparative Ct (2−ΔΔCt) method [43]. Data are presented as the mean±SD. The specific primers for each gene are shown in S1 Table.

Vector Construction and Arabidopsis Transformation

The coding regions of *GmCPD1*, *GmCPD2*, *GmCPD3* and *GmCPD4* with additional *Xba*I and *Sac*I restriction sites were PCR-amplified. The *Xba*I–*Sac*I flanked *GmCPDs* fragments were cloned into the *Xba*I–*Sac*I sites of pTF101.1-GFP vector, replacing GFP and generating pTF101.1-GmCPD1, pTF101.1-GmCPD2, pTF101.1-GmCPD3 and pTF101.1-GmCPD4. These resulting constructs were verified by sequencing and restriction analysis and transformed into *Agrobacterium tumefaciens* strain GV3101. The *Agrobacterium*-mediated flower infiltration transformation method [44] was used to introduce *GmCPDs* into *cpd-91* Arabidopsis mutant plants. T1 generation seeds were harvested and selected on antibiotic-containing MS plates with 10 mg/L glufosinate ammonium (Sigma, St. Louis, MO, USA). Positive plants were confirmed by PCR analysis and propagated to obtain the T3 generation.

Measurements and Statistical Analysis

All seedlings were axenically cultured on medium for light/dark analysis, BR treatment assays and leaf morphology analysis were scanned using an Epson perfection V700 photo scanner (Epson, Nagano, Japan). The images were analyzed using WinRHIZO Pro v.2009c software (Regent Instruments, Montreal, QC, Canada). For light/dark analysis, the hypocotyl lengths of 6-day-old seedlings were measured. Similarly, when the seedlings grown on medium with or without 24-epiBL in the BL treatment assay, the total root lengths of 10-day-old seedlings, hypocotyl length of 6-day-old seedlings and petiole length of 13-day-old seedlings were measured; the number of lateral roots of 10-day-old seedlings was also counted. For leaf morphology analysis, the first true leaves of 13-day-old seedlings were cut off at the bottom of the petioles and flattened on agar plates for scanning. Traits including petiole length, leaf area, length and width of the leaf blade were examined. Silique length and plant height were measured using a millimeter-graduated ruler. All measurements were repeated three times independently, and 30–50 seedlings were measured each time. Data are presented as the mean±SD and were subjected to Student's t test with a sample size of 30 to determine differences among the groups.

RESULTS

Cloning and Sequence Analysis of *GmCPD* genes in *Glycine max*

Four soybean *CPD* homologs (*GmCPDs*) were obtained from the soybean translated NCBI nucleotide database by a BLAST search using the amino acid sequence of Arabidopsis CPD (GenBank accession No. XP_002873219) as a query. These predicted genes were then assigned names based on their correspondence with AtCPD. The four potential homologous proteins, GmCPD1 (GenBank accession No. XP_003545232.1), GmCPD2 (GenBank accession No. XP_003519393.1), GmCPD3 (GenBank accession No. XP_003552845.1) and GmCPD4 (GenBank accession No. XP_003538460.1), are predicted to be between 473 and 480 amino acids in length and all belong to the cytochrome P450 (CYP) family.

The deduced amino acid sequences of the GmCPDs share 82–97% identity with each other and exhibit high similarity to the Arabidopsis CPD protein, with identities between 79% and 81% (Figure 1). GmCPD1 has the highest identity of 81% while GmCPD2 has the lowest. An alignment of GmCPDs with known CPDs from other species reveals identities of 81–87% for MtCPD1 of *Medicago truncatula*, 76–80% for PtCPD of *Populus trichocarpa*, 75–77% for CsCPD of *Cucumis sativus*, and 59–63% for OsCPD1 of monocot *Oryza sativa* (Figure 1). It is suggested that the amino acid sequences of CPDs are highly homologous across all species.

There are generally four structural domains in CPD proteins that exhibit catalytic features. The proline-rich region was thought to ensure the correct folding and proper orientation of the CPD protein. Domain A and domain B are involved in the dioxygen and steroid binding required for catalytic activity. The most characteristic P450 consensus sequence, the heme binding domain, is responsible for carbon monoxide binding ability [45,46]. As shown in Figure 1, all CPDs contain these characteristic domains, and their amino acid sequences are highly conserved. There are only two amino acid differences between AtCPD and GmCPDs in the proline-rich region, one amino acid difference in domain B and at most two amino acid differences in the heme-binding domain. As for domain A, AtCPD and GmCPDs share 100% amino acid sequence identity (Figure 1). Based on these findings,

GmCPDs bear a striking similarity to AtCPD in sequence and structure, a trait that might imply functional similarity.

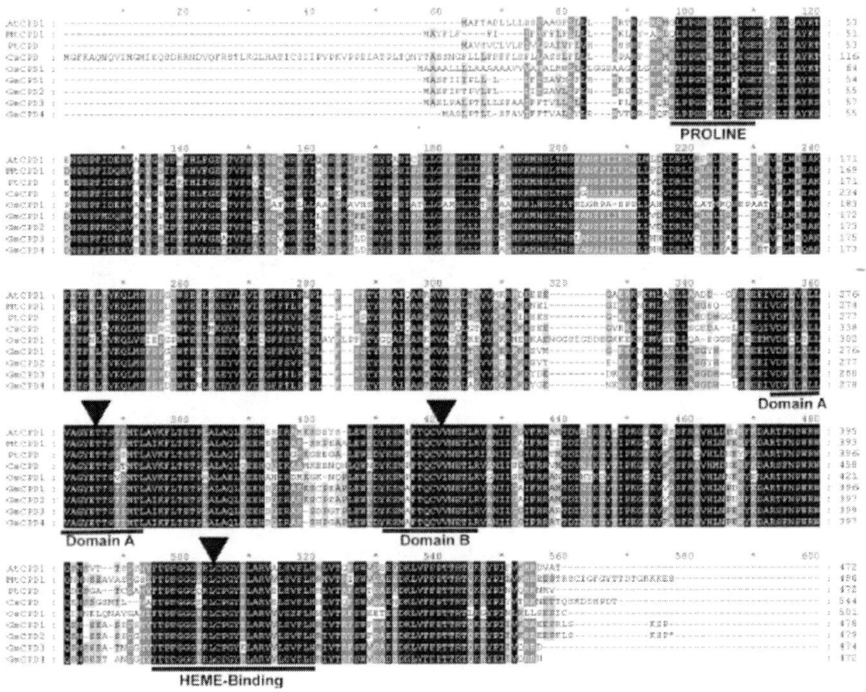

Fig 1. GmCPD amino acid sequence homology with CPDs from other species. Multiple alignments were performed by the Clustal W2 algorithm. Exact matches are boxed in black. Four domains found in cytochrome P450s: Proline-rich, Domain A (dioxygen-binding), Domain B (steroid-binding) and heme-binding are underlined. Arrowheads: Amino acid residues with important function. Accession numbers are as follows: AtCPD (XP_002873219), GmCPD1 (XP_003545232.1), GmCPD2 (XP_003519393.1), GmCPD3 (XP_003552845.1), GmCPD4 (XP_003538460.1), MtCPD1 (XP_003616626), PtCPD (XP_002311214), CsCPD (XP_004149251), OsCPD1 (NP_001066117).

Phylogenetic analysis was performed using the deduced amino acid sequences of GmCPD and a range of putative CPDs from higher plants. The tree is clearly divided into two major clades: one clade corresponds to monocots, while the other clade corresponds to dicots (Figure 2). The four GmCPDs all fall in the latter clade (Figure 2). GmCPD1 and GmCPD2 are clustered together with *Medicago truncatula* and *Cicer arietinum*, while

GmCPD3 and GmCPD4 branch off from the legume sub-clade (Figure 2). The four GmCPDs all cluster relatively closely with AtCPD (Figure 2), indicating that these proteins may have inherited more ancestral characteristics.

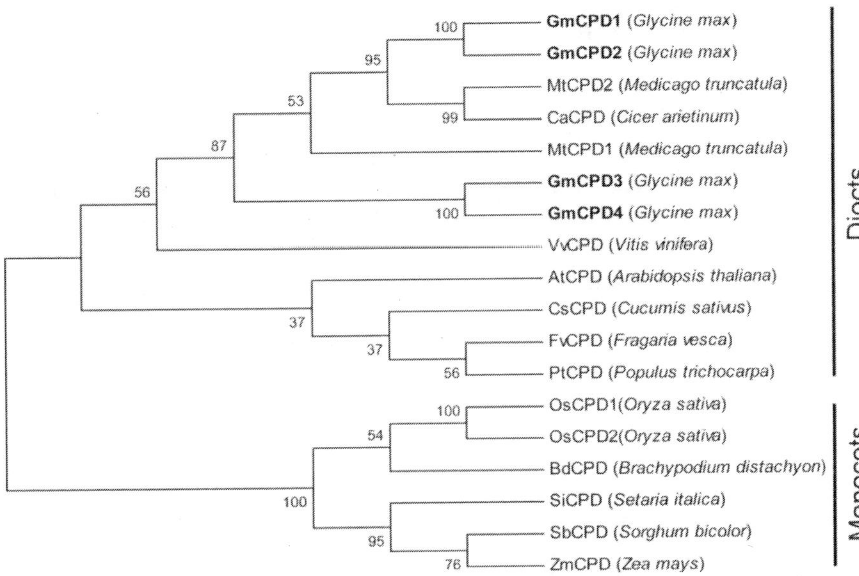

Fig 2. Phylogenetic tree of CPD proteins by neighbor-joining method using MEGA 5.02 software.

Accession numbers are as follows: AtCPD (XP_002873219), GmCPD1 (XP_003545232.1), GmCPD2 (XP_003519393.1), GmCPD3 (XP_003552 845.1), GmCPD4 (XP_003538460.1), MtCPD1(XP_003616626), MtCPD2(XP_ 003600878), CaCPD(XP_004490985), CsCPD (XP_004149 251), FvCPD (XP_004307639), PtCPD (XP_002311214), VvCPD (XP_0022 70553), OsCPD1 (NP_001066117), OsCPD2 (NP_001065721), BdCPD (XP_0035 78946), SiCPD (XP_004978643), SbCPD (XP_002450249), ZmCPD (NP_001140596).

Genomic location of each *GmCPD* was targeted on physical map of soybean (*Glycine max*) genome based on the information on SoyBase (http:// www.soy base.org) and Phytozome database (http:// phytozome. jgi.doe.gov). They are all located in separate chromosome: *GmCPD1* (Glyma.14g059900), *GmCPD2* (Glyma.02g256800), *GmCPD3* (Gly ma.18g028300) and *GmCPD4* (Glyma.11g228900) are located in Gm14 (B2), Gm02 (D1b), Gm18 (G) and Gm11 (B1), respectively (Figure 3). There were not many SSR markers around *GmCPDs*. Around *GmCPD1*, Sat_177 and Sat_264 are associated with the QTLs of flower number; Satt126 is associated with lodging and Sat_287 also relates to seed coat color (Figure 3). As for *GmCPD2*, Satt189, Satt350 and Satt546 are associated with the QTLs of first flower; Satt189 and Satt350 are associated with leaflet shape and leaf area respectively; Satt546 is associated with internode length; Sat_139, Satt546 and Satt172 are associated with the seed quality trait (Figure 3). Satt309, Satt356 and Satt570 locate closely to *GmCPD3*: Satt309 and Satt356 associated with the QTLs of pod maturity; Satt356 is linked with internode length; Satt570 is associated with seed protein, lateral root density and root width (Figure 3). Around the location of *GmCPD4*, Satt415 is associated with the internode length; Satt583 is associated with the length of reproductive stage; Sat_123 is associated with pod maturity and lodging; Sat_123, Satt583 and Sat_095 are all associated with seed weight (Figure 3). Above all, the four *GmCPD* homologous are associated with the QTLs related to main aspects of soybean development.

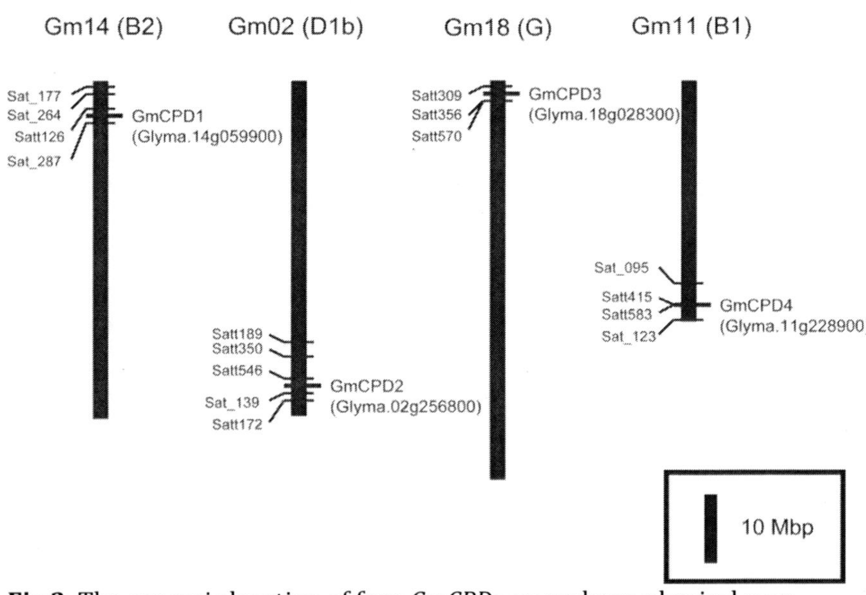

Fig 3. The genomic location of four *GmCPDs* on soybean physical map.

Distance along the vertical bars indicates the physical distance reported in the Soybase and Phytozome database. The SSR markers near to *GmCPDs* location were also labeled on the corresponding positions of chromosome.

Expression Patterns of *GmCPDs* in Soybean

Tissue-specific expression patterns of *GmCPDs* in soybean were systematically determined using RT-qPCR. These four *GmCPDs* are widely expressed in plant tissues but display different patterns. Although *GmCPD1*, *GmCPD2* and *GmCPD4* all showed higher expression levels in cotyledons and leaves, *GmCPD2* and *GmCPD4* had the highest level in cotyledons while *GmCPD1* had the highest level in leaves (Figure 4A). These results are consistent with the expression pattern of *CPD* in Arabidopsis [47]. However, *GmCPD3* is an exception, exhibiting the highest mRNA accumulation in young pods but very low concentrations in other tissues (Figure 4A). In addition, *GmCPD4* as well as *GmCPD1* and *GmCPD2* showed relatively high levels in young pods (Figure 4A). These results are consistent with the important roles proposed for BRs in processes such as fruit development and ripening [48,49].

The BR sensitivity of *GmCPDs* was also tested in soybean. *William 82* adult plants were treated with 24-epiBL, and the leaf samples were collected every half hour. As shown in the RT-qPCR results, the expression levels of the four *GmCPDs* fluctuated, but overall the expression levels tended to decrease. The expression patterns of *GmCPD1* and *GmCPD2* were nearly equivalent but distinct from *GmCPD3* and *GmCPD4* (Figure 4B). Following BR treatment, a sudden increase of *GmCPD1* and *GmCPD2* expression reached a maximum 0.5 h after treatment was initiated. *GmCPD1* and *GmCPD2* expression then sharply decreased, reaching a minimum approximately 1 h after treatment was started and subsequently increasing slightly to a plateau (Figure 4B). In contrast, the expression levels of *GmCPD3* and *GmCPD4* rapidly decreased following BR treatment, reaching a minimum at 1 h and then leveling off (Figure 4B). The above results indicate a highly sensitive response of *GmCPDs* to exogenous BR. Taken together with previous studies that show that *CPD* is feedback-inhibited by BR [13,47,50], our results corroborate the importance of *GmCPDs* in BR biosynthesis.

Complementation of an Arabidopsis CPD-Deficient Mutant phenotype by *GmCPDs* Expression

To test whether the *GmCPDs* can function in BR biosynthesis, the coding sequences of *GmCPD1*, *GmCPD2*, *GmCPD3* and *GmCPD*4 were placed under the control of a 35S constitutive promoter and introduced into a *cpd-91* mutant [13] of Arabidopsis. The goal was to evaluate whether the transgenes complement the mutant phenotype. Adult plants of *cpd-91*, a CYP90A1/CPD-deficient mutant, are small and dwarfed with rounded curled leaves. In contrast, the transgenic *GmCPD1*,

GmCPD2, *GmCPD3* and *GmCPD4 cpd-91* mutant lines were all similar to the wild type in size, showing a rescue of the *cpd-91* mutant adult phenotypes (Figure 5A). The RT-PCR results revealed that *GmCPD* genes can only be detected in the corresponding transgenic plants, indicating complementation by *GmCPDs* overexpression (Figure 5B).

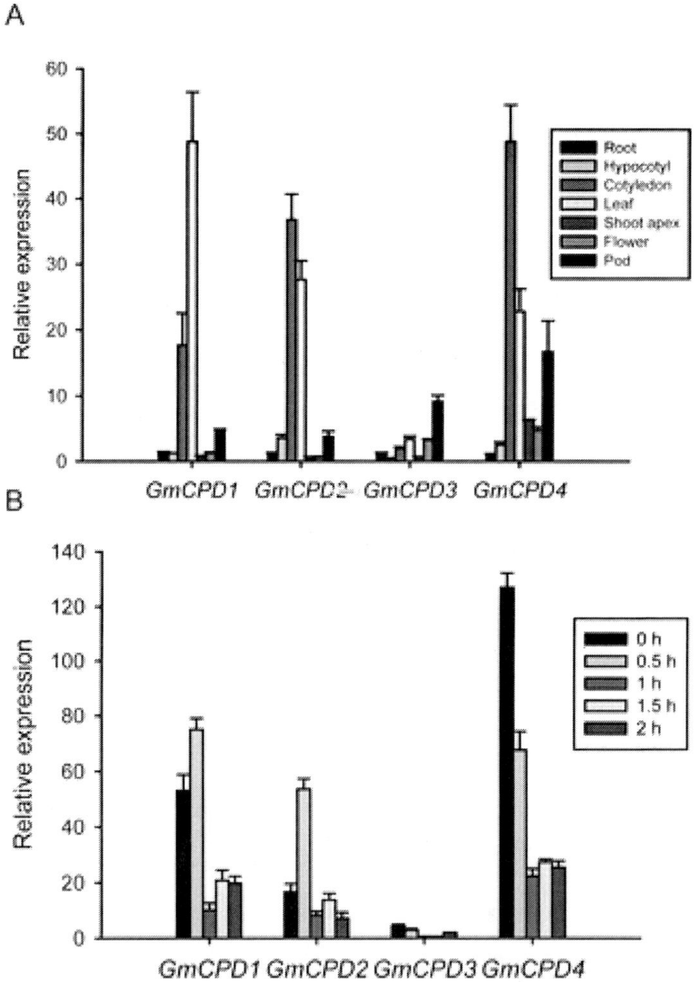

Fig 4. *GmCPDs* expression patterns in soybean. (**A**) Tissue-specific expression patterns of *GmCPDs* in soybean. The sampling time of each tissue is described in section 4 of the Materials and Methods (**B**) Inducible expression of four *GmCPD* genes in the soybean leaves under BR treatment. The relative expression levels are normalized to *GmG6PDH* (GenBank accession No. XM_003547631). The data represent the mean ± SD of three independent experiments.

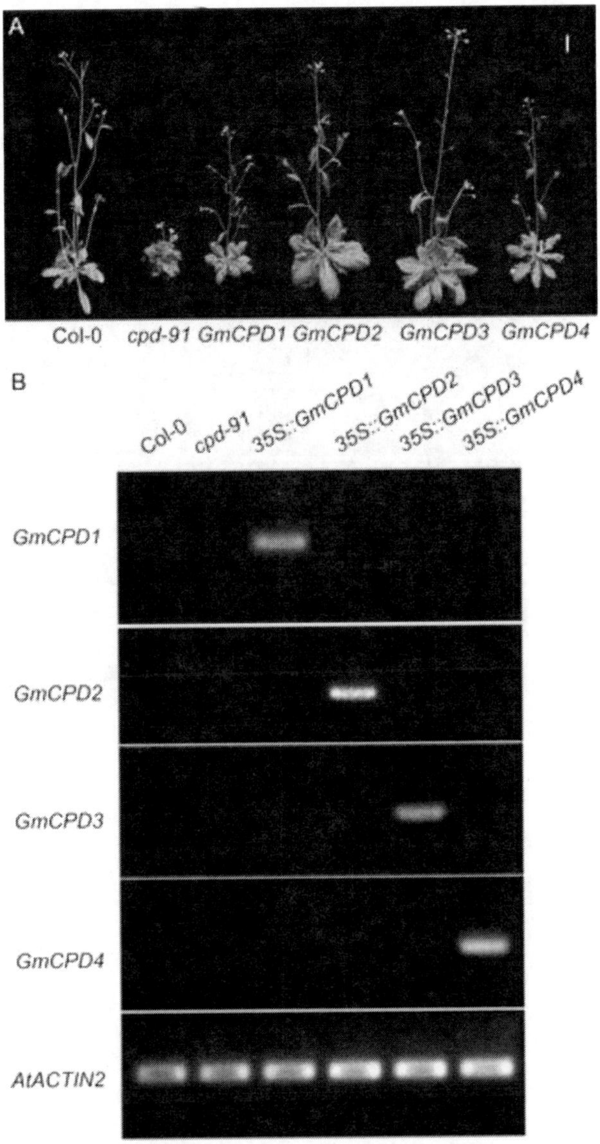

Fig 5. Four *GmCPDs* complement the phenotype of an Arabidopsis CPD-deficient mutant. (**A**) Phenotype comparison of adult plants. Bar = 10 mm; Col-0: wild type plants; *cpd-91*: CPD-deficient mutant; *GmCPD1*: 35S::*GmCPD1*; *GmCPD2*: 35S::*GmCPD2*; *GmCPD3*: 35S::*GmCPD3*; *GmCPD4*: 35S::*GmCPD4*; (**B**) RT-PCR analysis to detect *GmCPD* genes using specific primers described in S1 Table with *AtACTIN2* (GenBank accession No. AT3G18780) as a reference gene.

The leaf phenotypes of the transgenic lines all bear little resemblance to the *cpd-91* mutant, instead resembling the wild type phenotype (Figure 6A). In a quantitative comparison, the mutant retained minimum values of petiole length (Figure 6B), leaf area (Figure 6C) and length-width ratio (Figure 6D). The transgenic *GmCPD1*, *GmCPD2*, *GmCPD3* and *GmCPD4 cpd-91* lines were all similar to the wild type and were significantly different ($P < 0.01$) from the mutant (Figure 6B-D).

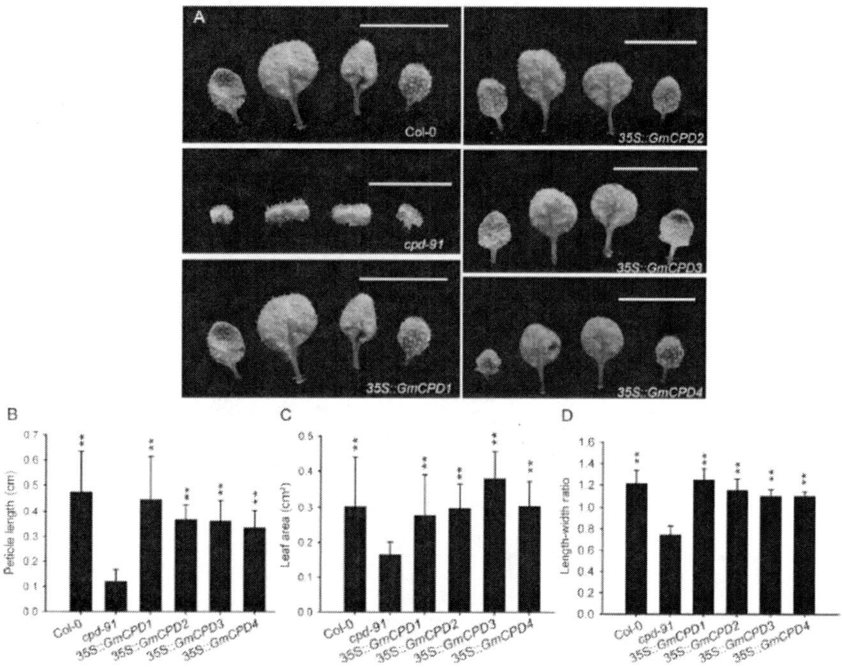

Fig 6. Comparison of leaf phenotype between the wild type (Col-0), CPD-deficient mutant (*cpd-91*) and transgenic plants.

(A) Detailed leaf morphologies of 13-day-old Arabidopsis plants, Bar = 10 mm; **(B-D)** Statistical analysis of leaf measurements: the petiole length (**B**), leaf area (**C**) and length-width ratio (**D**). The data represents the mean ± SD of three independent experiments. The asterisks indicate significant differences compared to the *cpd-91* mutant (**, $P < 0.01$ by the *t*-test).

Without the CYP90A/CPD gene, the morphological change in the roots was quite significant in the *cpd-91* mutant, with a small root length and undeveloped lateral roots (Figure 7A-C). Conversely, all transgenic lines exhibited developed root systems that were similar to the wild type plants

(Figure 7A). Student's t-tests indicated a significant difference ($P < 0.01$) between the transgenic lines and the mutant plants in root length and lateral root number (Figure 7B-C).

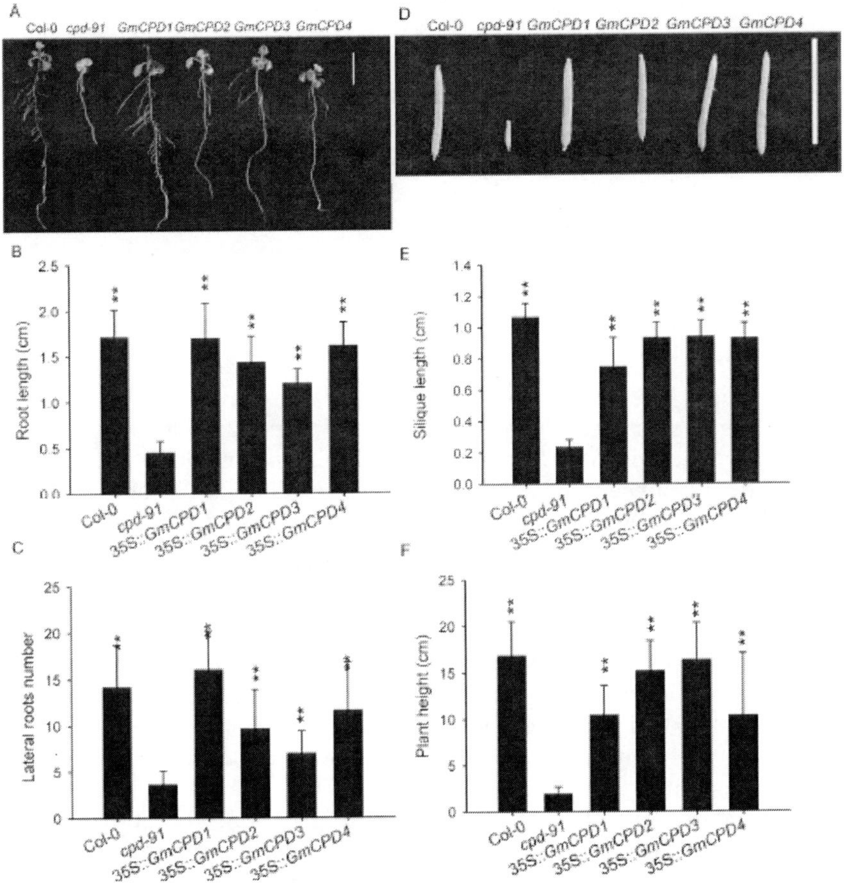

Fig 7. Phenotype comparison of root, silique and plant height between the wild type (Col-0), CPD-deficient mutant (*cpd-91*) and transgenic plants.

(**A**) Root phenotype of 13-day-old Arabidopsis plants. Col-0: wild type plants, *cpd-91*: CPD-deficient mutant, *GmCPD1*: 35S::*GmCPD1*, *GmCPD2*: 35S::*GmCPD2*, *GmCPD3*: 35S::*GmCPD3*, *GmCPD4*: 35S::*GmCPD4*, Bar = 10 mm; (**B and C**) Total root length (**B**) and the lateral root number (**C**) of plants shown in A; (**D**) Silique morphology comparison. Bar = 10 mm; (**E and F**) Measurement and statistical analysis of the silique length (**E**) and plant height (**F**). The data represents the mean ± SD of three independent experiments. The asterisks indicate significant differences compared to the *cpd-91* mutant (**, $P < 0.01$ by the t-test).

The most obvious complementation of the mutant is the rescue of dwarfness. Similar to other BR mutants, *cpd-91* mutant is severely dwarfed. In contrast, the plant height of every transgenic line was remarkably higher ($P < 0.01$) than that of the mutant and very similar to that of the wild type (Figure 7F).

The silique size of mutants is very small, only an average of 0.2 cm long and about twenty percent the length of Col-0 siliques (Figure 7D, E). Each transgenic line was extremely distinct ($P < 0.01$) from the non-transformed mutant in silique size and resembled the wild type (Figure 7E).

In conclusion, all four *GmCPDs* are functional and essential in leaf, root and plant type development.

GmCPD Homologs Restore BR Biosynthesis in Arabidopsis *cpd-91* mutants

To further confirm that the rescue of the *cpd-91* mutant phenotype is due to a restored BR biosynthesis pathway via *GmCPDs* transformation, we tested the BR responses of complemented Arabidopsis compared with untransformed *cpd-91* and wild type Col-0.

6-day-old seedlings grown in light and darkness were screened for hypocotyl elongation during skotomorphogenesis and photomorphogenesis. In the dark, *cpd-91* mutant seedlings underwent constitutive photomorphogenesis, exhibiting short hypocotyls and open cotyledons (Figure 8A, C). In contrast, the transgenic lines and wild type exhibited longer hypocotyls and closed apical hooks (Figure 8A, C). When grown in the light, *cpd-91* mutant seedlings exhibited shorter hypocotyls than the wild type (Figure 8B, D). This mutant phenotype was complemented by all four transgenes (Figure 8B, D). Student's *t*-tests indicate a significant difference ($P < 0.01$) in hypocotyl length between the transgenic lines and the mutant plants in both the light and darkness (Figure 8C, D).

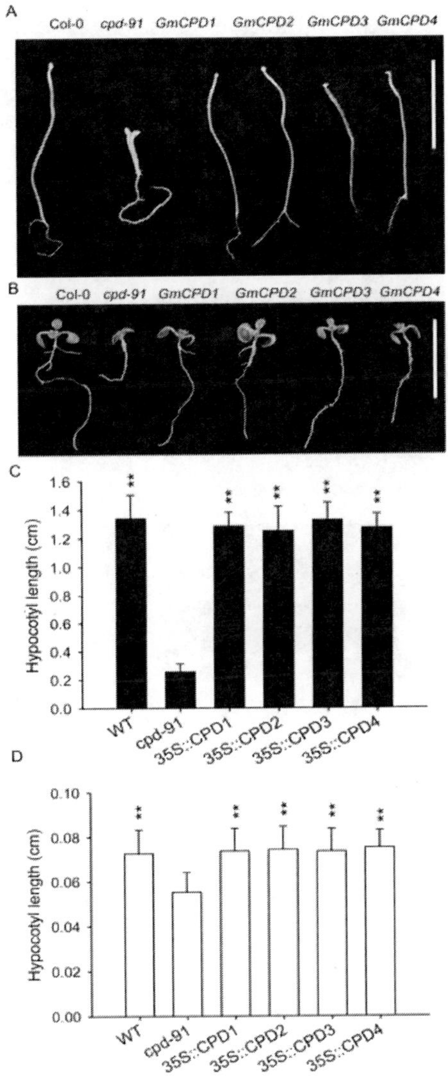

Fig 8. *GmCPDs* restore hypocotyl elongation of CPD-deficient Arabidopsis mutants in both light and darkness. (**A and B**) Morphologies of the five-day-old seedlings grown in darkness (**A**) and light (**B**). Col-0: wild type plants, *cpd-91*: CPD-deficient mutant, *GmCPD1*: 35S::*GmCPD1*, *GmCPD2*: 35S::*GmCPD2*, *GmCPD3*: 35S::*GmCPD3*, *GmCPD4*: 35S::*GmCPD4*, Bar = 10 mm; (**C and D**) Average hypocotyl length of seedlings in darkness (**C**) and light (**D**). The data represents the mean ± SD of three independent experiments. The asterisks indicate significant differences compared to the *cpd-91* mutant (**, $P < 0.01$ by the *t*-test).

In the root growth inhibition assay, 10-day-old complemented Arabidopsis, *cpd-91* mutants and the wild type were grown on 1/2 MS medium containing 100 nM 2,4-epibrassinolide (24-epiBL). All seedlings showed shortened roots in response to 24-epiBL but behaved differently in root shortening (Figure 9A-D). The transgenic lines and the wild type displayed greater shortening than *cpd-91* plants (Figure 9D), indicating a stronger response to BR.

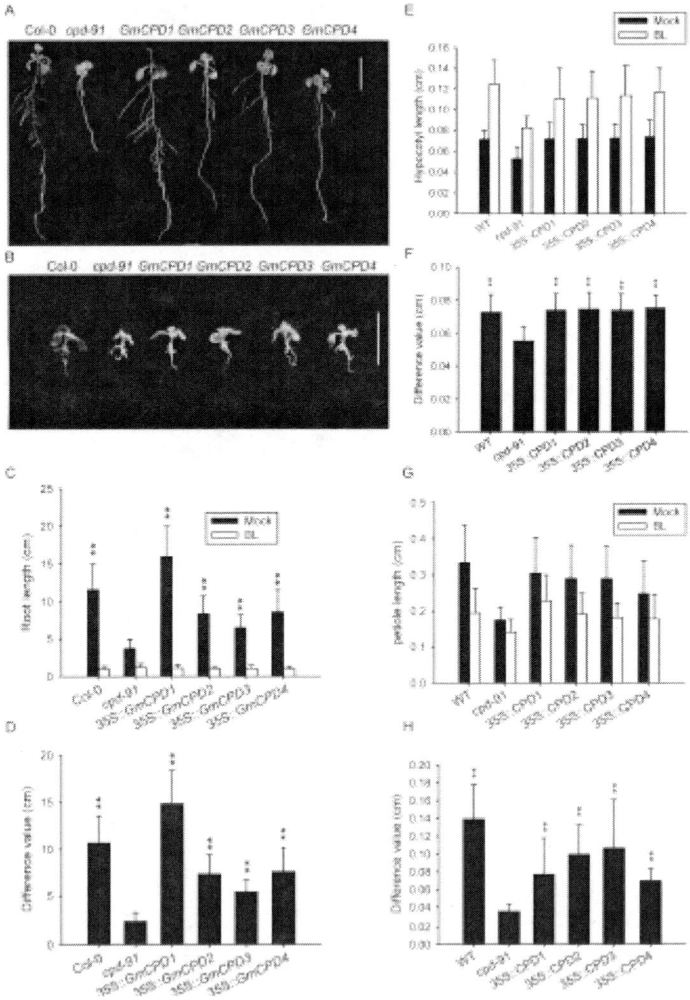

Fig 9. The variation in the length of root, hypocotyl and petiole showing different brassinosteroid (BR) responses in the wild type (Col-0), CPD-deficient mutant (*cpd-91*) and transgenic Arabidopsis plants.

(**A and B**) Phenotypes of the wild type, CPD-deficient mutant and transgenic plants in half-strength MS medium supplemented with (**B**) or without 100 nM 24-epiBL (**A**). Col-0: wild type plants, *cpd-91*: CPD-deficient mutant, *GmCPD1*: 35S::*GmCPD1*, *GmCPD2*: 35S::*GmCPD2*, *GmCPD3*: 35S::*GmCPD3*, *GmCPD4*: 35S::*GmCPD4*. Bar = 10 mm; (**C**) Measurements of the total root length of the seedlings in the root inhibition assay; (**D**) Root length shortening after BR treatment; (**E**) Hypocotyl length of 6-day-old seedling in medium with or without 24-epiBL; (**F**) The elongation of hypocotyl after BR treatment; (**G**) Petiole length of 13-day-old seedling with or without BR treatment; (**H**) Petiole length shortening after BR treatment. The data represents the mean ± SD of three independent experiments. The asterisks indicate significant differences compared to the *cpd-91* mutant (**, $P < 0.01$ by the *t*-test).

Additionally, under the treatment of 100 nM 24-epiBL, all the seedlings exhibited elongated hypocotyl and shortened petiole (Figure 9E-H). Compared to transgenic lines and Col-0 Arabidopsis, *cpd-91* mutant showed the shortest length of hypocotyl and petiole both in BL treatment and normal conditions (Figure 9E, G). The transgenic lines resembled the wild type and displayed greater hypocotyl elongation and petiole shortening than *cpd-91* mutant (Figure 9F, H).

Therefore, physiological response phenotypes of mutant plants are complemented by *GmCPDs* expression, suggesting a restored BR biosynthesis pathway in transgenic lines. This result further demonstrates that the *CPD* homologous genes in soybean, *GmCPD1*, *GmCPD2*, *GmCPD3* and *GmCPD4*, are functional in the BR pathway.

GmCPDs Are Involved in Floral Regulation of Arabidopsis

The above results show the phenotypic rescue of Arabidopsis CPD-deficient mutant by *GmCPDs* expression. In addition to rescuing morphology, overexpression of *GmCPDs* also complemented the delayed flowering of the *cpd-91* mutant. The observation that the *cpd-91* mutant flowered approximately 10 days later than the Col-0 wild type is in agreement with previous observations (Figure 10A, B). The transgenic plants transformed with any of the four *GmCPD* homologs all bloomed simultaneously with the wild type plant, much earlier than the *cpd-91* plants (Figure 10A, B).

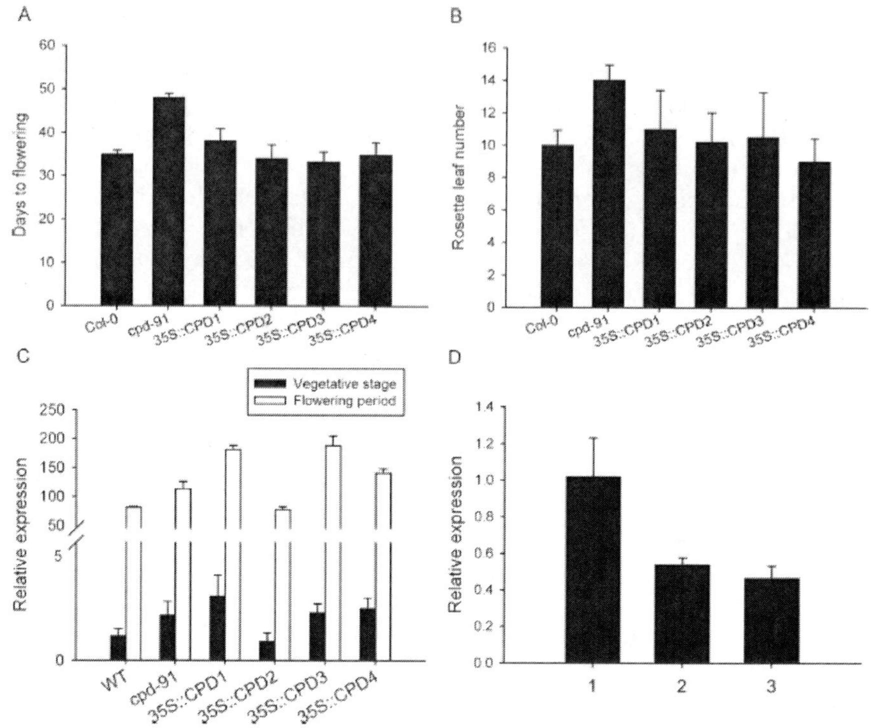

Fig 10. Complementation of late-flowering phenotype, expression pattern during flowering and effects on flowering related gene were indicated the roles of *CPD* in Arabidopsis flowering.

(**A and B**) Flowering time analysis among wild type (Col-0), CPD-deficient mutant (*cpd-91*) and transgenic Arabidopsis plants. Comparison of the days to flowering (**A**) and the rosette leaf number (**B**) at anthesis; (**C**) The expression pattern of Arabidaopsis *FLOWERING LOCUS T* (*AtFT*) in vegetative stage and flowering period; (**D**) The expression pattern of *AtCPD* during flowering. 1, 2 and 3 represent three developmental stages. 1: vegetative stage, two weeks after emergence; 2: flowering initiation, the time that inflorescence-bud just emerged; 3: flowering period, one week after flowering.

In order to investigate the roles of *CPD* in flowering regulation, the expression pattern of flowering integrating gene, *Flowering Locus T* (*FT*), was examined in the transgenic plants compared with non-transformed *cpd-91* and Col-0 wild-types (Figure 10C). *AtFT* acts as floral integrator of all four flowering pathways [51]. The *AtFT* product, which can move in long distance through the phloem to initiate flowering at the shoot apex, is a main determinant of the timing of flowering [52]. In our results, all the groups exhibited similar expression pattern that *AtFT* gene maintained at a very low level in the vegetative stage and expressed highly when flowered (Figure 10C). Except for that the *35S::GmCPD2* transgenic line showed similar level of *AtFT* expression to the wild type, the other three transgenic lines expressed diversely but all higher than *cpd-91* mutants (Figure 10C).

In addition, the expression pattern of *AtCPD* gene during flowering was also examined in Arabidopsis leaves that were collected in three developmental stages: vegetative growth (two-week-old), flowering initiation and flowering period (one week after beginning flowering). It is showed that the *AtCPD* transcripts were more abundant in vegetative stage, but decreased during flowering (Figure 10D).

The Potential Roles of *GmCPDs* in Soybean Flowering Regulation

To further study the roles of *GmCPDs* in flowering, *GmCPDs* transcript levels were tested in soybean, a typical short-day plant that can undergo flowering reversion. In a previous study by our lab, 13 days of SD treatment before transfer to an LD condition are enough for flowering reversion to occur in soybean var. *Zigongdongdou*. Based on this observation, an effective flowering reversion system was established. In this system, three developmental states, flowering, continuous vegetative growth and flowering reversion, can be observed in *Zigongdongdou* plants under different photoperiods (SD, LD, 13SD-LD). Genes related to photoperiodism and flower development are preferentially studied in this system. Accordingly, leaf samples were collected in each photoperiod, and the relative expression levels of *GmCPDs* were analyzed to investigate the potential roles of *GmCPDs* in flowering.

As shown in the results, all *GmCPDs* have the same expression pattern: expression was maintained at a much lower level in the LD condition (Figure 11A-D). Conversely, when treated in SD, *GmCPDs* expression levels were gradually elevated at first. Once SD treatment reached the 13th day, *GmCPDs* levels sharply increased to a maximum, then decreased suddenly under both the SD and LD conditions (Figure 11A-D). Obviously, these results suggest that *GmCPDs* expression is under photoperiod control and is upregulated by SD, a day length that induces flowering. Interestingly, the expression quantity of *GmCPDs* on the 13th day is around tenfold that of the 9th day and from nineteen to forty-five times that of the 19th day (Figure 11A-D). The peak on the 13th day is so sharp that we cannot help but wonder what happens on this day. It was found in a previous study that the apical meristem of *Zigongdongdou* begins to initiate floral primordia on the 13th day of SD treatment [39]. The above results suggest a certain relationship between *GmCPDs* and floral initiation through the photoperiod pathway.

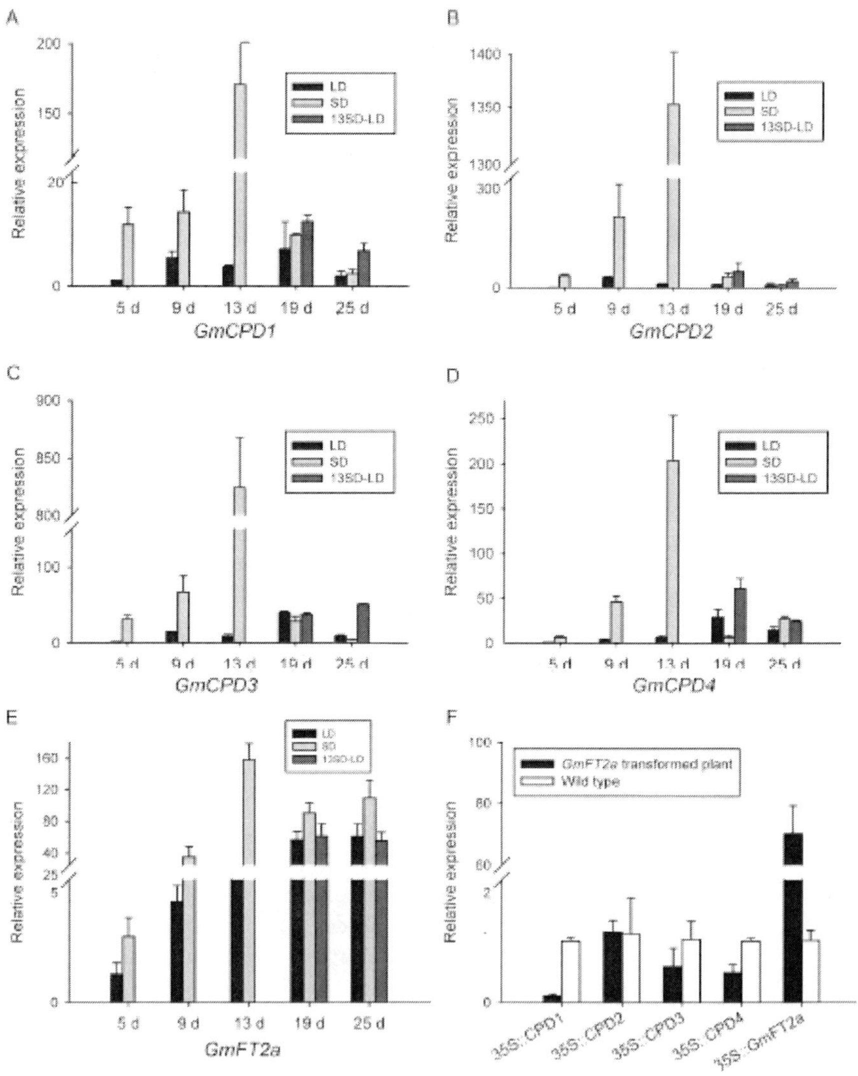

Fig 11. The expression pattern of *GmCPD* genes in the soybean flowering reversion system and *GmFT2a* transgenic soybean.

LD: long day treatment; SD: short day treatment; 13SD-LD: transfer to long day conditions after 13 days of short day treatment; leaf samples of *Zigongdongdou* plants under each photoperiod were collected and qRT-PCR was used to analyze the expression of *GmCPD1* (**A**), *GmCPD2* (**B**), *GmCPD3* (**C**), *GmCPD4* (**D**) and *GmFT2a* (**E**); (**F**) The expression of *GmCPDs* in *GmFT2a* transgenic soybean plants compare to non-transformed *Zigongdongdou*. The *x*-axis represents the days of treatment. Relative expression levels are normalized to *GmG6PDH* (GenBank accession No. XM_003547631). The data represent the mean ± SD of three independent experiments.

When plants were grown in LD after SD induction, the expression of *GmCPDs* all decreased but still higher than that of plants grown in either continuous LD or SD (Figure 11A-D). This result was not consistent with the expression pattern of *GmFT2a*, an integrator in photoperiod pathway. The expression of *GmFT2a* maintained in a rather low level in either LD or vegetative stage and raised around the 13th day in SD when flowering initiated (Figure 11E). When returned to the LD condition, *GmFT2a* expression decreased to the same level of that in the continuous LD treatment (Figure 11E). Unlike *GmFT2a*, the expression of *GmCPDs* had additive effect that the SD effects could be accumulated when turned into LD condition, suggesting the distinct roles of *GmCPDs* in flowering regulation.

Since there was no obvious effect on the pattern of *AtFT* expression in the absence of CPD (Figure 10C) and expression patterns between *GmFT2a* and *GmCPDs* in flowering reversion were different (Figure 11A-E), it might imply that no direct interaction between *GmFT2a* and *GmCPDs*. To text this, the expression of *GmCPDs* expression was examined in *GmFT2a* transgenic soybean and compared with the non-transformed *Zigongdongdou* (Figure 11F). The published data by our lab [53] have showed that one line of *GmFT2a* transgenic *Zigongdongdou* flowered approximately 20 days after emergence under non-inductive LD conditions. The expression level of *GmCPDs* in this line was found to be maintained in a quite low level and even decreased compared to the wild type on the occasion that *GmFT2a* expressed extremely high (Figure 11F). Therefore, the involvement of *GmCPDs* in flowering regulation may not be linked to the direct interaction with *GmFT2a*.

GmCPDs Expression in Soybean Varieties with Different Photoperiod Sensitivities

Soybean varieties are diverse in photoperiod sensitivity. *Zigongdongdou* is a photoperiod-sensitive late-flowering variety that only flowers under the SD condition. In contrast, the photoperiod-insensitive early-flowering variety *Heihe27* blooms approximately 25–27 days after emergence under both LD and SD conditions [54]. As it is shown in Figure 12A, at the 36th days after emergence, *Heihe 27* had already set pods while *Zigongdongdou* still underwent vegetative growth under the LD condition. These two typical varieties were chosen to evaluate the expression pattern of *GmCPDs* in soybean varieties with different photoperiod sensitivities.

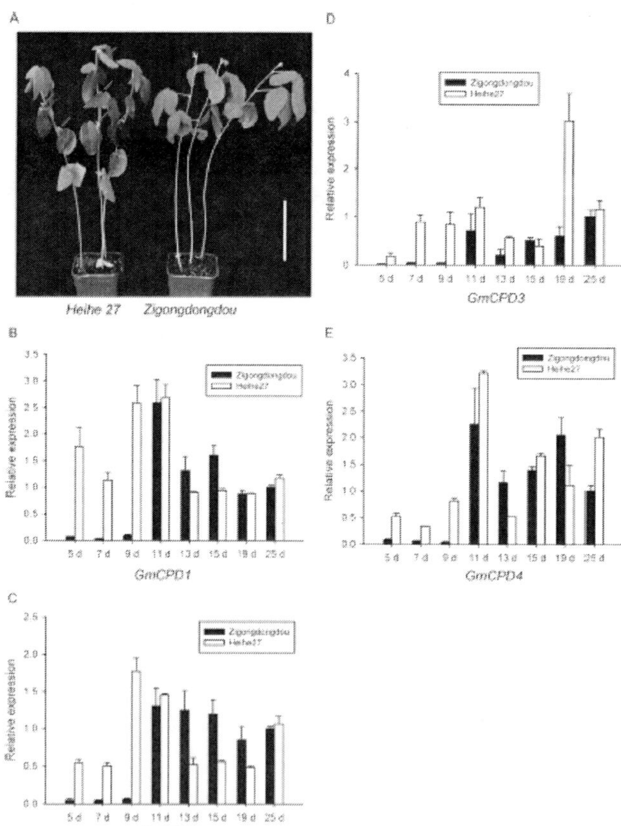

Fig 12. Expression patterns of *GmCPDs* in soybean varieties of different photoperiod sensitivities (*Zigongdongdou* and *Heihe27*) during LD treatment.

(**A**) The phenotype of early and late flowering soybean plants. *Heihe 27* on the left is a photoperiod-insensitive early-maturity variety; *Zigongdongdou* on the right is a photoperiod-sensitive late-maturity variety. The photo was taken at the 36th day after emergence under long day condition. *Heihe 27* had already produced one pod with length more than 2 cm and some younger pods (labeled by red arrows), while *Zigongdongdou* remained vegetative growth. Bar = 10 cm; Leaf samples from long day grown soybean plants were taken at the 5th, 7th, 9th, 11th, 13th, 15th, 19th and 25th day after cotyledon opening and labeled as 5 d, 7 d, 9 d, 11 d, 13 d, 15 d, 19 d and 25 d; qRT-PCR was carried out to determine the relative expression levels of *GmCPD1* (**B**), *GmCPD2* (**C**), *GmCPD3* (**D**) and *GmCPD4* (**E**) in each sample. Relative expression levels are normalized to *GmG6PDH* (GenBank accession No. XM_003547631); the data represent the mean ± SD of three independent experiments.

We screened leaf samples from *Zigongdongdou* and *Heihe27* after various days of LD treatment. In the 5th, 7th and 9th day after LD treatment (5 d, 7 d and 9 d), all *GmCPD* genes were expressed at very low levels in *Zigongdongdou* but at extremely high levels in *Heihe27* (Figure 12). After the 11th day, the transcript levels of *GmCPDs* were remarkably upregulated in *Zigongdongdou* but slightly decreased and maintained in *Heihe27* (Figure 12). The expression patterns of *GmCPD1*, *GmCPD2* and *GmCPD4* were nearly the same; the expression levels of these *GmCPDs* were obviously higher in *Heihe27* than *Zigongdongdou* from day 5 to day 11 d. From day 13 to day 19, the expression levels in *Zigongdongdou* were increased and higher than *Heihe27*, in which the levels were downregulated. At day 25, *Heihe27* had higher expression levels compared to *Zigongdongdou* (Figure 12A, B and D). As for the *GmCPD3* gene, the expression levels in *Heihe27* were always higher than *Zigongdongdou* except for day 15. *GmCPD3* was most highly expressed in *Heihe27* at day 19 (Figure 12C). However, the expressions of *GmCPDs* in *Zigongdongdou* under the LD condition in this experiment (Figure 12) have differences with the results shown in Figure 11A-D. This may due to the different culture temperature (described in section of Material and Methods) and sampling time. The leaf samples in this experiment were collected in the morning, while the samples in Figure 11 were collected in the afternoon. Since genes usually have different expression levels during the day, the results in the two experiments are not comparable. We only analyzed the expression differences of *GmCPDs* among *Zigongdongdou* and *Heihe 27* in this experiment that carried out in the same condition and sampled at the same time every day.

Leaves and cotyledons are the two main tissues in which *GmCPDs* are expressed (Figure 4A). Therefore, cotyledons were also collected from *Zigongdongdou* and *Heihe27* on the 3rd, 6th and 9th days after LD treatment. *GmCPD1, GmCPD2* and *GmCPD4* had similar expression patterns: their expression levels tended to decreased with time in *Zigongdongdou* but increased in *Heihe27*. Although levels in *Zigongdongdou* were higher on the 3rd day compared to *Heihe27*, the levels were much lower on the 6th and 9th days (Figure 13A, B and D). The expression pattern of *GmCPD3* was rather special: the expression levels of *GmCPD3* in both *Zigongdongdou* and *Heihe27* decreased each day, but the gene was still expressed more highly in *Heihe27* compared to *Zigongdongdou* (Figure 13C).

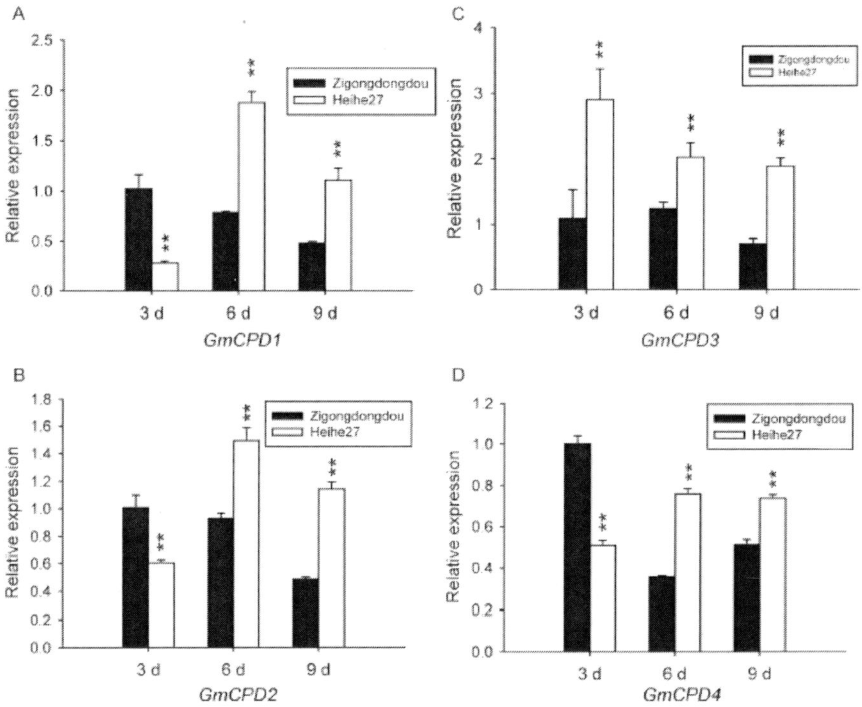

Fig 13. Comparison of *GmCPDs* expression in the cotyledons of *Zigongdongdou* and *Heihe27*, two soybean varieties with distinct photoperiod sensitivities.

Relative expression levels of *GmCPD1* (**A**), *GmCPD2* (**B**), *GmCPD3* (**C**) and *GmCPD4* (**D**) were analyzed by qRT-PCR on the 3rd day (3 d), 6th day (6 d) and 9th day (9 d) after cotyledon opening under the LD condition. Relative expression levels are normalized to *GmG6PDH* (GenBank accession No. XM_003547631). The data represent the mean ± SD of three independent experiments, and asterisks indicate significant differences compared to the *Zigongdongdou* plants (**, $P < 0.01$ by the *t*-test).

Regardless of whether leaves or cotyledons were measured, all *GmCPDs* exhibited more vigorous expression (much higher level) in *Heihe27* compared to *Zigongdongdou*, especially in the early days of LD treatment. The flowering of *Heihe27* is less regulated by photoperiod and can be initiated by the LD non-inducible day length. Compared to the strict short day flowering variety *Zigongdongdou*, the expression of *GmCPDs* is increased in *Heihe27*. In our opinion, differences in florescence between varieties is not only determined by the flowering regulation pathway but also by how each variety has been prepared for flowering; such preparation is affected by many factors, which may include *GmCPDs*.

DISCUSSION

BR Intermediate Products Catalyzed by GmCPD Move Long Distances

In the tissue-specific expression assay, the observation that *GmCPDs* are expressed mainly in cotyledons and leaves (Figure 4A) is consistent with the expression pattern of *CPD* in Arabidopsis [47] but does not coincide with the distribution pattern of bioactive BR [55,56]. The bioactive BR levels in vegetative tissues are much lower [30,56,57], with the highest levels generally occurring in reproductive organs [48,49,56,58], where BR can easily perform its intended function due to its lack of transport [30,55]. It has also been reported that the transcript levels of most BR biosynthesis genes are generally higher in tissues with high BR levels [48,49,57,59]; *CPD* obviously is an exception. Because CYP90A/CPD encoded by the *CPD* gene catalyzes an early step of BR synthesis [19], long-distance movements are required for BR intermediate products to finish synthesis where untransported bioactive BR are accumulated. With this assumption, the paradox that *GmCPDs* transcript levels are not higher in tissues with high BR levels is readily explained.

However, long-distance transformation is so costly that we wonder if the higher expression of *GmCPDs* in vegetative tissues holds further meanings. One possibility is because *CPD* is under light-dependent diurnal regulation primarily mediated by phytochrome signaling [23], leaves and cotyledons, where phytochrome collects, are preferred. In addition, recent work has revealed that BR plays a controlling role in the assembly and function of the photosynthetic apparatus. Moreover, severe thermal instability of oxygen yields has been observed in *cpd* mutants [60], suggesting the potential role of *CPD* in photosynthesis. All of these intriguing hypotheses are worthy of further investigation.

Universality and Characteristics of GmCPDs Compared with AtCPD

The homologous sequences of *CPD* in soybean have not been isolated until the current study. Strong similarities were found between *GmCPDs* and *AtCPD* in many aspects. First, GmCPDs and AtCPD bear high identities in amino acid sequence and structure. Second, *GmCPD1*, *GmCPD2* and *GmCPD4* were most highly expressed in leaves and cotyledons, consistent with the *AtCPD* expression pattern. Most importantly, transformation of *GmCPD* genes into an Arabidopsis CPD-deficient mutant restored the BR biosynthesis pathway and complemented the mutant phenotype with respect to root development, leaf expansion, plant type architecture and flowering regulation, suggesting functional similarity between *GmCPDs* and *AtCPD*.

In addition, *GmCPDs* exhibit some special characteristics in soybean. One is that *GmCPD3* only expresses highly in the young pods of soybean plants. The other is the potential role of *GmCPDs* in soybean flowering regulation. We scanned the entire developmental stage of soybean in a flowering reversion system and found that *GmCPDs* were under photoperiod control. The highest *GmCPD* transcript levels were observed on the 13th day under SD treatment, when the floral meristem initiated. Additionally, *GmCPDs* expressed distinctly in soybean varieties with different photoperiod sensitivities, with insensitive varieties exhibiting higher expression levels especially in the early stages of development. The late flowering phenotype of the *cpd* mutant indicated an essential role of *CPD* in flowering regulation, but the expression patterns of *GmCPDs* in soybean suggested a contributing role of *GmCPDs* in the early stages of flowering development.

Furthermore, all the four *GmCPDs* may perform individual roles and cooperate to regulate flowering. The genomic locations of *GmCPD1* and *GmCPD2* were associated to the QTLs of flower number and the time of the first flower (Figure 3). Taken into account that *GmCPD2* with the lowest identity of *AtCPD* was not influenced in the transcription level by highly expressed *GmFT2a* while other *GmCPD* homologs decreased in expression (Figure 11F), *GmCPD1* with the highest identity to *AtCPD* is more likely to play the major role in flowering regulation. Additionally, analysis of SSR markers around *GmCPD3* and *GmCPD4* suggested their association with QTLs of pod maturity and seed quality traits (Figure 3). This result, taken together with *GmCPD3* specifically expressing in young pods (Figure 4A), was rather indicated that *GmCPD3* and *GmCPD4* may involve in post-flowering development and fruit ripping. Considering their behavior in flowering regulation, *GmCPD3* and *GmCPD4* are possible to contribute in the whole reproductive stage. Especially *GmCPD4*, bearing similar pattern with *GmCPD1* and *GmCPD2* in flowering regulation, may be the most versatile among this *GmCPD* genes.

GmCPDs Act as Participants in Flowering Regulation

Our study confirmed previous observations that *cpd* mutants exhibit a prolonged vegetative phase and delayed flowering (Figure 10A, B) [12,34]. This phenotype can be rescued by overexpression of any of the *GmCPDs* we isolated (Figure 10 A, B). It is therefore clear that *GmCPDs* are associated with flowering. *CPD* has been reported to interact with genes involved in the circadian clock [23,35], the upstream of *FT* in photoperiod pathway. However, in the analysis of *AtFT* expression in wild type, *cpd-91* mutant and mutant with *GmCPDs* transformation, no obvious difference in expression pattern was found (Figure 10C). In Col-0 Arabidopsis, the expression level of *AtCPD* was higher in vegetative stages and decreased after flowering (Figure 10D). Therefore, *GmCPDs* may participate in flowering induction. Considering that there was no evidence of changes in flowering time when exogenous BR was applied, thus, *GmCPD* is not the trigger of flowering, acting as a participant rather than a decider.

This hypothesis was illustrated by our analysis of *GmCPD* expression patterns in a flowering reversion system (Figure 11). The striking observations were that expression of *GmCPDs* is under photoperiod control and is upregulated sharply on the 13th day of SD treatment. The 13thday of SD treatment (13SD) is rather special. In a previous study by our lab, Xiaomei Li *et al* investigated the morphological and anatomical changes that occur during flowering reversion of *Zigongdongdou* [39]. At day 13 under SD condition, the apical meristem began to initiate floral primordia inside the newly formed bracts. Before day 13, the apical meristem retained its vegetative status, and the floral primordia only appeared in the axils of newly formed trifoliolates. The same result also shown by Cunxiang Wu *et al* (Figure 9) and Hongbo Sun *et al* (Figure 6) [40,53]; although the lateral floral meristems appeared at SD7, inflorescence differentiation was initiated at the shoot apices at SD13, indicated by the formation of floral meristems and primordia. Logically, these results highlight the potential role of *GmCPDs* in the floral transition of apical meristem.

One possible explanation of the delayed flowering in *cpd* mutants is that floral meristem formation is retarded in the absence of the *CPD* gene, resulting in prolonged flower development manifested as a flowering time delay. How *CPD* participates in floral meristem initiation has not been reported up to now, but the highest level of endogenous BRs and the highest expression of the BR-biosynthesis genes, *DWF4*, *BR6ox1* and *BR6ox2*, have been observed in the apical shoots of Arabidopsis [57]. The effects of BR in cell elongation and cell wall modification is reported to be of vital importance for shoot apical meristem (SAM) function and inflorescence architecture in rice[61]. Further study on the relationship between *CPD* and the shoot apex meristem switch is needed. The new roles of *CPD* in plant development await uncovering.

REFERENCES

1. Michaels SD (2009) Flowering time regulation produces much fruit. Current opinion in plant biology 12: 75–80. pmid:18938104
2. Achard P, Baghour M, Chapple A, Hedden P, Van Der Straeten D, et al. (2007) The plant stress hormone ethylene controls floral transition via DELLA-dependent regulation of floral meristem-identity genes. Proceedings of the National Academy of Sciences 104: 6484–6489. pmid:17389366
3. Lee JH, Yoo SJ, Park SH, Hwang I, Lee JS, et al. (2007) Role of SVP in the control of flowering time by ambient temperature in Arabidopsis. Genes & development 21: 397–402.
4. Wollenberg AC, Strasser B, Cerdán PD, Amasino RM (2008) Acceleration of flowering during shade avoidance in Arabidopsis alters the balance between FLOWERING LOCUS C-mediated repression and photoperiodic induction of flowering. Plant physiology 148: 1681–1694. pmid:18790998
5. Kotchoni SO, Larrimore KE, Mukherjee M, Kempinski CF, Barth C (2009) Alterations in the endogenous ascorbic acid content affect flowering time in Arabidopsis. Plant physiology 149: 803–815. pmid:19028878
6. Clouse SD, Sasse JM (1998) BRASSINOSTEROIDS: Essential Regulators of Plant Growth and Development. Annu Rev Plant physiology Plant Mol Biol 49: 427–451. pmid:15012241
7. Alabadi D, Blazquez MA, Carbonell J, Ferrandiz C, Perez-Amador MA (2009) Instructive roles for hormones in plant development. Int J Dev Biol 53: 1597–1608. pmid:19247940
8. Yang CJ, Zhang C, Lu YN, Jin JQ, Wang XL (2011) The mechanisms of brassinosteroids' action: from signal transduction to plant development. Mol Plant 4: 588–600. pmid:21471332
9. Chory J, Nagpal P, Peto CA (1991) Phenotypic and Genetic Analysis of det2, a New Mutant That Affects Light-Regulated Seedling Development in Arabidopsis. Plant Cell 3: 445–459. pmid:12324600
10. Azpiroz R, Wu Y, LoCascio JC, Feldmann KA (1998) An Arabidopsis brassinosteroid-dependent mutant is blocked in cell elongation. The Plant Cell 10: 219–230. pmid:9490745
11. Li J, Chory J (1997) A putative leucine-rich repeat receptor kinase involved in brassinosteroid signal transduction. Cell 90: 929–938. pmid:9298904
12. Domagalska MA, Schomburg FM, Amasino RM, Vierstra RD, Nagy F, et al. (2007) Attenuation of brassinosteroid signaling enhances FLC expression and delays flowering. Development 134: 2841–2850. pmid:17611230
13. Szekeres M, Nemeth K, KonczKalman Z, Mathur J, Kauschmann A, et al. (1996) Brassinosteroids rescue the deficiency of CYP90, a cytochrome P450, controlling cell elongation and de-etiolation in arabidopsis. Cell 85: 171–182. pmid:8612270
14. Xu Y, Zhang X, Li Q, Cheng Z, Lou H, et al. (2014) BdBRD1, a brassinosteroid C-6 oxidase homolog in Brachypodium distachyon L., is required for multiple organ development. Plant Physiology and Biochemistry.
15. Higgins JA, Bailey PC, Laurie DA (2010) Comparative genomics of flowering time pathways using Brachypodium distachyon as a model for the temperate grasses. PLoS ONE 5: e10065. pmid:20419097

16. Zhang Y, Li B, Xu Y, Li H, Li S, et al. (2013) The cyclophilin CYP20-2 modulates the conformation of BRASSINAZOLE-RESISTANT1, which binds the promoter of FLOWERING LOCUS D to regulate flowering in Arabidopsis. The Plant Cell 25: 2504–2521. pmid:23897924
17. Kauschmann A, Jessop A, Koncz C, Szekeres M, Willmitzer L, et al. (1996) Genetic evidence for an essential role of brassinosteroids in plant development. Plant Journal 9: 701–713.
18. Ohnishi T, Szatmari A-M, Watanabe B, Fujita S, Bancos S, et al. (2006) C-23 hydroxylation by Arabidopsis CYP90C1 and CYP90D1 reveals a novel shortcut in brassinosteroid biosynthesis. The Plant Cell 18: 3275–3288. pmid:17138693
19. Ohnishi T, Godza B, Watanabe B, Fujioka S, Hategan L, et al. (2012) CYP90A1/CPD, a brassinosteroid biosynthetic cytochrome P450 of Arabidopsis, catalyzes C-3 oxidation. J Biol Chem 287: 31551–31560. pmid:22822057
20. Divi UK, Krishna P (2009) Brassinosteroid: a biotechnological target for enhancing crop yield and stress tolerance. N Biotechnol 26: 131–136. pmid:19631770
21. Li J, Li Y, Chen S, An L (2010) Involvement of brassinosteroid signals in the floral-induction network of Arabidopsis. J Exp Bot 61: 4221–4230. pmid:20685730
22. Noguchi T, Fujioka S, Choe S, Takatsuto S, Tax FE, et al. (2000) Biosynthetic pathways of brassinolide in Arabidopsis. Plant physiologyogy 124: 201–209. pmid:10982435
23. Bancos S, Szatmari AM, Castle J, Kozma-Bognar L, Shibata K, et al. (2006) Diurnal regulation of the brassinosteroid-biosynthetic CPD gene in arabidopsis. Plant Physiology 141: 299–309. pmid:16531479
24. Toyomasu T, Kawaide H, Mitsuhashi W, Inoue Y, Kamiya Y (1998) Phytochrome regulates gibberellin biosynthesis during germination of photoblastic lettuce seeds. Plant Physiology 118: 1517–1523. pmid:9847128
25. Yamaguchi S, Smith MW, Brown RGS, Kamiya Y, Sun TP (1998) Phytochrome regulation and differential expression of gibberellin 3 beta-hydroxylase genes in germinating Arabidopsis seeds. Plant Cell 10: 2115–2126. pmid:9836749
26. Ait-Ali T, Frances S, Weller JL, Reid JB, Kendrick RE, et al. (1999) Regulation of gibberellin 20-oxidase and gibberellin 3beta-hydroxylase transcript accumulation during De-etiolation of pea seedlings. Plant physiology 121: 783–791. pmid:10557226
27. Jackson SD, James PE, Carrera E, Prat S, Thomas B (2000) Regulation of transcript levels of a potato gibberellin 20-oxidase gene by light and phytochrome B. Plant Physiology 124: 423–430. pmid:10982455
28. Harmer SL, Kay SA (2005) Positive and negative factors confer phase-specific circadian regulation of transcription in Arabidopsis. Plant Cell 17: 1926–1940. pmid:15923346
29. Green RM, Tobin EM (1999) Loss of the circadian clock-associated protein I in Arabidopsis results in altered clock-regulated gene expression. Proceedings of the National Academy of Sciences of the United States of America 96: 4176–4179. pmid:10097183
30. Symons GM, Reid JB (2004) Brassinosteroids do not undergo long-distance transport in pea. Implications for the regulation of endogenous brassinosteroid levels. Plant physiology 135: 2196–2206. pmid:15299131
31. McClung CR (2006) Plant circadian rhythms. Plant Cell 18: 792–803. pmid:16595397
32. Suárez-López P, Wheatley K, Robson F, Onouchi H, Valverde F, et al. (2001) CONSTANS mediates between the circadian clock and the control of flowering in Arabidopsis. Nature 410: 1116–1120. pmid:11323677
33. Yanovsky MJ, Kay SA (2002) Molecular basis of seasonal time measurement in Arabidopsis. Nature 419: 308–312. pmid:12239570
34. Li J, Chory J (1997) A putative leucine-rich repeat receptor kinase involved in brassinosteroid signal transduction. Cell 90: 929–938. pmid:9298904

35. Hanano S, Domagalska MA, Nagy F, Davis SJ (2006) Multiple phytohormones influence distinct parameters of the plant circadian clock. Genes to Cells 11: 1381–1392. pmid:17121545
36. Battey N, Lyndon R (1990) Reversion of flowering. The botanical review 56: 162–189.
37. Han T, Gai J, Wang J, Zhou D (1997) Discovery of flowering reversion in soybean plants. Acta Agronomica Sinina 24: 168–171.
38. Washburn CF, Thomas JF (2000) Reversion of flowering in Glycine max (Fabaceae). American Journal of Botany 87: 1425–1438. pmid:11034918
39. Li X-M, Wu C-X, Ma Q-B, Zhang S, Li C-L, et al. (2005) Morphology and anatomy of the differentiation of flower buds and the process of flowering reversion in soybean cv. Zigongdongdou. Acta Agronomica Sinina 31: 1437–1442.
40. Wu CX, Ma QB, Yam KM, Cheung MY, Xu YY, et al. (2006) In situ expression of the GmNMH7 gene is photoperiod-dependent in a unique soybean (Glycine max [L.] Merr.) flowering reversion system. Planta 223: 725–735. pmid:16208488
41. Gai J, Wang Y (2001) A study on the varietal eco-regions of soybeans in China. Scientia Agricultura Sinica 34: 139–145.
42. Curtis DF, Tanner JW, Luzzi BM, Hume DJ (2000) Agronomic and phenological differences of soybean isolines differing in maturity and growth habit. Crop Science 40: 1624–1629.
43. Zhen J, Cun-xiang W, Miao W, Hong-bo S, Wen-sheng H, et al. (2011) Effects of leaf number of stock or scion in graft union on scion growth and development of soybean. Acta Agronomica Sinina 37: 650–660.
44. Clough SJ, Bent AF (1998) Floral dip: a simplified method for Agrobacterium-mediated transformation of Arabidopsis thaliana. Plant Journal 16: 735–743. pmid:10069079
45. Koch BM, Sibbesen O, Halkier BA, Svendsen I, Moller BL (1995) The Primary Sequence of Cytochrome P450tyr, the Multifunctional N-Hydroxylase Catalyzing the Conversion of L-Tyrosine to P-Hydroxyphenylacetaldehyde Oxime in the Biosynthesis of the Cyanogenic Glucoside Dhurrin in Sorghum-Bicolor (L) Moench. Archives of Biochemistry and Biophysics 323: 177–186. pmid:7487064
46. Liu S, Ju J, Xia G (2014) Identification of the flavonoid 3′-hydroxylase and flavonoid 3′,5′-hydroxylase genes from Antarctic moss and their regulation during abiotic stress. Gene 543: 145–152. pmid:24631264
47. Mathur J, Molnar G, Fujioka S, Takatsuto S, Sakurai A, et al. (1998) Transcription of the Arabidopsis CPD gene, encoding a steroidogenic cytochrome P450, is negatively controlled by brassinosteroids. Plant Journal 14: 593–602. pmid:9675902
48. Montoya T, Nomura T, Yokota T, Farrar K, Harrison K, et al. (2005) Patterns of Dwarf expression and brassinosteroid accumulation in tomato reveal the importance of brassinosteroid synthesis during fruit development. Plant Journal 42: 262–269. pmid:15807787
49. Symons GM, Davies C, Shavrukov Y, Dry IB, Reid JB, et al. (2006) Grapes on steroids. Brassinosteroids are involved in grape berry ripening. Plant physiology 140: 150–158. pmid:16361521
50. Kim HB, Kwon M, Ryu H, Fujioka S, Takatsuto S, et al. (2006) The regulation of DWARF4 expression is likely a critical mechanism in maintaining the homeostasis of bioactive brassinosteroids in Arabidopsis. Plant Physiology 140: 548–557. pmid:16407451
51. Wellmer F, Riechmann JL (2010) Gene networks controlling the initiation of flower development. Trends in Genetics 26: 519–527. pmid:20947199
52. Corbesier L, Vincent C, Jang S, Fornara F, Fan Q, et al. (2007) FT protein movement contributes to long-distance signaling in floral induction of Arabidopsis. science 316: 1030–1033. pmid:17446353
53. Sun H, Jia Z, Cao D, Jiang B, Wu C, et al. (2011) GmFT2a, a soybean homolog of FLOWERING LOCUS T, is involved in flowering transition and maintenance. PLoS One 6: e29238. pmid:22195028

54. Zhen J, Cun-xiang W, Miao W, Hong-bo S, Wen-sheng H, et al. (2011) Effects of leaf number of stock or scion in graft union on scion growth and development of soybean. Acta Agronomica Sinina 37: 650–660.
55. Symons GM, Ross JJ, Jager CE, Reid JB (2008) Brassinosteroid transport. J Exp Bot 59: 17–24. pmid:17709326
56. Bajguz A, Tretyn A (2003) The chemical characteristic and distribution of brassinosteroids in plants. Phytochemistry 62: 1027–1046. pmid:12591256
57. Shimada Y, Goda H, Nakamura A, Takatsuto S, Fujioka S, et al. (2003) Organ-specific expression of brassinosteroid-biosynthetic genes and distribution of endogenous brassinosteroids in Arabidopsis. Plant physiology 131: 287–297. pmid:12529536
58. Nomura T, Ueno M, Yamada Y, Takatsuto S, Takeuchi Y, et al. (2007) Roles of brassinosteroids and related mRNAs in pea seed growth and germination. Plant physiology 143: 1680–1688. pmid:17322340
59. Bancos S, Nomura T, Sato T, Molnar G, Bishop GJ, et al. (2002) Regulation of transcript levels of the Arabidopsis cytochrome p450 genes involved in brassinosteroid biosynthesis. Plant physiology 130: 504–513. pmid:12226529
60. Krumova S, Zhiponova M, Dankov K, Velikova V, Balashev K, et al. (2013) Brassinosteroids regulate the thylakoid membrane architecture and the photosystem II function. Journal of Photochemistry and Photobiology B-Biology 126: 97–104. pmid:23911861
61. Tsuda K, Kurata N, Ohyanagi H, Hake S (2014) Genome-wide study of KNOX regulatory network reveals brassinosteroid catabolic genes important for shoot meristem function in rice. The Plant Cell 26: 3488–3500. pmid:25194027

CHAPTER 15

Transcriptome Analysis of Differentially Expressed Genes Relevant to Variegation in Peach Flowers

Yingnan Chen[1], Yan Mao[1], Hailin Liu[1], Faxin Yu[2], Shuxian Li[1], Tongming Yin[1]

[1] *The Southern Modern Forestry Collaborative Innovation Center, Nanjing Forestry University, Nanjing, Jiangsu, China,*

[2] *Institute of Biology and Resources, Jiangxi Academy of Sciences, Nanchang, Jiangxi, China*

ABSTRACT

Background

Variegation in flower color is commonly observed in many plant species and also occurs on ornamental peaches (*Prunus persica* f. *versicolor* [Sieb.] Voss). Variegated plants are highly valuable in the floricultural market. To gain a global perspective on genes differentially expressed in variegated peach flowers, we performed large-scale transcriptome sequencing of white and red petals separately collected from a variegated peach tree.

Results

A total of 1,556,597 high-quality reads were obtained, with an average read length of 445 bp. The ESTs were assembled into 16,530 contigs and 42,050 singletons. The resulting unigenes covered about 60% of total predicted genes in the peach genome. These unigenes were further subjected to functional annotation and biochemical pathway analysis. Digital expression analysis identified a total of 514 genes differentially expressed between red and white flower petals. Since peach flower coloration is determined by the expression and regulation of structural genes relevant to flavonoid biosynthesis, a detailed examination detected four key structural genes, including *C4H*, *CHS*, *CHI* and *F3H*, expressed at a significantly higher level in red than in white petal. Except for the structural genes, we also detected 11 differentially expressed regulatory genes relating to flavonoid biosynthesis. Using the differentially expressed structural genes as the test objects, we validated the digital expression results by using quantitative real-time PCR, and the differential expression of *C4H*, *CHS* and *F3H* were confirmed.

Conclusion

In this study, we generated a large EST collection from flower petals of a variegated peach. By digital expression analysis, we identified an informative list of candidate genes associated with variegation in peach flowers, which offered a unique opportunity to uncover the genetic mechanisms underlying flower color variegation.

BACKGROUND

Flower color is of paramount importance in plant biology [1]. Three major groups of pigments–betalains, carotenoids and flavonoids–are responsible for the attractive natural display of flower colors [2], [3]. Flavonoids, particularly anthocyanidins, are the most common flower pigments, and contribute to a wide range of colors, from pale yellow to red, purple and blue [4]. To date, most enzymes involved in the anthocyanin biosynthetic pathway have been identified in various plant species [5]–[7]. All anthocyanidins are derived from a general phenylpropanoid pathway that converts the aromatic amino acid phenylalanine to anthocyanidins through a series of enzyme-catalyzed reactions [8]–[12]. During the past few decades, much of the molecular information available on the regulation and biosynthesis of flower pigments has been derived from studies performed in model systems such as *Zea mays* L. (maize) [13], [14], *Arabidopsis* [6], [15], petunia [16], [17] and snapdragon [3], [15]. An increasing number of non-classical plants, however, are providing unique insights into molecular mechanisms involved in flower pigment formation, leading to further understanding of how flower color varies among wild species [18]–[20].

Ornamental peach, a member of the *Rosaceae* family, is an important horticultural tree. *Prunus persica* f. *versicolor* (Sieb.) Voss, one of the main varieties, is characterized by the presence of chimeric flowers (Figure 1). Because variegation in flowers often attracts consumer attention, variegated plants are generally of high value in the ornamental market. This unstable phenotype has been observed in natural populations of petunia, snapdragon, morning glory, azalea and other plant species [21]. Flower variegation is usually due to the presence of a group of colored cells descended from a single ancestral cell in which a somatic mutation from the recessive white to the pigmented revertant allele has occurred. Somatic mutation frequency and timing during petal development determine variegation patterns [22].

Figure 1. Sampling stage of variegated flower petals.

Molecular mechanisms of flower variegation have been investigated in various plant species [22]–[25]. Some of this variegation is caused by transposable element insertion into structural genes associated with the anthocyanin synthetic pathway [26]–[28]. Because of the complexity of the flavonoid biosynthetic pathway, however, the exact genetic mechanisms underlying the generation of flower pigmentation chimeras may differ among species. Chaparro et al. [29] studied variegation in anthocyanin production in both vegetative and reproductive tissues of the peach cultivar Pillar; they verified that this phenotype is heritable, although the degree of variegation differed according to the genetic background of outcross progeny. The genetic mechanisms in peach responsible for this unstable phenotype have not yet been elucidated. The whole peach genome has been sequenced [30], and abundant transcriptome sequences are available to the public [31]. The reference genome and large EST collection enhance our ability to align sequences, identify genes, and characterize transcriptomes, thereby facilitating identification of the genetic basis of variegated pigmentation in peach.

To understand the mechanism of variegated pigmentation, detection of differentially expressed genes from different-colored flowers is essential. Transcriptome sequencing is an efficient way to measure transcriptome composition and uncover differentially expressed genes [32]–[34]. Many studies using high-throughput next-generation sequencing technology have surveyed the complex transcriptomes of various plants including *Arabidopsis thaliana* [35], *Digitalis purpurea*[36], *Carthamus tinctorius* [37], *Persea americana* [38] and *Salix suchowensis* [39]. In this study, transcriptomes of white and red flower petals sampled from a *Prunus persica* f. *versicolor* individual were sequenced using a 454 GS-FLX sequencer. By analyzing the data with various bioinformatics tools, we aimed to discover candidate genes involved in peach flower variegation.

MATERIALS AND METHODS

Plant Material

We separately collected petals of expanded but unflushed white and red variegated flower buds (Figure 1) from a tree of *Prunus persica* f. *versicolor* in Nanjing Lovers Garden, Jiangsu, China in March 2012. Petals were immediately frozen in liquid nitrogen and stored at −80°C until RNA extraction. The field studies did not involve any endangered or protected species, and sample collection was authorized by the administration office of Nanjing Lovers Garden.

RNA Extraction and cDNA Synthesis

Total RNA was extracted separately from white and red petals using the CTAB method [39]. RNA integrity was confirmed by 1% agarose gel electrophoresis. After digestion with DNase (Takara) at 37°C for 30 min to remove DNA residues, RNA concentration was determined using a Nanodrop spectrophotometer (Thermo). mRNA was then purified from total RNA using an Oligotex mRNA midi kit (Qiagen), with its quality assessed using an Agilent Technologies 2100 Bioanalyzer (Agilent). cDNA synthesis was performed using the 454 cDNA amplification technique with a cDNA Synthesis System kit (Roche) following the manufacturer's protocol.

Sequencing Library Construction and 454 Sequencing

Sequencing libraries were separately constructed for white and red petals using a Rapid Library Prep kit (Roche). Quality of sequencing libraries was checked using an Agilent 2100 Bioanalyzer. Approximately 2.1 million beads per flower color were separately loaded onto two sections of a pico-titer plate. A sequencing run was carried out on a Roche 454 GS FLX sequencer at Nanjing Forestry University. All ESTs in this study were deposited in NCBI with an accession number SRR1037160.

EST Sequence Processing and Assembly

Raw 454 sequence files in SFF format were base-called using the Pyrobayes base caller [40]. In addition, 78,689 peach ESTs were downloaded from GenBank in February 2012. The 454 sequencing data and GenBank ESTs were processed with GS FLX v2.0.01 software (454 Life Sciences, Roche) to remove low-quality and adaptor sequences. To remove possible contamination, the resulting high-quality 454 and GenBank sequences were screened against the NCBI UniVec database, *E. coli* genome sequences, and peach ribosomal RNA and chloroplast genome sequences. Sequences shorter than 50 bp were discarded before assembly. Finally, the processed 454 and GenBank sequences were assembled into putative transcripts (including contigs and singletons) using the 454 assembly program Newbler v2.7 with the following overlap detection settings: seed step, 12; seed length, 16; seed count, 1; minimum overlap length, 40; minimum overlap identity, 95%; alignment identity score, 2; alignment difference score, −3. We parsed the 454 ReadStatus.txt file to identify singletons, which were unassembled reads. The contig and singleton files were used to generate a unigene file.

Mapping Unigenes to Peach Genome Predicted Genes

The generated unigenes were aligned to predicted genes of the peach genome using BLAST [41]. Based on the International Peach Genome Initiative (IPGI; http://www.rosaceae.org/species/prunus /prunus_persica), there are 27,864 predicted genes in the peach genome. Because the number of obtained unigenes was much greater than the number of predicted genes, it is possible that different unigenes were segments of the same predicted gene. In this study, unigenes mapping to the same predicted gene were integrated into one unique gene, which was used to collect transcript count information for detection of differentially expressed genes in white and red flower petals.

Gene Annotation and Pathway Prediction

Unigene annotation was performed by BLASTX analysis [42] against peach protein (http://www.phytozome.com/peach.php; v2.0), NCBI non-redundant protein (nr) and UniProt databases using a cutoff E-value of 10^{-5}. Gene Ontology (GO) terms were assigned to each unigene based on GO annotations of its corresponding homologs in the UniProt database [43], and, using interpro2go and pfam2go mapping files available on the GO website (http://www.geneontology.org), their corresponding InterPro and Pfam domains. GO mapping results were further plotted by uploading the GO list file to the Web Gene Ontology Annotation Plot (WEGO) website (http://wego.genomics.org.cn/cgi-bin/wego/index.pl). The detailed annotation was then used to retrieve keywords for identification of flower pigmentation-related genes.

The KEGG Automatic Annotation Server (KAAS; http://www.genome.jp/tools/kaas/) was employed to perform metabolic pathway mapping [44], [45]. KAAS assigned each peach gene a KEGG Orthology (KO) number, which was then used for mapping to a KEGG reference metabolic pathway.

Differentially Expressed Genes between White and Red Petals

Following cDNA sequence assembly and gene prediction, transcript count information was collected for sequences corresponding to each predicted gene associated with a flower petal color. To obtain relative expression levels in each sample, transcript counts were normalized to the total number of produced transcripts per sample. Significance of gene differential expression level was assessed using R [46], χ^2 and Fisher exact tests as implemented in the publicly available web tool IDEG6 (http://telethon.bio.unipd.it/bioinfo/IDEG6/) [47]. A gene was considered to be differentially expressed when results from the above tests were all significant at a level of $P \leq 0.0001$.

Quantitative Real-time PCR Analysis

Total RNA was extracted from red and white flower petals as described above. Approximately 2 μg of total RNA per sample was treated with

DNaseI (Takara), and then subjected to reverse transcript to cDNA using reverse transcription system (Promega). The quantitative real-time PCR (qRT-PCR) was performed using Applied Biosystems 7500 Real-time PCR system (Applied Biosystems) with SYBR Premix Ex Taq (Takara). Each reaction contained 2 μL the first-strand cDNA as template, in a total volume of 20 μL reaction mixture. The amplification program was performed as 95°C 30 s followed by 95°C for 5 s and 60°C for 34 s (40 cycles). Gene-specific primers, shown in Table S3, were used for detecting the relative quantification of each gene. qRT-PCR expression levels were compared based on the mean of three independent experimental repeats. Calculation of relative expression level was performed using the $2^{-\Delta\Delta CT}$ method [48]. *TEF2* was used as an internal control for normalization [49].

RESULTS

Transcriptome Sequencing and Assembly

A half 454 GS-FLX run was performed for each sample, resulting in the generation of 1,556,684 reads. After quality control, 1,556,597 reads with an average length of l, 445 bp remained: 837,041 from white petals and 719,556 from red petals (Table 1). These reads, together with 76,822 high-quality GenBank ESTs, were subjected to assembly analysis. Sequence assembly yielded 58,580 unigenes comprising 16,530 contigs and 42,050 singletons. Contigs and singletons had average lengths of 1,555 bp and 364 bp, respectively (Table 2). A plot of unigene length distribution revealed that most unigenes (82.8%) were longer than 300 bp (Figure 2). From our EST collection, we were able to identify a number of highly expressed unigenes in peach flowers (Figure 3). Approximately 12,589 unigenes were assembled from more than 10 EST reads; these unigenes (~21% of all unigenes) corresponded to ~95% of total EST reads.

Table 1. Statistics for peach ESTs generated by 454 GS-FLX sequencing.

	Red petals	White petals	Total
Number of reads	719,556	837,041	1,556,597
Average read length (bp)	443.3	448	445.8
Total bases (bp)	318,983,491	374,962,455	693,945,946
Number of reads in contigs	675,990	783,159	1,459,149
Number of reads as singletons	28,444	35,141	63,585

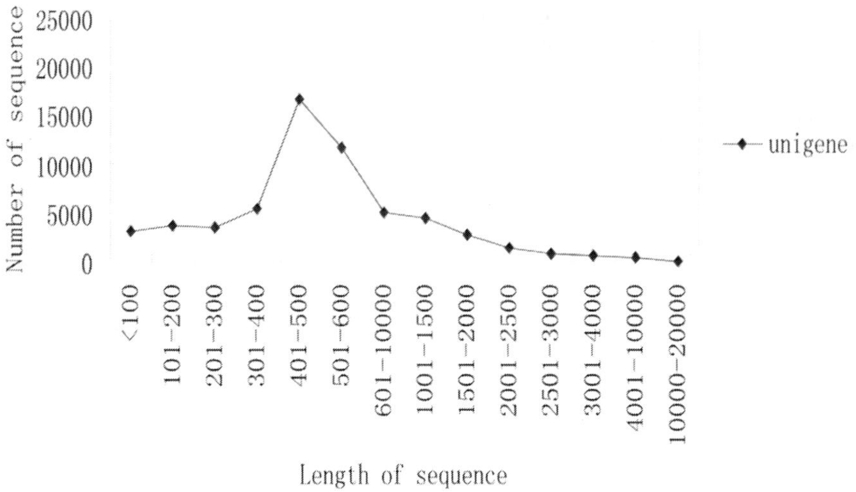

Figure 2. Length distributions of peach unigenes.

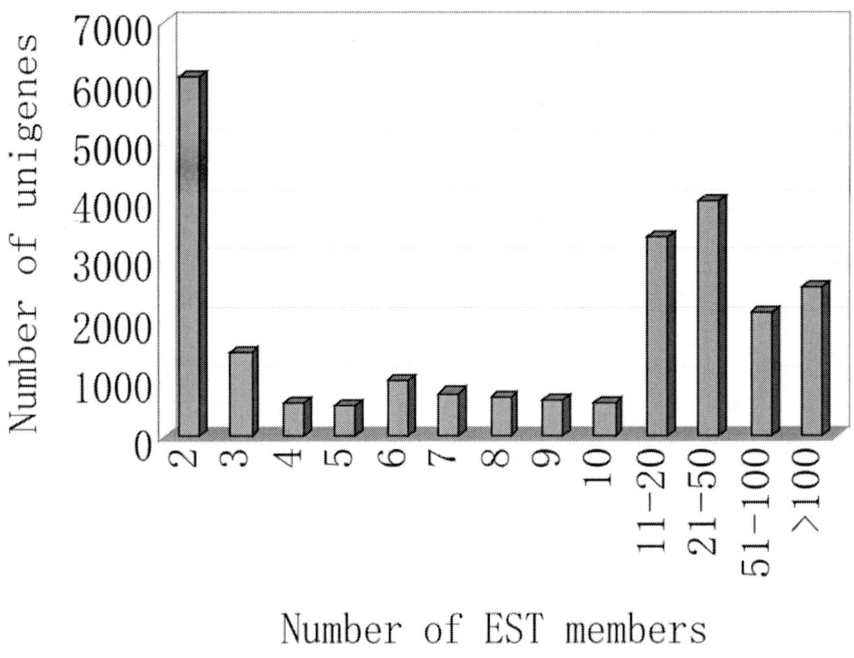

Figure 3. Distribution of ESTs in peach unigenes.

Table 2. Statistics for assembled peach unigenes.

	Singleton	Contig	Unigene(cluster)
Number of sequences	42,050	16,530	58,580
Average read length (bp)	364	1,555	699
Total bases (bp)	26,469,716	27,763,950	40,933,465
Number of unigenes mapped to peach genome predicted genes	25,042	15,645	40,687
Number of unigenes unmapped to peach genome predicted genes	17008	885	17893

doi:10.1371/journal.pone.0090842.t002

Mapping Unigenes to Predicted Genes in the Peach Genome

Using the peach genome (v1.0; IPGI), approximately 70% (40,687) of obtained unigenes could be mapped to predicted genes of *P. persica*. Among these, 15,645 were contigs (95% of all contigs) and 25,042 were singletons (60% of singletons). The mapped unigenes were aligned to 16,733 predicted genes, corresponding to approximately 60% of the 27,864 predicted genes in the peach genome. The remaining 17,893 unigenes could not be mapped to any predicted genes. Among the unmappable unigenes, 885 were contigs and 17,008 were singletons (Table 2). We further examined the unmappable unigenes, and found that about 20.6% had no significant matches with plant sequences. In total, alignment of unigenes to predicted *P. persica* genes generated 34,626 unique genes: 16,733 identified as peach genome predicted genes and 17,893 unmappable unigenes. When unigenes were mapped to different scaffolds of the peach genome (http://www.rosaceae.org/species/prunus/prunus_persica), the number of mapped unigenes was found to be highly correlated with the number of predicted genes on each scaffold (paired t-test, R^2=0.992) (Figure 4). The number of expressed genes was thus proportional to the number of predicted genes allocated to each scaffold.

Roman numerals I–VIII correspond to the eight large scaffolds of the peach genome (v1.0; IPGI), and are ordered along the x-axis with the number of mapped unigenes on each scaffold. IX+ refers to all remaining small scaffolds of the peach genome.

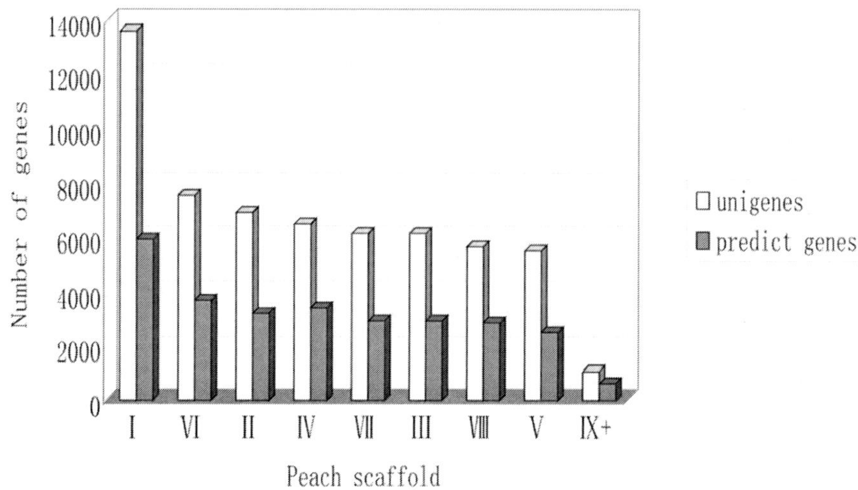

Figure 4. Distribution of unigenes and predicted genes allocated to each peach genome scaffold.

Functional Annotation of Peach Transcriptomes

Based on alignments of unigenes to peach genome predicted genes, 34,626 unique genes were identified. To infer putative functions of these unique genes, we blasted their sequences against the GenBank nr database using a significance cutoff of $E \leq 10^{-5}$. The analysis indicated that 21,663 unique genes (63%) had significant matches to the nr database, among which 15,846 were peach genome predicted genes (94.7% of the 16,733 unigene-matched predicted genes) and 5,817 were unmappable unigenes (32% of all unmappable unigenes). Most of the unmappable unigenes were singletons (95.1%), which were much shorter than contigs (364 bp vs. 1,555 bp on average). In general, the longer the sequence is, the greater the chance of annotation, and thus the more number of GO terms that can be recovered [50]. This suggests that the inability to assign putative functions to most unmappable unigenes was due to the absence of conserved functional domains in short sequences.

Gene Ontology (GO) describes gene products in terms of their associated molecular functions, biological processes and cellular components. To assign putative functional roles to the obtained unique genes, GO terms were assigned based on sequence similarities to known GO-annotated proteins (and their InterPro and Pfam domains) in the UniProt database. As a result, 10,485 unique genes were assigned at least one GO term. Among GO terms, 896 were in the biological process category, 606 in the molecular function category, and 607 in the cellular component category (Figure 5). These unique genes were further displayed using a set of GO slims, which are a list of high-level GO terms providing a broad overview of ontology content (http://www.geneontology.org/GO.slims.shtml). Figure 5 displays the functional classification of peach unique genes into plant-specific GO slims within cellular component, molecular function and biological process categories. Genes involved in cell, cell part, binding, catalytic, cellular process and metabolic process categories were the highest-represented groups, indicating that flower buds were undergoing rapid growth and carrying out intensive metabolic activities. In the biological process category, it is noteworthy that genes involved in pigmentation were also highly represented, indicating active pigmentation activities.

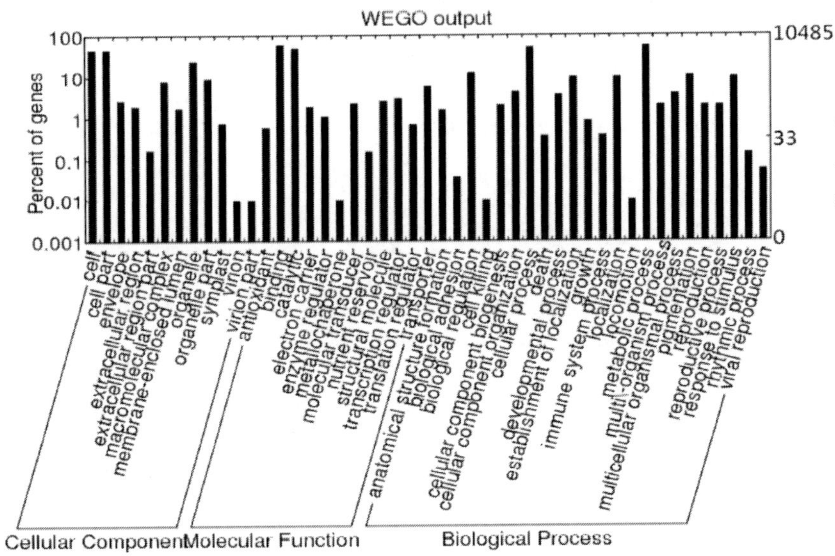

Figure 5. Number of peach unigenes in each functional category.

Peach unigenes were classified into different functional groups based on a set of plant-specific GO slims within cellular component, molecular function and biological process categories.

Biochemical Pathways

KEGG, an alternative functional gene annotation system, performs assignments based on Enzyme Commission (EC) numbers of genes associated with biochemical pathways. To further demonstrate the usefulness of the generated peach ESTs for discovering flower pigmentation-related genes, we identified biochemical pathways represented by our EST collection. Annotations of peach unique genes were fed into the Pathway Tools program (KAAS; http://www. genome.jp/tools/kaas/). This process predicted 229 pathways represented by a total of 5,345 unique genes. Of these KEGG-annotated genes, 2,425 (45.4%) were involved in metabolism, 1,186 (22.2%) in genetic information processing, 346 (6.5%) in environmental information processing, 569 (10.6%) in cellular processes and 792 (14.8%) in organism systems (Table S1). Among genes involved in metabolism, 14 genes were found to encode key enzymes involved in flavonoid biosynthesis. We mapped and highlighted these genes onto the "flavonoid biosynthesis" pathway (Figure S1), which demonstrated that most of the key enzymes in this pathway were covered by our sequence data. We also detected three genes involved in flavone and flavonol biosynthesis; they were mapped and highlighted onto the "flavone and flavonol biosynthesis" pathway (Figure S2). In contrast to the flavonoid biosynthesis pathway, only a small proportion of flavone and flavonol biosynthetic pathway key enzymes were represented by our sequence data.

Identification of Differentially Expressed Genes in White and Red Flower Petals

In this study, we obtained 16,530 contigs and 42,050 singletons, among which 15,645 contigs and 25,042 singletons were mapped to 16,733 peach predicted genes. There were 855 unmapped contigs and 17,008 unmapped singletons. The purpose of this study was to uncover genes differentially expressed between flower petal colors. The 17,008 unmapped singletons were less closely related to peach genome predicted genes and were not present in sufficient quantities for statistical analysis [50]; they were consequently excluded from the differentially expressed gene analysis. Number of reads was obtained for each gene using a custom PERL script. The digital expression profiling analysis identified 514 genes differentially expressed between red and white flower petals, with $P<0.0001$ for all employed statistics (Table S2); 367 of these genes showed significantly higher expression in red flower petals, whereas 147 exhibited significantly higher expression in white petals.

In peach, the flavonoid biosynthetic pathway leads to production of colored pigments. We therefore specifically examined expression of flavonoid structural genes, and identified four differentially expressed genes, all highly expressed in red petals. These genes encode the enzymes chalcone and stilbene synthase (CHS), chalcone-flavanone isomerase (CHI), cinnamate-4-hydroxylase (C4H) and flavanone 3-hydroxylase (F3H). Functionally, C4H catalyzes the incorporation of a 4′-hydroxyl group into cinnamate during p-coumarate formation. CHS condenses one molecule of p-coumaroyl-CoA with three molecules of malonyl-CoA to produce the chalcone tetrahydroxychalcone. (Chalcone is the precursor for all classes of flavonoids, including flavones, flavonols, flavandiols, flavan-4-ols, condensed tannins, isoflavonoids and anthocyanins.) CHI is responsible for the conversion of tetrahydroxychalcone to naringenin, and F3H catalyzes the formation of DHK from naringenin (Figure S3). DHK can be further hydroxylated to form dihydroquercetin (DHQ) and dihydromyricetin (DHM). DHK, DHQ and DHM subsequently lead synthetic branches producing pelargonidin-based (orange to red), cyanidin-based (red to magenta) and delphinidin-based (purple) pigments, respectively. Interestingly, the four differentially expressed genes all mapped to early committed steps on the flavonoid biosynthetic pathway before the formation of DHK (Figure S3). From 454 sequencing, absolute read numbers of *C4H*, *CHS*, *CHI* and *F3H* in red vs. white flower petals were 146:54, 4050:1448, 122:14 and 26:1, respectively. After normalization, expression levels of *C4H*, *CHS*, *CHI* and *F3H* were 3.1-, 3.3-, 10.1- and 30.2-fold higher, respectively, in red flower petals than in white ones.

In the flavonoid biosynthetic pathway, transcription levels of flavonoid biosynthesis genes are regulated by various TFs. TFs related to flavonoid biosynthesis can be divided into three classes: MYB, basic-Helix-Loop-Helix (bHLH) and WD40 [15], [51], [52]. Apart from the structural genes, we further examined expression levels of these three TF classes. We detected 11 TFs differentially expressed between red and white petals, including 3 bHLHs, 4 MYBs and 4 WD40s. Nine of these TFs were highly expressed in red petals, and two were highly expressed in white petals (Table 3). These DNA-binding proteins interact with promoter regions of target genes and regulate the initiation rate of mRNA synthesis.

Table 3. Differentially expressed structural and regulatory genes related to flavonoid pathways in variegated peach flowers.

ID	Function distribution	Highly expressed tissue
ppa025745m	Chalcone and stilbene synthase (CHS) family protein	red
ppa011276m	Chalcone-flavanone isomerase (CHI) family protein	red
ppa004544m	cinnamate-4-hydroxylase (C4H)	red
ppa007636m	flavanone 3-hydroxylase (F3H)	red
ppa009757m	basic helix-loop-helix (bHLH) DNA-binding superfamily protein	red
ppa003543m	basic helix-loop-helix (bHLH) DNA-binding superfamily protein	red
ppa013751m	ILI1 binding bHLH 1	red
ppa010069m	myb domain protein 113	red
ppa011751m	myb domain protein 24	red
ppa010277m	myb domain protein 4	red
ppa007222m	myb-like transcription factor family protein	white
ppa005673m	Transducin family protein/WD-4 repeat family protein	red
ppa005800m	Transducin family protein/WD-4 repeat family protein	red
ppa002072m	Transducin/WD4 repeat-like superfamily protein	red
ppa014936m	Transducin/WD4 repeat-like superfamily protein	white

Using the differentially expressed structural genes as the test objects, we further validated the digital expression profiling by qRT-PCR technology. Statistically, *C4H*, *CHS*, and *F3H* expressed significantly higher in red flower petals than in white ones, with $p<0.05$ (Figure 6). Thus, the significant difference in expression level of these three structural genes was confirmed by both of the analytical techniques. Whereas the expression level of *CHI* was not statistically different between colors. The comparative transcription level of this gene was found to be only slightly higher in red than in white petals by qRT-PCR analysis (Figure 6). Nevertheless, in general, the qRT-PCR results were in rough accordance with the electronic data of gene expression analysis.

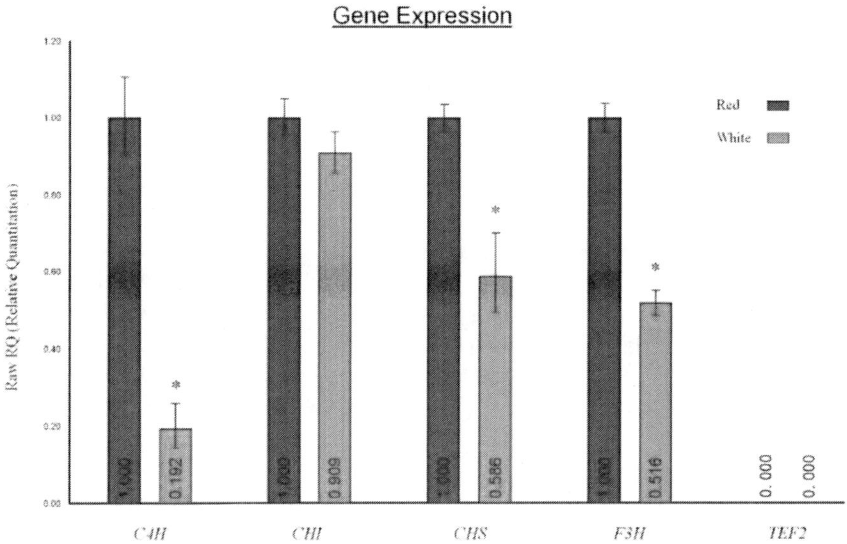

Figure 6. qRT-PCR analysis of *C4H*, *CHS*, *CHI* and *F3H* gene expression in white and red flower petals.

TEF2 was used as an internal control for normalization. qRT-PCR data calculated with the $2^{-\Delta\Delta Ct}$ method [48]. The expression level of each gene in red flower petals was arbitrarily set as 1 and its corresponding transcript level in white flower petals was calibrated against red one. Error bars represent standard error for three independent experimental replicates. *, $p<0.05$, as determined by one-way ANOVA.

DISCUSSION

Approximately 60% of total predicted genes in the peach genome were covered by the unigenes obtained in this study. In *Arabidopsis*, approximately 55–67% of genes are expressed in a single tissue based on microarray analysis [53]. In human and mouse, around 60–70% of genes are expressed in a specific tissue [54]. Transcriptomic studies have revealed that about 64%, 66% and 68% of genesare expressed in whole flower tissues of cucumber [55], willow [39] and peach [31], respectively. In our study, we sequenced genes expressed in flower petals, not entire flowers. Our EST collection, designed to capture the majority of genes expressed in peach flower petals, would thus be expected to encompass fewer expressed genes. Because more transcript sequences were generated in peach (1.5 million) than in willow (1.2 million) [39] or cucumber (0.35 million) [55], the sequence depth of genes covered by our EST dataset should therefore be higher. We thus obtained a reliable dataset to explore differentially expressed genes relevant to variegation in peach flowers. Most differentially expressed genes that have currently been revealed by transcriptome sequencing are derived from studies performed with tissues of different genotypes [39], [55], [56]. In contrast, ESTs generated in our study are collections of expressed genes of the same genotype. Our approach is thus appropriate for identification of differentially expressed genes relevant to the focal phenotype.

The primary pigments related to flower color are anthocyanins, which contribute to a variety of colors, such as red, pink and blue [4]. Biochemical pathway mapping revealed that most of the structural genes in the anthocyanin biosynthetic pathway were covered by our sequence data (Figure S1), and we detected four structural genes highly expressed in red petals. In addition to anthocyanins, some co-pigments, such as flavones and flavonols, can change flower color hues. Only a few structural genes (Figure S2) involved in flavone and flavonol biosynthesis were captured, however, and none of them were differentially expressed between red and white petals. These results suggest that peach flower variegation is linked mainly to anthocyanin biosynthesis, with flavones and flavonols having a limited role.

With respect to enzymatic genes involved in the anthocyanin biosynthetic pathway, *C4H*, *CHS* and *F3H* were confirmed to express at a significantly higher level in red petals than in white petals by both the transcriptome profiling and qRT-PCR analysis. These genes were associated with committed steps before the formation of DHK (Figure S3). Because red flower color in peach is related to the synthesis of pelargonidin-based (orange to red) and cyanidin-based (red to magenta) pigments, our results suggest that the low *C4H*, *CHS* and *F3H* expression levels in white petals reduce DHK formation, thereby inhibiting pelargonidin and cyanidin production. In contrast, the high expression levels of these genes observed in red petals ensure sufficient anthocyanin yields to make flowers red. In this experiment, none of the structural genes downstream of DHK (Figure S3) were found to be significantly differentially expressed between white and red petals. It can thus be concluded that variegation in peach flowers is related primarily to structural genes upstream of DHK in the anthocyanin synthetic pathway, and is less likely affected by genes downstream.

In nature, white is the most abundantly occurring flower color, and can also be artificially produced from colored flowers. The first successful reversion to white flowers was achieved by suppressing *CHS* in tobacco and petunia [57]. Suppression of structural genes in the anthocyanin synthetic pathway has subsequently proven useful for modifying colored flowers to white. Studies have shown that transgenic plants carrying single gene constructs of *CHS*[58], [59] or *F3H* [60] can all exhibit white or faint-colored flowers.

Final anthocyanin concentrations in plant cells are not determined solely by structural gene expression levels; some regulatory genes are clearly also involved in control of flavonoid biosynthesis gene expression. These regulatory genes, typically specific TFs, influence anthocyanin biosynthesis intensity and pattern and generally control expression of many different structural genes [21]. Three classes of TFs–bHLH, MYB and WD40–have been found to be related to flavonoid biosynthesis [15], [51], [52]. In maize, an MYB-related protein and a bHLH-containing protein have been shown to interact to activate genes in the anthocyanin biosynthetic pathway [61]. In this study, we detected 11 differentially expressed TFs from the three flavonoid biosynthesis-related classes. The high *C4H*, *CHS* and *F3H* expression levels observed in red peach flower petals may be due to regulation by one or more of these TFs. Although the exact regulatory TFs remain unknown, this study provides an informative list of candidates.

In peach flowers, different variegation patterns are observed: some occur among different flowers on the same branch, while others occur within the same flower (Figure S4). Much genetic evidence supports the hypothesis that flower variegation is caused by transposable elements inserted into structural genes or regulatory elements related to anthocyanin synthesis [26]–[28]. The different transcript levels detected in this study for structural genes and TFs might be due to insertions of transposable elements. Examination of candidate DNA sequences for transposable element insertions is needed to determine the mechanisms of peach flower variegation. Although transposable elements cannot be detected based solely on transcriptome sequencing, our study has provided some novel insights into the molecular mechanisms underlying variegation in peach flowers.

CONCLUSIONS

Flower color is one of the most attractive characteristics of ornamental plants, and variegated plants are highly valuable in the floricultural market. In this study, we generated a large EST collection from flower petals of a variegated peach. Based on the digital expression analysis, a total of 514 genes were identified to differentially express between red and white flower petals. The red and white petals were collected from the same tree, and they were in the same developmental stage. Moreover, flower coloration was specifically connected with the flavonoid biosynthetic pathway. All these conditions enabled us to narrow down the considerable number of differentially expressed genes to a small number of candidates, which warranted further investigation. Finally, three key structural genes in the anthocyanin biosynthesis pathway were confirmed to express significantly different between colors, and we also detected 11 differentially expressed TFs related to anthocyanin biosynthesis. Our results provide critical information for uncovering candidate genes associated with variegation in peach flowers. We believe this transcriptome dataset will continue to provide unique insights into the molecular mechanisms controlling variegated flower pigmentation, and will eventually help the molecular engineering of variegated plants.

REFERENCES

1. Kevan PG, Baker HG (1983) Insects as flower visitors and pollinators. Annu Rev Entomol 28: 407–53.
2. Harborne JB, Williams CA (2000) Advances in flavonoid research since 1992. Phytochemistry 55: 481–504.
3. Grotewold EA (2006) The genetics and biochemistry of floral pigments Rev. Plant Biol 57: 761–80.
4. Goto T, Kondo T (1991) Structure and molecular stacking of anthocyanins–flower color variation. Angew Chem Int Ed Engl 30: 17–33.
5. Holton TA, Cornish EC (1995) Genetics and biochemistry of anthocyanin biosynthesis. Plant Cell 7: 1071–1083.
6. Winkel-Shirley B (2001) Flavonoid biosynthesis: a colorful model for genetics, biochemistry, cell biology, and biotechnology. Plant Physiol 126: 485–493.
7. Davies KM (2009) Modifying anthocyanin production in flowers. In Anthocyanins. Edited by Winefield C, Davies K, Gould K. New York, Springer, 49–80.
8. Kroon J, Souer E, de Graaff A, Xue Y, Mol J, et al. (1994) Cloning and structural analysis of the anthocyanin pigmentation locus Rt of Petunia hybrida: characterization of insertion sequences in two mutant alleles. Plant J 5: 69–80.
9. Ronchi A, Petroni K, Tonelli C (1995) The reduced expression of endogenous duplications (REED) in the maize R gene family is mediated by DNA methylation. EMBO J 14: 5318–5328.

10. Fujiwara H, Tanaka Y, Fukui Y, Nakao M, Ashikari T, et al. (1997) Anthocyanin 5-aromatic acyltransferase from Gentiana triflora: purification, characterization and its role in anthocyanin biosynthesis. Eur J Biochem 249: 45–51.
11. Yoshida K, Toyama Y, Kameda K, Kondo T (2000) Contribution of each caffeoyl residue of the pigment molecule of gentiodelphin to blue color development. Phytochemistry 54: 85–92.
12. Yabuya T, Yamaguchi M, Fukui Y, Katoh K, Imayama T, et al. (2001) Characterization of anthocyanin p-coumaroyltransferase in flowers of Iris ensata. Plant Sci 160: 499–503.
13. Bruce W, Folkerts O, Garnaat C, Crasta O, Roth B, et al. (2000) Expression profiling of the maize flavonoid pathway genes controlled by estradiol-inducible transcription factors CRC and P. Plant Cell. 12: 65–80.
14. Morohashi K, Casas MI, Falcone Ferreyra ML, Mejía-Guerra MK, Pourcel L, et al. (2012) A genome-wide regulatory framework identifies maize pericarp color1 controlled genes. Plant Cell 24: 2745–2764.
15. Mol J, Grotewold E (1998) KoesR (1998) How genes paint flowers and seeds. Trends Plant Sci 3: 212–217.
16. Forkmann G, de Vlaming P, Spribille R, Wiering H, Schram AW (1986) Genetic and biochemical studies on the conversion of dihydroflavonols to flavonols and flavones of Petunia hybrida. Z Naturforsch C: Biosci 41c: 179–186.
17. Dixon RA, Lamb CJ, Masoud S, Sewalt VJ, Paiva NL (1996) Metabolic engineering: prospects for crop improvement through the genetic manipulation of phenylpropanoid biosynthesis and defense responses-a review. Gene 179: 61–71.
18. Suzuki S, Nishihara M, Nakatsuka T, Misawa N, Ogiwara I, et al. (2007) Flower color alteration in Lotus japonicus by modification of the carotenoid biosynthetic pathway. Plant Cell Rep 26: 951–959.
19. Davies KM, Albert NW, Schwinn KE (2012) From landing lights to mimicry: the molecular regulation of flower colouration and mechanisms for pigmentation patterning. Funct Plant Biol 39: 619–638.
20. Dubois A, Carrere S, Raymond O, Pouvreau B, Cottret L, et al. (2012) Transcriptome database resource and gene expression atlas for the rose. BMC Genomics 13: 638.
21. To KY, Wang CK (2006) Molecular breeding of flower color. In Floriculture, Ornamental and Plant Biotechnology: Advances and Topical Issues Volume I. Edited by Teixeira da Silva JA. Isleworth, England, 300–310.
22. Iida S, Hirota T, Morisaki T, Marumoto T, Hara T, et al. (2004) Tumor suppressor WARTS ensures genomic integrity by regulating both mitotic progression and G1 tetraploidy checkpoint function. Oncogene 23: 5266–5274.
23. van Houwelingen A, Souer E, Spelt K, Kloos D, Mol J, et al. (1998) Analysis of flower pigmentation mutants generated by random transposon mutagenesis in Petunia hybrida. Plant J 13: 39–50.
24. Iida S, Hoshino A, Johzuka-Hisatomi Y, Habu Y, Inagaki Y (1999) Floricultural traits and transposable elements in the Japanese and common morning glories. Ann NY Acad Sci 18: 265–74.
25. Liu D, Galli M, Crawford NM (2001) Engineering variegated floral patterns in tobacco plants using the Arabidopsis transposable element Tag1. Plant Cell Physiol 42: 419–423.
26. Inagaki Y, Hisatomi Y, Suzuki T, Kasahara K, Iida S (1994) Isolation of a Suppressor-mutator/Enhancer-like, transposable element, Tpn1, from Japanese morning glory bearing variegated flowers. Plant Cell 6: 375–383.
27. Hoshino A, Inagaki Y, Iida S (1995) Structural analysis of Tpnl, a transposable element from Japanese morning glory bearing variegated flowers. Mol Gen Genet 247: 114–117.
28. Quattrocchio F, Wing J, van der Woude K, Souer E, deVetten N, et al. (1999) Molecular analysis of the anthocyanin2 gene of petunia and its role in the evolution of flower color. Plant Cell 11: 1433–1444.

29. Chaparro JX, Werner D, Whetten R, O'Malley D (1995) Characterization of an unstable anthocyanin phenotype and estimation of somatic mutation rates in peach. J. Hered 86: 186–193.
30. Verde I, Abbott AG, Scalabrin S, Jung S, Shu SQ, et al. (2013) The high-quality draft genome of peach (Prunus persica) identifies unique patterns of genetic diversity, domestication and genome evolution. Nat Genet 45: 487–494.
31. Wang L, Zhao S, Gu C, Zhou Y, Ma JJ, et al. (2013) Deep RNA-Seq uncovers the peach transcriptome landscape. Plant Mol Biol 83: 365–377.
32. Mardis ER (2008) The impact of next-generation sequencing technology on genetics. Trends Genet 24: 133–141.
33. Morozova O, Marra MA (2008) Applications of next-generation sequencing technologies in functional genomics. Genomics 92: 255–264.
34. Parchman TL, Geist KS, Grahnen JA, Benkman CW, Buerkle CA (2010) Transcriptome sequencing in an ecologically important tree species: assembly, annotation, and marker discovery. BMC Genomics 11: 180.
35. Zhang X, Byrnes JK, Gal TS, Li WH, Borevitz JO (2008) Whole genome transcriptome polymorphisms in Arabidopsis thaliana. Genome Biol 9: R165.
36. Wu B, Li Y, Yan HX, Ma YM, Luo HM, et al. (2012) Comprehensive transcriptome analysis reveals novel genes involved in cardiac glycoside biosynthesis and mlncRNAs associated with secondary metabolism and stress response in Digitalis purpurea. BMC Genomics 13: 15.
37. Huang LL, Yang X, Sun P, Tong W, Hu SQ (2012) The first illumina-based de novo transcriptome sequencing and analysis of safflower flowers. PLoS ONE 7: e38653. doi 0.1371/journal.pone.0038653.
38. Mahomed W, van den Berg N (2011) EST sequencing and gene expression profiling of defence–related genes from Persea americana infected with Phytophthora cinnamomi. BMC Plant Biol 11: 167.
39. Liu J, Yin T, Ye N, Chen Y, Yin T, et al. (2013) Transcriptome analysis of the differentially expressed Genes in the Male and Female Shrub Willows (Salix suchowensis). PLoS ONE 8: e60181
40. Quinlan RJ, Tobin JL, Beales PL (2008) Chapter 5 modeling ciliopathies: primary cilia in development and disease. Curr Top Dev Biol 84: 249–310.
41. Altschul SF, Gish W, Miller W, Meyers EW, Lipman DJ (1990) Basic local alignment search tool. J Mol Biol 215: 403–410.
42. Altschul SF, Madden TL, Schaffer AA, Zhang J, Zhang Z, et al. (1997) Gapped BLAST and PSI-BLAST: a new generation of protein database search programs. Nucleic Acids Res 25: 3389–3402.
43. The UniProt Consortium (2008) The Universal Protein Resource (UniProt). Nucleic Acids Res 36: D190–D195.
44. Aoki-Kinoshita KF, Kanehisa M (2007) Gene annotation and pathway mapping in KEGG. Methods Mol Biol 396: 71–91.
45. Moriya Y, Itoh M, Okuda S, Yoshizawa AC, Kanehisa M (2007) KAAS: an automatic genome annotation and pathway reconstruction server. Nucleic Acids Res 35: W182–W185.
46. Stekel DJ, Git Y, Falciani F (2000) The comparison of gene expression from multiple cDNA Libraries. Genome Res 10: 2055–2061.
47. Romualdi C, Bortoluzzi S, D'Alessi F, Danieli GA (2003) IDEG6: a web tool for detection of differentially expressed genes in multiple tag sampling experiments. Physiol Genomics 12: 159–62.
48. Livak KJ, Schmittgen TD (2001) Analysis of Relative Gene Expression Data Using Real-Time Quantitative PCR and the 2–ΔΔCT Method. Methods 25: 402–408.
49. Tong ZG, Gao ZH, Wang F, Zhou J, Zhang Z (2009) Selection of reliable reference genes for gene expression studies in peach using real-time PCR. BMC Mol Biol 10: 71.
50. Zhang Z, Wang Y, Wang S, Liu J, Warren W, et al. (2011) Transcriptome analysis of female and male Xiphophorus maculatus Jp 163 A. PLoS ONE. 6: e18379

51. Broun P (2005) Transcriptional control of flavonoid biosynthesis: a complex network of conserved regulators involved in multiple aspects of differentiation in multiple aspects of in differentiation Arabidopsis. Curr Opin Plant Biol 8: 272–279.
52. Morita Y, Saitoh M, Hoshino A, Nitasaka E, Iida S (2006) Isolation of cDNAs for R2R3-MYB, bHLH and WDR transcriptional regulators and identification of c and ca mutations conferring white flowers in the Japanese morning glory. Plant Cell Physiol 47: 457–470.
53. Schmid M, Davison TS, Henz SR, Pape UJ, Demar M, et al. (2005) A gene expression map of Arabidopsis thaliana development. Nat Genet 37: 501–506.
54. Ramskold D, Wang ET, Burge CB, Sandberg R (2009) An abundance of ubiquitously expressed genes revealed by tissue transcriptome sequence data. PLoS Comput Biol 5: e1000598
55. Guo SG, Zheng Y, Joung JG, Liu SQ, Zhang ZH, et al. (2010) Transcriptome sequencing and comparative analysis of cucumber flowers with different sex types. BMC Genomics 11: 384.
56. Zhu ML, Zheng XC, Shu QY, Li H, Zhong PX, et al. (2012) Relationship between the composition of flavonoids and flower colors variation in tropical water lily (Nymphaea) cultivars. PloS ONE 7: e34335
57. van der Krol AR, Lenting PE, Veenstra J, van der Meer IM, Koes RE, et al. (1988) An antisense chalcone synthase gene in transgenic plants inhibits flower pigmentation. Nature 333: 866–869.
58. Napoli C, Lemieux C, Jorgensen R (1990) Introduction of a chimeric chalcone synthase gene into petunia results in reversible co-suppression of homologous genes in trans. Plant Cell 2: 279–289.
59. Gutterson N (1995) Anthocyanin biosynthetic genes and their application to flower color modification through sense suppression. Hort Sci 30: 964–966.
60. Zuker A, Tzfira T, Ben-Meir H, Ovadis M, Shklarman E, et al. (2002) Modification of flower color and fragrance by antisense suppression of the flavanone 3-hydroxylase gene. Mol Breeding 9: 33–41.
61. Schwinn K, Venail J, Shang Y, Mackay S, Alm V, et al. (2006) A small family of MYB-regulatory genes controls floral pigmentation intensity and patterning in the genus Antirrhinum. Plant Cell 18: 831–851.

CHAPTRER 16

Combination of *Cyclamen Persicum* Mill. Floral Gene Promoters and Chimeric Repressors for the Modification of Ornamental Traits in *Torenia Fournieri*Lind.

Ichiro Kasajima[1,2,] Norihiro Ohtsubo[1,3] and Katsutomo Sasaki[1]

[1]*Institute of Vegetable and Floriculture Science, National Agriculture and Food Research Organization (NARO), Ibaraki 305-8519, Japan.*

[2]*Present address: Department of Agriculture, Iwate University, Iwate 020-8550, Japan.*

[3]*Present address: Graduate School of Life and Environmental Sciences, Kyoto Prefectural University, Kyoto 606-8522, Japan.*

ABSTRACT

Although chimeric repressors such as the Arabidopsis TCP3 repressor are known to have significant effects on flower morphology and color, their cellular-level effects on flower petals are not understood. The promoter sequences of the genes expressed in the flowers of cyclamen, a representative potted flower grown during the winter season, are also unknown. Here, we isolated eight promoters from cyclamen genes that are reportedly expressed in the petals. These promoters were then fused to four chimeric repressors and introduced into the model flower torenia to screen for effective combinations of promoters and repressors for flower breeding. As expected, some of the constructs altered flower phenotypes upon transformation. We further analyzed the effects of chimeric repressors at the cellular level. We observed that complicated petal and leaf serrations were accompanied by excessive vascular branching. Dichromatism in purple anthocyanin was inferred to result in bluish flowers, and imbalanced cell proliferation appeared to result in epinastic flowers. Thus, the genetic constructs and phenotypic changes described in this report will benefit the future breeding and characterization of ornamental flowers.

INTRODUCTION

The breeding of new flower cultivars with various phenotypes, such as different petal colors and shapes, is key to promoting flower use. Straightforward destruction of endogenous genes, introduction of new metabolic pathways from different species through transformation, and random mutagenesis using ion beam irradiation have been effective in generating double flowers as well as altering flower color, pigmentation patterns, and shapes (for example, refs 1,2,3,4,5). In addition to these approaches, the use of chimeric repressors to confer different characteristics upon flowers in a dominant genetic manner has also been tested.

Chimeric repressors are artificially generated by combining transcription activators with peptides known as repression domains. In plants, class II ethylene-responsive element-binding factor and TFIIIA-type zinc finger repressors, including SUPERMAN, share the 'EAR motif'.[6] Short peptides that include this motif function as repression domains, changing transcription activators into transcription suppressors. The most frequently used and strongest repression domain is 'SRDX,' which is a leucine-rich peptide consisting of 12 amino acid residues (LDLDLELRLGFA). This peptide was obtained by altering the native sequence of the repression domain of SUPERMAN.[7,8] The wide range of flower phenotypes generated to date using chimeric repressors is listed in the FioreDB database (http://www.cres-t.org/fiore/public_db/index.shtml; refs 9,10).

Due to the ease of transformation compared with other ornamental crops, torenia (*Torenia fournieri* Lind.) is used as a model system for transgenic flower breeding. Various chimeric repressors of Arabidopsis (*Arabidopsis thaliana* (L.) Heynh.) transcription factors have been surveyed in torenia.[11] According to the comparison of all such chimeric repressors examined in torenia thus far, repressors of *TCP* (Teosinte branched1, Cycloidea and Pcf) and *SEP3* (SEPallata3) are among those causing the clearest phenotypic changes.[12,13,14,15] Endogenous promoters of torenia have been used to drive the Arabidopsis TCP3 repressor to change torenia flower phenotypes specifically and in various ways.[16] Thus, variation in promoters and chimeric repressors promotes the breeding of a variety of transgenic flowers. However, despite the publication of several relevant reports, the physiological effects of chimeric repressors linked to phenotypic changes in transgenic flowers, such as petal serration, have not been examined in ornamental flowers.

In the present study, we first isolated promoters from cyclamen (*Cyclamen persicum* Mill.), which is a representative potted winter-flowering plant. We identified eight promoter sequences of genes known to be expressed in cyclamen petals. We then cloned these promoters and used them to drive four different chimeric repressors in torenia. Examination of the transgenic torenia flowers harboring these genetic constructs allowed us to characterize the effects of these cyclamen promoters and chimeric repressors. Further analyses of the transgenic lines suggested morphological or biochemical effects linked to the phenotypic changes in the flowers.

MATERIALS AND METHODS

Isolation of cyclamen promoters

The full-length cDNA sequences of *CpAP1, CpPI, CpAP3A, CpAP3B, CpSEP2, CpSEP3, CpFLC, CpCHS, CpDFR* and *CpOMT* were previously reported (Table 1). Three gene-specific primers (GSP1, GSP2 and GSP3; Supplementary Table S1) were designed in reverse orientation corresponding to the 5′-untranslated region (UTR) or the first exon of each cDNA sequence, and additional gene-specific primers were designed for some of the genes to extend the promoter sequences. Genomic DNA was extracted from cyclamen (*Cyclamen persicum* Mill. cultivar 'Mini Loyal Purple') leaves using 'PVPP buffer' containing 4 MNaCl and was purified via ultracentrifugation.[17,18] The promoter sequences were amplified by adaptor PCR with a 'mismatched DS4 adaptor' (ref. 19; Supplementary Table S1), which potentially has a better capability to amplify target sequences. Briefly, genomic DNA was digested with eight different restriction enzymes (*Dra*I, *Eco*RV, *Sca*I, *Ssp*I, *Hin*dIII, *Kpn*I, *Sac*I and *Spe*I) and then ligated to adaptors and amplified by nested PCR using gene-specific primers and adaptor primers (DS4-AP1, DS4-AP2 or DS4-AP3; Supplementary Table S1). PCR was performed with KOD Plus Neo polymerase (Toyobo, Osaka, Japan) for 35 cycles, with an annealing temperature of 68 °C and an extension time of 5 min. The amplified fragments were recovered from the agarose gels, TA-cloned into plasmids, and sequenced. The promoters were then amplified from the genomic DNA using cloning primers (CP-F and CP-R; Supplementary Table S1) in which *Hin*dIII (5′- AAGCTT-3′) or *Bam*HI (5′- GGATCC-3′) sites were attached at either end for subcloning (underlined in Supplementary Table S1). Amplification was performed with KOD Plus Neo polymerase for 45 cycles, with an annealing temperature of 68 °C and an extension time of 5 min. The accession numbers and lengths of the isolated promoters are shown in Table 1.

Table 1. Properties of cyclamen petal genes (*CPGs*) and their promoters.

Gene	*MADS class*	*INSDC[a] accession number*		*Promoter length[b]*
		Gene	*Promoter*	*(bp)*
CpAP1	A	AB600230.1	LC176494	1596
CpPI	B	AB600234.2	LC176500	2030
CpAP3A	B	AB600231.1	LC176495	1179
CpAP3B	B	AB600232.1	LC176496	1554
CpSEP2 (*CpMADS1*)[c]	E	AB600235.1	LC176501	1863
CpSEP3 (*CpMADS2*)[c]	E	AB600236.1	—	—
CpFLC (*CpMADS3*)[c]	—	AB600233	—	—
CpCHS[d]	—	DD354767	LC176497	1555
CpDFR[d]	—	DD354768	LC176498	1528
CpOMT[d]	—	DD248507	LC176499	1655

[a]International Nucleotide Sequence Database Collaboration (collaboration between DDBJ, EMBL-EBI, and NCBI; http://www.insdc.org/). [b]Lengths of the isolated sequences, upstream of the start codon (ATG). [c]The original names for *CpSEP2*, *CpSEP3*, and *CpFLC* were *CpMADS1*, *CpMADS2* and *CpMADS3*, respectively.[25] [d]These genes are expressed in cyclamen petals, as their cDNA sequences were isolated from cyclamen petals (Japan patents 2006-115861 and 2005-312388; https://www.j-platpat.inpit.go.jp/web/tokujitsu/tkbs/TKBS_GM101_Top.action, Japanese).

Subcloning

The protein-coding regions of *AtTCP3*, *CpTCP1B*, *AtSEP3*, and *CpSEP3* (International Nucleotide Sequence Database Collaboration (INSDC) accession numbers NM_104201.3, LC186052, NM_180622.2 and AB600236.1, respectively) were cloned into the *Bam*HI–*Sma*I site of the p35SSRDXG vector20 to produce *35S*p:*CR* vectors. Single-nucleotide (silent) mutations were introduced into the *AtSEP3* and *CpSEP3* sequences to eliminate restriction sites without changing the encoded protein sequences. Cyclamen promoter fragments were introduced into the *Hind*III-*Bam*HI sites of *35S*p:*CR* vectors or into the pG2-35S-GUS vector to produce 32 *CPG*p:*CR* vectors and four *CPG*p:*GUS* vectors, respectively. The regions corresponding to each transgene were transferred into the binary vector pBCKH (for hygromycin selection of transgenic plants) or pBCKK (for kanamycin selection of transgenic plants; ref. 20) using the Gateway system (Thermo Fisher Scientific, Waltham, MA, USA).

Plant materials, plant growth conditions and transformation

In vitro cloned seedlings of torenia (*Torenia fournieri* Lind. cultivar 'Crown Violet' and the hybrid torenia cultivar 'Summerwave Blue') were maintained on sterilized media in transparent plastic boxes at 25 °C under white fluorescent light (photosynthetic photon flux density (PPFD)=150 μmol m^{-2} s^{-1}) with a 16-h/8-h light/dark photoperiod. For transformation, bulk (mixture) or individual plant expression vectors carrying the appropriate constructs were introduced into *Agrobacterium tumefaciens* strain EHA105 via electroporation, and the *Agrobacterium* were subsequently infiltrated into leaf discs as previously described.21 Briefly, thousands of 5-mm leaf discs were prepared from *in vitro* cultures, inoculated with *Agrobacterium*, then co-cultured on media without antibiotics, and transformed calli were generated on selection media containing antibiotics. One or two transgenic lines were typically obtained from 100 leaf discs. The number of total leaf discs was ~8000. Regenerated plant cultures were maintained on media containing trehalose, and the media were renewed every 2 months.22 Transformants were then transferred to pots and grown in a sunlit greenhouse.

The transformants generated with bulk plasmids were examined via nested PCR to determine the inserted constructs (Supplementary Figure S4). The first round of PCR was performed with 'Whole-F1' and 'Whole-R1' primers.

The first PCR product (diluted 1/1000 in solution) was then amplified using a combination of the 'Whole-F2' primer, individual promoter-specific primers (PSP; Supplementary Table S1), and the mixture of all repressor-specific primers (RSP; Supplementary Table S1). Both the first and second rounds of PCR were performed with KOD Plus Neo polymerase for 40 cycles, with an annealing temperature of 55 °C and an extension time of 2 min. The repressor genes (*AtTCP3-RD*, *CpTCP1B-RD*, *AtSEP3-RD* or *CpSEP3-RD*) introduced into the transformants were finally determined based on the PCR product sizes (approximately 650 bp for *CpTCP1B-RD*, 500 bp for *CpSEP3-RD*, 400 bp for *AtSEP3-RD*, and 200 bp for *AtTCP3-RD*). The whole-F2 primer generates large DNA fragments when there is no construct corresponding to the target promoter sequence. Inclusion of the whole-F2 primer in the second PCR is necessary to reduce non-specific amplification. This scheme was repeated for all *CPG* promoters that may have been introduced into the target line. Tetraploid transformants and those transformants harboring multiple constructs were not used for phenotypic analyses. As described in the 'Results' section, 95 transgenic crown violet lines possessing single constructs were generated via transformation with individual constructs (59 lines) or bulk constructs (36 lines); 52 transgenic crown violet lines were generated with a bulk plasmid of 8 constructs (*CpOMT*p and *CpSEP2*p constructs, inserts not determined); and 12 transgenic summerwave blue lines were generated with the *CpAP3B*p-*CpTCP1B-RD* construct.

GUS staining

GUS staining was performed with T_1 (first generation) transgenic plants (Figure 2). Plant tissues were pretreated with 90% acetone and placed under vacuum for 30 min in 100 mM sodium phosphate buffer (pH 7.0) containing 0.1% Triton X-100, 0.5 mg 5-bromo-4-chloro-3-indolyl-β-D-glucuronide (X-Gluc), 10 mM ethylenediaminetetraacetic acid (EDTA), 5 mM potassium ferricyanide and 5 mM potassium ferrocyanide. The tissues were then incubated in this solution at 37 °C in the dark overnight. The stained tissues were washed with 70% ethanol, treated with ethanol-acetic acid (6:1, v/v) overnight, washed again with 70% ethanol, and treated with chloral hydrate solution (chloral hydrate:glycerol:sterile water=8 g:1 mL:2 mL) overnight. Images were obtained using a single-lens reflex digital camera with a white backlight to observe whole staining or with a stereoscopic microscope (MZ16 FA, Leica, Wetzlar, Germany) to observe staining in small areas.

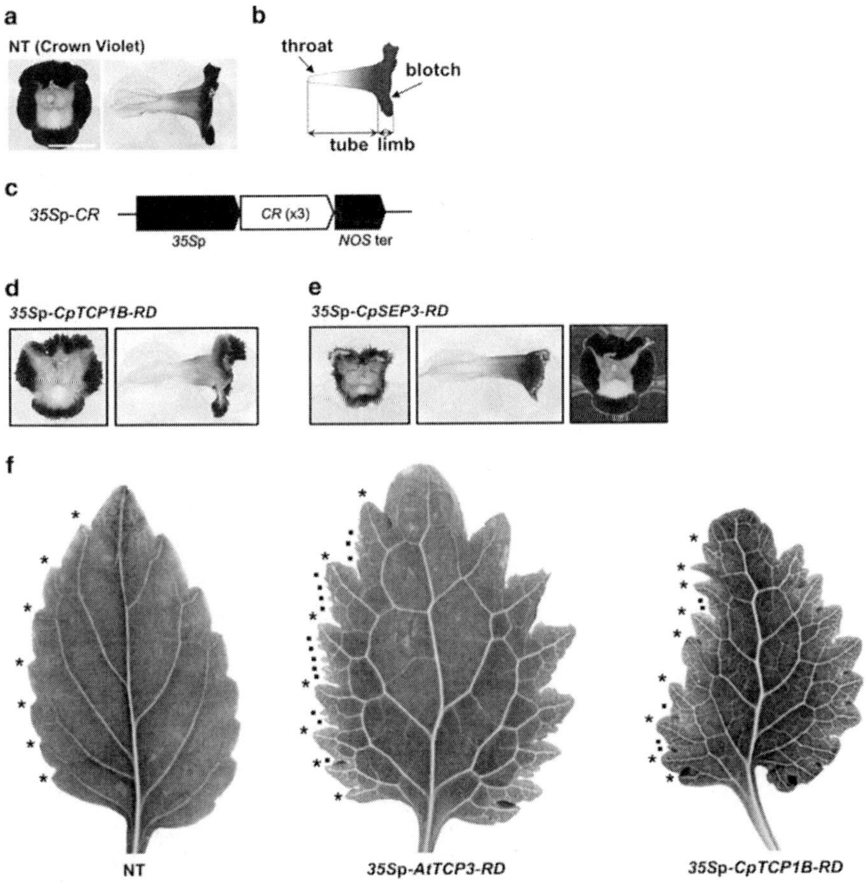

Figure 1. Structure of torenia flowers and the phenotypes of *35S*p-*CR* lines. (**a**) A non-transgenic (NT) Crown Violet flower. Bar=1 cm. (**b**) Schematic representation of the flower structure. (**c**) The genetic constructs introduced. *35S*p was combined with three chimeric repressor (*CR*) genes (*CpTCP1B-RD, CpSEP3-RD* or *AtTCP3-RD*). (**d**) A transgenic line expressing *35S*p-*CpTCP1B-RD*. (**e**) Two transgenic lines expressing *35S*p-*CpSEP3-RD*. (**f**) Leaf silhouette and vascular structure of the non-transgenic line, the *35S*p-*AtTCP3-RD* line, and the *35S*p-*CpTCP1B-RD* line. The primary serrations are indicated with asterisks, and the secondary serrations are indicated with dots only on the left side of the leaves.

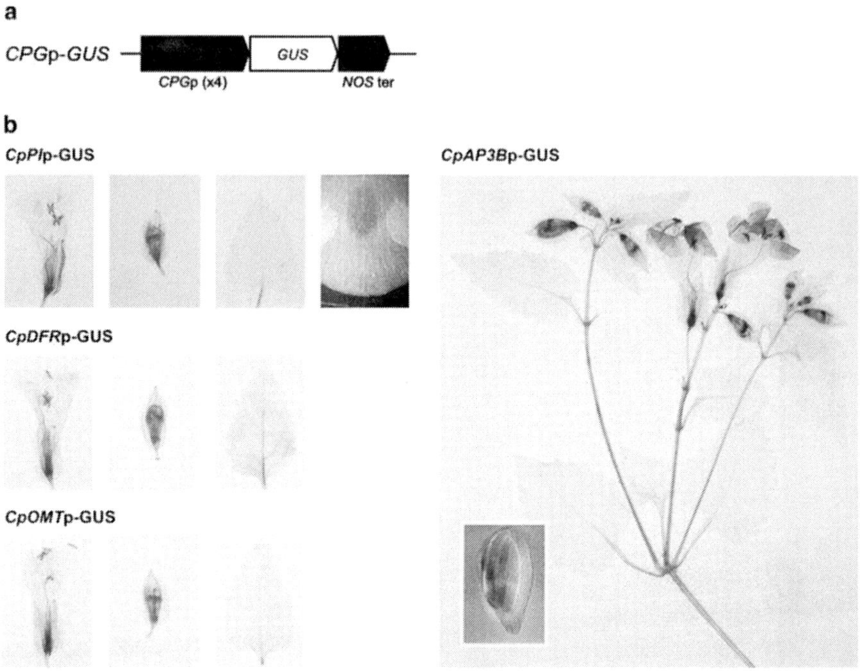

Figure 2. *CPG*p-*GUS* lines. (**a**) The genetic constructs introduced. Four cyclamen petal gene (*CPG*) promoters (*CpPI*p, *CpDFR*p, *CpOMT*p, and *CpAP3B*p) were used to drive *GUS* gene expression. '*NOS* ter' represents the *NOPALINE SYNTHASE* terminator. (**b**) GUS staining of the *CpPI*p-*GUS* line (left to right: flower, floral bud, leaf, and blotch), the *CpDFR*p-*GUS* line (flower, floral bud and leaf), the *CpOMT*p-*GUS* line (flower, floral bud and leaf), and the *CpAP3B*p-*GUS* line (inset shows the anther excised from floral bud).

Vascular structure

Leaves were imaged using a single-lens reflex digital camera with a white backlight. The resulting digital images were retouched using Adobe Photoshop software (Adobe Systems Inc., San Jose, CA, USA) by converting the color image to a black and white image and then adjusting the brightness and contrast to make the vasculature clear (Figure 1f). Petals were treated with ethanol-acetic acid overnight and then placed in chloral hydrate solution overnight. The vasculature was observed under dark-field conditions with a stereoscopic microscope (MZ16 FA, Leica; Figure 6b).

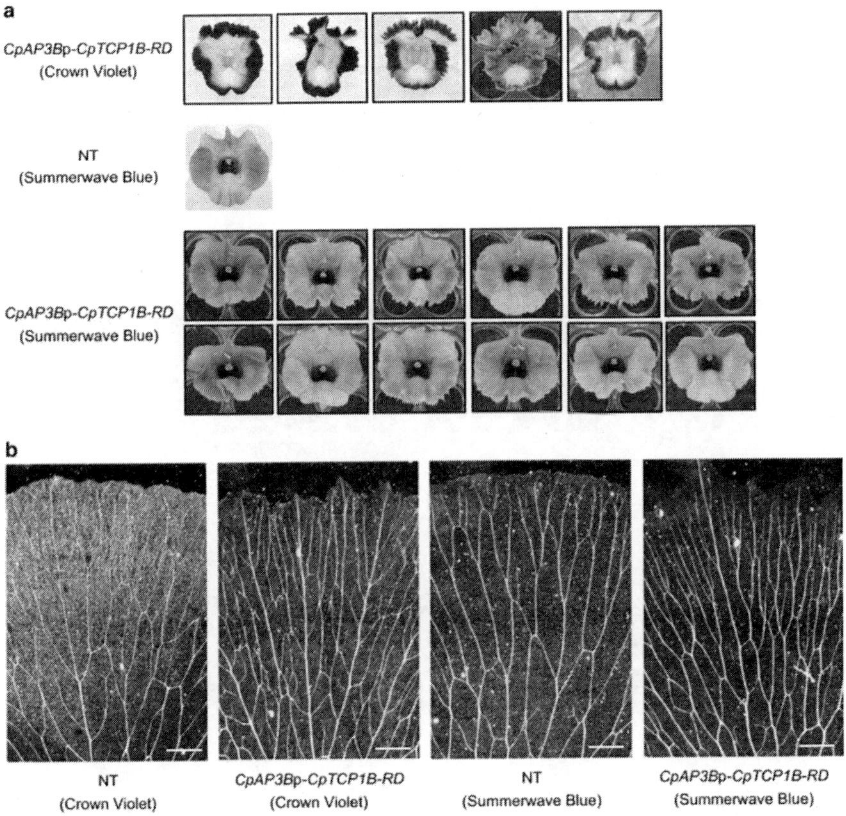

Figure 6. Phenotypes of *CpAP3B*p-*CpTCP1B-RD* lines. (**a**) Five *CpAP3B*p-*CpTCP1B-RD* lines of 'Crown Violet,' the non-transgenic cultivar 'Summerwave Blue,' and twelve *CpAP3B*p-*CpTCP1B-RD* lines of 'Summerwave Blue.' (**b**) Vascular structures in the petals of non-transgenic 'Crown Violet,' a *CpAP3B*p-*CpTCP1B-RD* line of 'Crown Violet,' non-transgenic 'Summerwave Blue,' and a *CpAP3B*p-*CpTCP1B-RD* line of 'Summerwave Blue.' Bars=1 mm.

Petal cross-sections and scanning electron microscopy

To prepare petal cross-sections, agar with an isotonic solute was produced (2% agar, 500 mM trehalose, 5 mM $CaCl_2$, 0.2% Tween-20). Side lobes excised from open flowers were sliced at a thickness of 55 μm with a vibrating microslicer (DTK-1000, Dosaka EM, Kyoto, Japan) and embedded in the agar. The slices were subsequently observed with a light microscope (AX70, Olympus, Tokyo, Japan; Figure 8e). The adaxial and abaxial surfaces of side lobes excised from open flowers were also observed with a scanning electron microscope (VE-7800, Keyence, Osaka, Japan; Figure 8b).

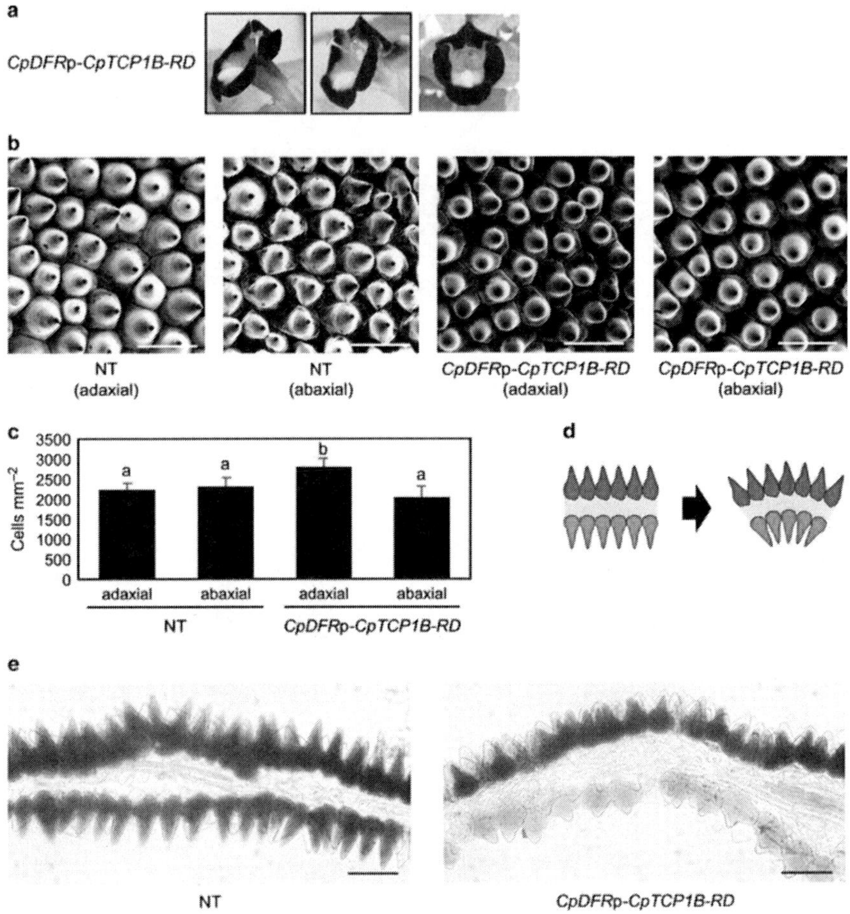

Figure 8. Microscopy analysis of *CpDFR*p-*CpTCP1B-RD* lines. (**a**) Phenotypes of three *CpDFR*p-*CpTCP1B-RD* lines. (**b**) Conical cells of the petals imaged via scanning electron microscopy on the adaxial/abaxial sides of the non-transgenic line and curled *CpDFR*p-*CpTCP1B-RD* line. (**c**) Conical cell density on petal surfaces. The data represent the mean and standard deviation (n=4). The values labeled with different letters (**a** or **b**) are significantly different from each other by Student's t-test (P<0.05). (**d**) Schematic representation of the effect of differential cell densities. (**e**) Cross-section of the lobe of the non-transgenic line and curled *CpDFR*p-*CpTCP1B-RD* line. Bars=50 μm in all images.

Colorimetry

The brightness of eight different parts of the same flowers was measured in images of flowers using Adobe Photoshop software (Figures 5b and 7b). The surface reflectance of the adaxial side of the side lobes was measured with an ultraviolet/visible (UV/VIS) spectrophotometer (UV-2450, Shimadzu, Kyoto, Japan; Figure 7c). RGB color under flat white light was calculated using a standardized CIE 1931 RGB color matching function. Hue, lightness, saturation, and coordinates in a round diagram were also calculated following previous reports (refs 23,24; Figure 7d). The recently developed round diagram is based on the RGB color system, which distributes hue angles and saturation values equally around a circle, consistent with our perception of color.24

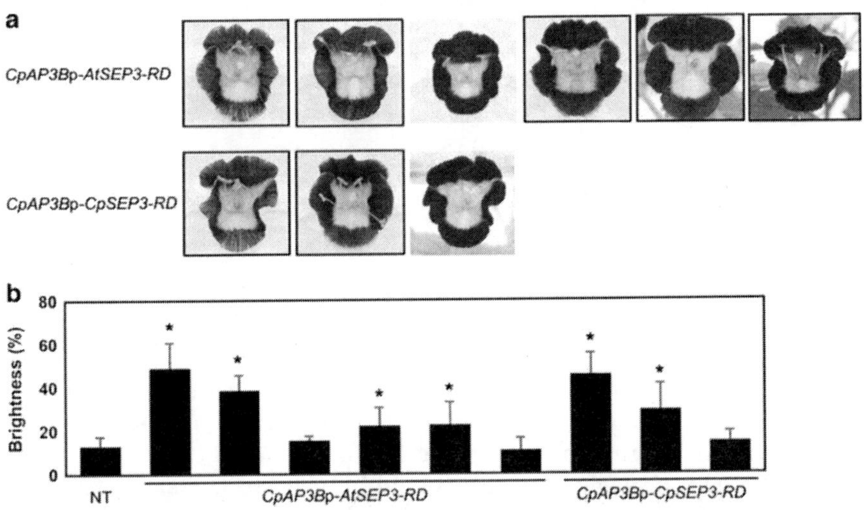

Figure 5. Comparison between the effects of *AtSEP3-RD* and *CpSEP3-RD*. (**a**) Phenotypes of six *CpAP3B*p-*AtSEP3-RD* lines and three *CpAP3B*p-*CpSEP3-RD* lines. (**b**) Brightness of the lobes measured in the non-transgenic line, *CpAP3B*p-*AtSEP3-RD* lines, and *CpAP3B*p-*CpSEP3-RD* lines. The data represent the mean and standard deviation (n=8). Asterisks indicate significant differences from the non-transgenic line by Student's t-test (P<0.05). The data for transgenic lines correspond to flowers shown in (**a**) in the same order.

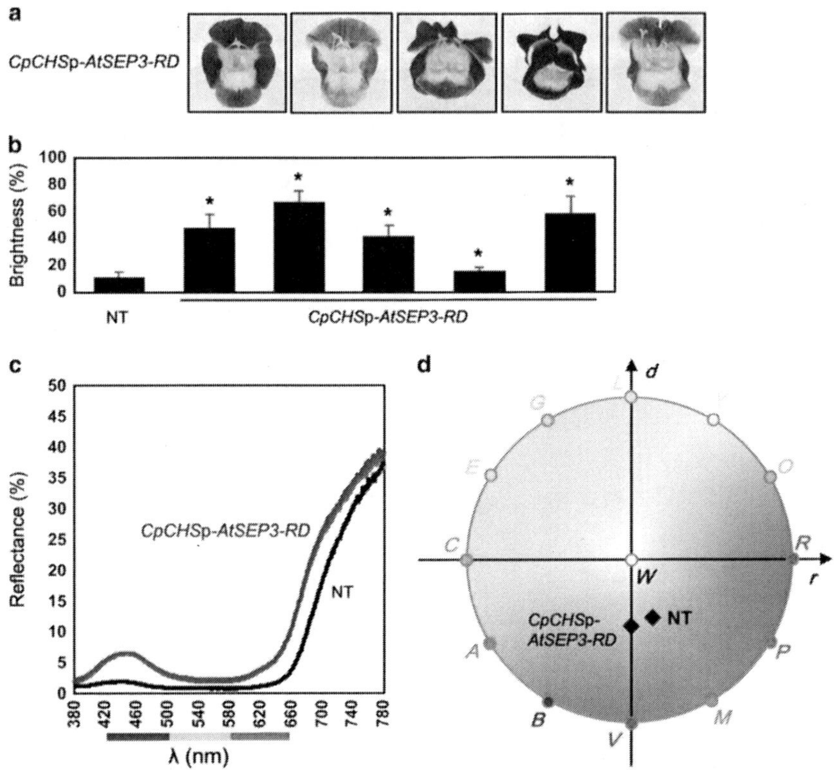

Figure 7. Colorimetric analysis of *CpCHS*p-*AtSEP3-RD* lines. (**a**) Phenotypes of five *CpCHS*p-*AtSEP3-RD* lines. (**b**) Brightness of the lobes measured in the non-transgenic line and *CpCHS*p-*AtSEP3-RD* lines. The data represent the mean and standard deviation (n=8). Asterisks indicate significant differences from the non-transgenic line by Student's t-test ($P<0.05$). The data for the transgenic lines correspond to the flowers shown in (**a**) in the same order. (**c**) Reflectance spectra of visible light on the lobes of the non-transgenic line (black and gray lines) and of a light-colored *CpCHS*p-*AtSEP3-RD* line (violet and purple lines). The measurements were performed in two replications with different flowers; each replication produced similar results, and the lines (black and gray, or violet and purple) overlap with each other. The blue, green, and red bars at the bottom of the graph indicate the approximate color of light at each wavelength. (**d**) Colors (hue and saturation) of the non-transgenic line and a *CpCHS*p-*AtSEP3-RD* line plotted on a *round* color diagram.

RESULTS

Isolation of cyclamen petal gene promoters

To express the chimeric repressors under the control of a variety of cyclamen promoters, we isolated 1.5-kb or longer upstream sequences (promoter sequences) of cyclamen petal genes (referred to as *CPGs* hereafter). Eleven homeotic MADS-box genes have been isolated from cyclamen, including *CpAP1*, *CpPI*, *CpAP3A*, *CpAP3B*, *CpSEP2*, *CpSEP3*, and *CpFLC*, which are all expressed in the petals (ref. 25; Table 1). In addition to these seven *CPGs*, we targeted three genes involved in anthocyanin pigmentation (*CpCHS*, *CpDFR* and *CpOMT*), which are also expressed in the petals (see footnote of Table 1). Here, the abbreviations *AP*, *PI*, *FLC*, *CHS*, *DFR* and *OMT* represent apetala, pistillata, flowering locus c, chalcone synthase, dihydroflavonol 4-reductase, and O-methyltransferase, respectively.

After adaptor PCR analysis, we successfully isolated the promoter sequences of 8 of the 10 target *CPGs* (*CpAP1*, *CpPI*, *CpAP3A*, *CpAP3B*, *CpSEP2*, *CpCHS*, *CpDFR*, and *CpOMT*). These promoter sequences are abbreviated as *CpAP1*p, *CpPI*p, *CpAP3A*p, *CpAP3B*p, *CpSEP2*p, *CpCHS*p, *CpDFR*p, and *CpOMT*p, respectively. All isolated promoter sequences (*CPG*p sequences), except for *CpAP3A*p (1179 bp), were longer than 1.5 kb, ranging from 1528 to 2030 bp (Table 1).

In this study, the transformed material was the torenia cultivar 'Crown Violet' (Figure 1a,b) unless otherwise indicated. Torenia presents flowers with compound petals (gamopetalous corolla). The basal part of the corolla (referred to as the 'throat' hereafter) is yellow, while the middle part is light purple. The basal and middle parts of the corolla are collectively referred to as the 'tube.' The top part of the corolla (the 'limb') is dark purple and is divided into four parts ('lobes'). The lobe margins are rounded, and there is a yellow spot on the basal lobe of the corolla, which is referred to as the 'blotch.' The dark purple color of the limbs is caused by anthocyanin pigments.26

To examine promoter activity, we generated one transgenic torenia line expressing the *GUS* (β-glucuronidase) gene under four representative *CPG* promoters (*CpPI*p, *CpDFR*p, *CpOMT*p and *CpAP3B*p; Figure 2a). These transgenic plants were stained with a buffer containing 5-bromo-4-chloro-3-indolyl-β-D-glucuronide (X-Gluc), which is catabolized by the GUS enzyme to form a blue pigment. Blue pigmentation was observed only in the blotch in the *CpPI*p-*GUS* line (Figure 2b), indicating that *CpPI*p primarily drives gene expression in blotch cells. Weak staining was observed in the limbs of open flowers of the *CpDFR*p-*GUS* line (Figure 2b), while no staining was observed in the *CpOMT*p-*GUS* line (Figure 2b). The *CpAP3B*p-*GUS* line exhibited the greatest density of blue pigmentation among the four promoters examined (Figure 2b). Thus, strong staining was observed in the limbs of open flowers and in the anthers of floral buds, with weaker staining observed in the tubes of open flowers and in the basal part of the stem.

Preparation of chimeric repressors

In this study, we used four chimeric repressors to modify torenia flowers. In addition to *AtTCP3-RD* and *AtSEP3-RD*, whose effects have been previously reported in torenia,[13,14] a gene homologous to *AtSEP3* and a putative allele of *CpTCP1* from cyclamen were isolated, and corresponding chimeric repressors (*CpSEP3-RD* and *CpTCP1B-RD*) were prepared. Expression vectors in which *CpTCP1B-RD* or *CpSEP3-RD* was driven by the Cauliflower mosaic virus 35S promoter (*35S*p) were used for functional characterization of these repressors, as was the expression vector of *AtTCP3-RD* for comparison (Figure 1c).

We generated one transgenic torenia line harboring the *35S*p-*CpTCP1B-RD* construct (Figure 1d) and two transgenic torenia lines harboring the *35S*p-*CpSEP3-RD* construct (Figure 1e). The line expressing *35S*p-*CpTCP1B-RD* possessed serrated petals. One of the lines expressing *35S*p-*CpSEP3-RD* exhibited jaggy petals, which were more pointed than the serrated petals. Compared with previous reports on *35S*p-*AtTCP3-RD* and *35S*p-*AtSEP3-RD*, *35S*p-*CpTCP1B-RD* showed similar effects to *35S*p-*AtTCP3-RD*, while *35S*p-*CpSEP3-RD* showed similar effects to *35S*p-*AtSEP3-RD* in transgenic torenia petals.

Torenia lines expressing *35S*p-*AtTCP3-RD* or *35S*p-*CpTCP1B-RD* produce petals and leaves with deeper and more complicated serrations. That is, non-transgenic leaves exhibit only 'primary' serrations, but the transgenic lines exhibit small 'secondary' serrations formed on the primary serrations as well as deeper primary serrations (Figure 1f). In our observations, we noted that the transgenic lines with more complicated serrations displayed more vascular branching in the leaves compared with the leaves of the non-transgenic line (Figure 1f). Vascular branching was particularly promoted at the leaf margins of these transgenic lines.

Expression of chimeric repressors under the control of cyclamen petal gene promoters

We prepared 32 different expression vectors using different combinations of the eight *CPG*p sequences and the four chimeric repressors (*CPG*p-*CR* vectors; Figure 3a). Torenia plants were transformed with each vector separately, and 59 lines were generated from these transformations. An additional 36 lines were generated through bulk transformation, whose T-DNA inserts were determined via PCR. Figure 3b shows representative lines with the most clearly modified flower phenotypes associated with each construct. Thirteen constructs resulted in phenotypes that differed from those of the non-transgenic line. *CpAP1*p generated flowers with jaggy margins when driving *AtTCP3-RD* or *CpTCP1B-RD*. *CpPI*p generated flowers without a blotch or with a smaller blotch (blotchless flowers) when driving *AtSEP3-RD* or *CpSEP3-RD*. *CpAP3B*p generated serrated and wavy flowers when driving *AtTCP3-RD* or *CpTCP1B-RD*. *CpAP3B*p also generated blotchless and striped flowers when driving *AtSEP3-RD* or *CpSEP3-RD*. *CpCHS*p generated 'trumpet-shaped' flowers with outer-curling limbs when driving *AtTCP3-RD* and 'tube' flowers without limbs when driving *CpTCP1B-RD*. *CpCHS*p generated blotchless and light-colored flowers when driving *AtSEP3-RD* or *CpSEP3-RD*. *CpDFR*p generated trumpet-shaped flowers when driving *CpTCP1B-RD*. *CpAP3A*p, *CpSEP2*p and *CpOMT*p combined with four repressors did not generate any phenotypic changes.

The absence of phenotypic changes associated with the constructs driven by *CpSEP2*p or *CpOMT*p was supported by the finding that a population of an additional 52 lines transformed with a bulk plasmid consisting of eight constructs driven by these promoters did not show any phenotypic changes (Supplementary Figure S1). Although it was not determined which construct was introduced into each line, eight constructs containing *CpSEP2p* or *CpOMTp* were expected to be randomly introduced into these lines. Excluding this population, one to seven transgenic lines per 32 constructs (approximately three lines, on average) were analyzed (Supplementary Figure S2). Approximately half of the lines were modified from the non-transgenic line when the 13 constructs were introduced, and the phenotypes were similar to each other when the transgenic lines possessed the same construct. It is possible that the remaining 19 constructs may produce phenotypic changes if larger populations of transgenic plants are generated. The average rate of phenotypically modified transgenic lines in the *CPG*p-*CR* population was ~30% for *CpTCP1B-RD*, *AtSEP3-RD* and *CpSEP3-RD*, while the rate for *AtTCP3-RD* was less than 20% (Figure 3c).

Comparison between the chimeric repressors derived from Arabidopsis and cyclamen

AtTCP3-RD and *CpTCP1B-RD* caused similar phenotypic changes in transgenic torenia flowers, although both the degree and the rate of phenotypic changes appeared to be greater in association with *CpTCP1B-RD*. However, *AtSEP3-RD* and *CpSEP3-RD* caused similar phenotypic changes and showed a similar degree and rate of phenotypic changes (Figures 3b and Supplementary Figure S2). These trends were reconfirmed through an analysis of representative constructs.

The *CpAP1*p-*AtTCP3-RD* and *CpAP1*p-*CpTCP1B-RD* constructs were chosen for comparison of the effects of *AtTCP3-RD* and *CpTCP1B-RD*. Both constructs generated serrations at the lobe margins, but the degree of serration tended to be greater in *CpAP1*p-*CpTCP1B-RD* lines (Figure 4). The *CpAP3B*p-*AtSEP3-RD* and *CpAP3B*p-*CpSEP3-RD* constructs were chosen for a comparison of the effects of *AtSEP3-RD* and *CpSEP3-RD*. Both constructs decreased the purple coloration of the lobes and the yellow coloration of the blotch (Figure 5a). The degree of phenotypic changes was quantified based on the brightness of the lobes (Figure 5b); when the purple coloration is reduced by the chimeric repressors to a greater extent, the lobes become brighter. The distribution of lobe brightness in the transgenic lines was similar between *AtSEP3-RD* and *CpSEP3-RD*, suggesting that *AtSEP3-RD* and *CpSEP3-RD* have similar effects in torenia.

Microscopic analysis of serrated flowers

Although the beauty of flowers will always be subjective depending on personal preferences, the flowers among these lines showing the greatest degree of modification are the serrated, wavy (and light-colored) flowers associated with *CpAP3B*p-*CpTCP1B-RD*, the blotchless, light-colored flowers associated with *CpCHS*p-*AtSEP3-RD*, and the trumpet-shaped flowers associated with *CpDFR*p-*CpTCP1B-RD*. In addition, all flowers on individual plants showed similar, consistent phenotypes. Examples of plants producing these three types of flowers are shown in Supplementary Figure S3. The flowers generated by these three constructs were further analyzed.

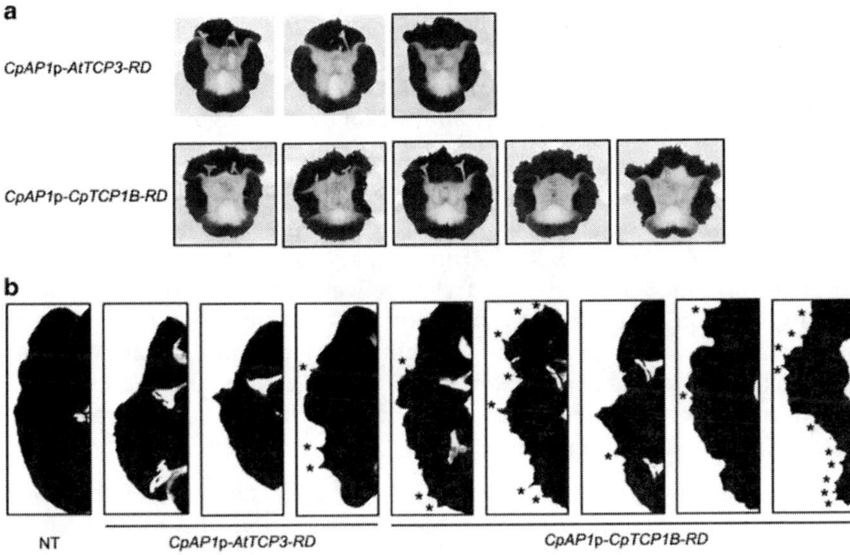

Figure 4. Comparison between the effects of *AtTCP3-RD* and *CpTCP1B-RD*. (**a**) Phenotypes of three *CpAP1*p-*AtTCP3-RD* lines and five *CpAP1*p-*CpTCP1B-RD* lines. (**b**) Silhouettes of the top lobes of the non-transgenic line, *CpAP1*p-*AtTCP3-RD* lines, and *CpAP1*p-*CpTCP1B-RD* lines. Serrations are indicated by asterisks. The silhouettes of the transgenic lines correspond to the flowers shown in (**a**) in the same order.

Phenotypic changes were observed to varying degrees in all five *CpAP3B*p-*CpTCP1B-RD* lines of the 'Crown Violet' cultivar (Figure 6a). We also generated 12 *CpAP3B*p-*CpTCP1B-RD* lines of the torenia cultivar 'Summerwave Blue,' all of which exhibited serrations and/or slight reductions in purple coloration (Figure 6a). We observed vascular structures on the serrated side lobes of *CpAP3B*p-*CpTCP1B-RD* lines (Figure 6b). Vascular branching was enhanced by the introduction of the *CpAP3B*p-*CpTCP1B-RD* construct in both the crown violet and summerwave blue genetic backgrounds.

Colorimetric analysis of light-colored flowers

Transgenic lines harboring the *CpCHS*p-*AtSEP3-RD* construct produced blotchless and light-colored flowers (Figure 7a). The lobe colors of all five lines were brighter than those of the non-transgenic line (Figure 7b). Although it may not be clear from the digital images (due to a problem with color regeneration from the digital camera), the light-colored transgenic flowers tended to be slightly more bluish than the non-transgenic flowers. To examine this phenomenon more precisely, we analyzed the side lobes of one of the *CpCHS*p-*AtSEP3-RD* lines following rigorous colorimetric procedures.

First, we measured the reflectance spectra of visible wavelengths on the lobes with a spectrophotometer (Figure 7c). Wavelengths in the ranges of 420–500, 500–580 and 580–660 nm roughly correspond to blue, green and red coloration, respectively.24 The reflectance recorded in the non-transgenic line consisted of a medium peak in the blue range (~440 nm), low values in the green range, and a large slope in the red range. Reflectance was enhanced for all wavelengths in the *CpCHS*p-*AtSEP3-RD* line, but the positions of the blue peak and the red slope were similar to those in the non-transgenic line.

We transformed the visible light spectra of the reflectance of the lobes to RGB color using an 'RGB color matching function.' In general, color is composed of three factors: hue (for example, red, yellow, green and purple), lightness (brightness), and saturation (vividness). In colorimetry, hue is expressed as the hue angle (0–360°), while lightness and saturation are expressed as ratios (0–100%). The non-transgenic line and the *CpCHS*p-*AtSEP3-RD* line exhibited hue angles (H^{RGB} value) of 290° and 269°, lightness values (L^{RGB}) of 4% and 12%, and saturation values (S^{RGB2}) of 37% and 40%, respectively. The lobe colors were then plotted on a '*round* diagram' (Figure 7d).

Microscopic observation of trumpet-shaped flowers

Trumpet-shaped flowers were generated in two of the three *CpDFR*p-*CpTCP1B-RD* lines (Figure 8a). We examined the effect of this construct on the generation of trumpet-shaped flowers by performing microscopic observations of the side lobes of both the non-transgenic line and a trumpet-shaped *CpDFR*p-*CpTCP1B-RD* line.

We directly observed the density of conical cells on the adaxial (upper) and abaxial (lower) surfaces of the side lobes in the non-transgenic line and the *CpDFR*p-*CpTCP1B-RD* line through scanning electron microscopy (Figure 8b). The conical cell density was significantly greater on the adaxial surface than on the abaxial surface of the *CpDFR*p-*CpTCP1B-RD* line (Figure 8c). Figure 8d illustrates a possible geometric effect of increased conical cell density on the adaxial surface of the lobes, while Figure 8e shows cross-sections of fresh lobes prepared from both the non-transgenic and *CpDFR*p-*CpTCP1B-RD* lines.

DISCUSSION

Activities of *CPG* promoters in torenia and their relationship with flower phenotypes

The sizes of the isolated promoter sequences (all longer than 1.5 kb, except for *CpAP3A*p) are expected to be sufficient for the expression of downstream genes because conserved domains are found within 1.0 kb of promoter sequences for dicot MADS-box genes.27 In addition, floral gene promoters ranging from 1394 bp to 2083 bp derived from Arabidopsis or torenia have been successfully used previously for gene expression in torenia petals.16 Different combinations of *CPG* promoters and chimeric repressors generated different types of flowers in torenia, although some specific combinations did not result in any phenotypic changes (Figure 3). The observed phenotypic changes in the petals fell into three categories: changes in lobe morphology (e.g., serrated lobes), changes in the yellow coloration of the blotch, and changes in the purple coloration of the limbs. Figure 9 shows the phenotypic changes in these categories caused by different combinations of *CPG* promoters and chimeric repressors (*TCP-RD* or *SEP3-RD*). For example, because transgenic lines harboring *CpPI*p-*SEP3-RD* constructs exhibited a blotchless phenotype, they were considered modified in category 'Y' (yellow) but not in categories 'L' (lobes) and 'P' (purple). In contrast, transgenic lines harboring *CpAP3B*p-*TCP-RD* constructs were modified in categories 'L' and 'P,' but these lines always exhibited normal blotch colors.

		[Promoter]				
		*CpAP1*p (L)	*CpPI*p (Y)	*CpAP3B*p (L,Y,P)	*CpCHS*p (L,Y,P)	*CpDFR*p (L)
[Chimeric repressor]	TCP-RD (L,P)	L	-	L,P	L,P	L
	SEP3-RD (Y,P)	-	Y	Y,P	Y,P	-

Figure 9. Relationship between *CPG* promoters, chimeric repressors, and the categories of flower phenotypes. The phenotypes of transgenic torenia flowers generated by a combination of five *CPG* promoters (*CpAP1*p, *CpPI*p, *CpAP3B*p, *CpCHS*p or *CpDFR*p) and two types of chimeric repressors (*TCP-RD* or *SEP3-RD*) were categorized as 'L' (lobe morphology), 'Y' (yellow color

of the blotch), or 'P' (purple color of the petal). The potential effects of the *CPG* promoters and chimeric repressors, as deduced from the flower phenotypes, are also indicated in parentheses.

This categorization of the observed phenotypic changes in the transgenic lines allowed us to deduce the capacity of *CPG* promoters and chimeric repressors to modify these respective categories of flower phenotypes. Thus, *TCP* repressors can modify L-type and P-type phenotypes, while *SEP3* repressors can modify Y-type and P-type phenotypes. *CpAP1*p and *CpDFR*p can modify L-type phenotypes; *CpPI*p can modify Y-type phenotypes; and *CpAP3*Bp and *CpCHS*p can modify all three categories. Each phenotype category is modified only when both the promoter and repressor have the ability to modify it. This model, based on the differential activities of both *CPG* promoters and chimeric repressors, would reasonably explain the variable flower phenotypes generated by different constructs.

The tissue-specific activities of the four promoters (Figure 2) are also consistent with this model. *CpPI*p activity was specifically detected in the blotch, which explains why *CpPI*p can modify only the Y-type phenotype and why the flower throat remained yellow in the lines possessing the *CpPI*p-*SEP3-RD* construct. In contrast, *CpAP3B*p activity was observed in the whole petal, explaining why this promoter can modify all three phenotypic categories. *CpDFR*p was only weakly active in the limbs, and this promoter specifically modified L-type phenotypes. L-type phenotypes may be modified by TCP repressors more easily compared with P-type phenotypes. No activity of *CpOMT*p was detected in our experiments, and this promoter accordingly did not cause any changes in flower phenotypes. As *CpAP3A*p or *CpSEP2*p also did not cause phenotypic changes, these two promoters may not be active in torenia flowers.

The observed activity of *CpAP3B*p in the petals and stamens is in accord with its phylogenetic classification as a class-B MADS-box gene, although no phenotypic changes were observed in the stamens in this study. The blotch-specific activity of *CpPI*p was unexpected, considering its classification as a class-B *MADS-box* gene, and the GUS staining associated with *CpDFR*p in this study appeared weaker than the GUS staining associated with the torenia *DFR* promoter in a previous study.16 Torenia may be lacking in *trans* regulators that properly recognize and strongly drive the transgenic cis regulators *CpPI*p and *CpDFR*p. Due to their specific expression patterns, the activities of these promoters may result in a lack of phenotypic changes in transgenic torenia in some cases and unexpected flower phenotypes in others.

Role of TCP transcription factors in the regulation of cell proliferation

TCP transcription factors are known to regulate cell proliferation.28Among the *TCP* family genes, *AtTCP3* and *CpTCP1* are close homologs of the *CINCINNATA* gene (*AmCIN*) of snapdragon (*Antirrhinum majus*; ref. 15). A mutation in the *AmCIN* gene causes crinkly (curly) leaves, particularly at the margins, and arrests the growth of the lobes in snapdragon,29 while a snapdragon mutant harboring mutations in the *TCP* genes *DICHOTOMA* and *CYCLOIDEA* exhibits a rounded corolla.30Thus, TCP has various effects on flower morphogenesis.

The curly lobes observed in the *CpDFR*p-*CpTCP1B-RD* line (Figure 8a) were consistent with the mutant phenotype of the *AmCIN* gene in snapdragon. Outer curling of the lobes (that is, epinasty) is caused by an imbalance in cell proliferation and/or cell expansion between the adaxial (inner) and abaxial (outer) surfaces of the lobes. Microscopic analysis revealed an increased density of conical cells only on the adaxial surface of the *CpDFR*p-*CpTCP1B-RD* line (Figure 8c), indicating that *CpTCP1B-RD* accelerated cell proliferation on the adaxial surface of transgenic plants. The conical cells of the *CpDFR*p-*CpTCP1B-RD* line also tended to be flatter than those of the non-transgenic line (Figure 8e), similar to what was observed in a snapdragon *AmCIN* gene mutant.31

In addition to the petal curvature caused by *CpDFR*p-*CpTCP1B-RD*, the *TCP* repressors also generated complicated serrations in leaves and petals (Figures 1 and 6). The effects of TCP repressors on serrations can be divided into two categories. First, TCP repressors cause deeper primary serrations. This effect is consistent with the CUC2 (CUp-shaped Cotyledon 2)-mediated formation and regulation of leaf serrations.32 Indeed, *CUC2* expression is elevated in Arabidopsis plants with mutations in TCP genes.33 Second, TCP repressors generate small secondary serrations in leaves (Figure 1f) and possibly in petals (Figure 6b). The positions of the secondary serrations appear to overlap with the positions at which additional vasculature is formed in transgenic lines. This positive and possibly simultaneous regulation of the formation of secondary serrations and vascular branching should be independent of the CUC2 pathway because CUC2 is known to neither induce secondary serrations nor regulate vascular branching.32

A mutation in the *FRILL1* gene also causes frilled flowers in Arabidopsis,34 and the vascular pattern is altered in such flowers.35 The frilled phenotype of this mutant is attributed to the differential regulation of endoreduplication,36 a phenomenon in which some diploid (2C) cells are converted to tetraploid (4C), octaploid (8C) or hexadecaploid (16C) cells with an enlarged cell size. Cell ploidy analyses have shown that endoreduplication frequently occurs in vegetative cells of Arabidopsis (a diploid) (for example see refs. 37,38). However, there is no sign of endoreduplication in diploid torenia leaves; i.e., there is only a single peak corresponding to diploid (2C) cells (Kasajima et al., unpublished). Although we do not have sufficient data to reach a definite conclusion, vascular branching, rather than endoreduplication, may be linked to the serrations observed in torenia.

Phenotypic changes caused by SEP3 repressors are not necessarily strong, but the reduction of purple coloration may be a sign of the conversion of petals to sepals. This possibility could be examined using gene expression analysis in future studies.

Why do transgenic plants with *CpCHS*p-*SEP3-RD* constructs produce bluish flowers?

*CpCHS*p-*SEP3-RD* constructs generated blotchless, light-colored flowers (Figure 3). The characteristics of these flowers lie partly in the simple balance of colors, in the absence of a yellow blotch, and in lighter lobe colors. The lobe coloration of these lines also appeared somewhat bluish when directly observed with the naked eye, which was confirmed by colorimetric analysis (Figure 7). Judging from the obtained colorimetric values, the non-transgenic flowers were magenta (red-purple), while the *CpCHS*p-*AtSEP3-RD* flowers were violet (blue-purple) and three times lighter. The purple petal color of torenia is caused by anthocyanins.26

The color of purple flowers is affected by changes in the spectra of incident light. Therefore, the abovementioned flower colors (magenta and violet) were calculated under a hypothetical 'flat' white light, which mimics the solar spectrum, with an even distribution of spectral power across all visible wavelengths (380–780 nm). The color of purple flowers, including that of purple torenia flowers, becomes bluish under white light-emitting diode (LED) light and reddish under incandescent light due to the colorimetric phenomenon known as the 'alexandrite effect' or 'color change'.24 This may be the principal reason that the color (hue) of a given purple flower appears different between images.

The observed difference in the hues between the non-transgenic line and the *CpCHS*p-*AtSEP3-RD* line may have been caused by a colorimetric phenomenon other than changes in the chemical structure of anthocyanins. In a previous report, a bluish torenia line (cultivar 'Crown Violet') was generated by suppressing the *DFR* gene.26,39Suppression of the *DFR* gene modified co-pigment accumulation and shifted the spectral pattern of light absorbance to the right (that is, to longer wavelengths). However, there was no obvious shift or changes in the shape of the reflectance spectrum in the *CpCHS*p-*AtSEP3-RD* line (Figure 7c), except that the entire spectrum was magnified to approximately three times that of the non-transgenic line. Although we did not measure anthocyanin concentrations in this study, we deduced that 'dichromatism' causes the hue to change in the presence of different concentrations of purple anthocyanins. Strong red reflectance will cause the flower color to be reddish at high anthocyanin concentrations in the non-transgenic line, whereas intermediate blue reflectance will be enhanced in the presence of lower concentrations of purple anthocyanins in the *CpCHS*p-*AtSEP3-RD* line, causing the flower to become bluish. The exponential relationship between light transmittance and pigment concentrations, as described by the Beer-Lambert law, causes this kind of alteration in the balance between the intensities of different wavelengths at different pigment concentrations.23 The concentration and chemical structure of anthocyanins in the petals of *CpCHS*p-*AtSEP3-RD* lines should be examined in the future to obtain further insight into the color of these flowers.

CONCLUSION

In this study, we newly isolated 8 promoters of genes expressed in cyclamen petals and prepared 32 constructs. We then identified 13 genetic constructs that modify flower phenotypes, including *CpCHS*p-*AtSEP3-RD*, *CpDFR*p-*CpTCP1B-RD*, and *CpAP3B*p-*CpTCP1B-RD*, by 'screening' the constructs in the model flower torenia. These three constructs generated pale bluish petals, trumpet-like petals, and serrated petals, respectively. We also examined the cellular effects underlying these phenotypic changes and deduced that the changes are linked to dichromatism, imbalanced cell proliferation, and excessive vascular branching, respectively. Taken together, the information and genetic resources generated in this study will benefit future molecular flower breeding efforts and further characterization of these valuable transgenic plants.

REFERENCES

1. Katsumoto Y, Fukuchi-Mizutani M, Fukui Y, Brugliera F, Holton TA, Karan M et al. Engineering of the rose flavonoid biosynthetic pathway successfully generated blue-hued flowers accumulating delphinidin. Plant Cell Physiol 2007; 48: 1589–1600.
2. Klocko AL, Borejsza-Wysocka E, Brunner AM, Shevchenko O, Aldwinckle H, Strauss SH. Tarnsgenic suppression of AGAMOUS genes in apple reduces fertility and increases floral attractiveness. PLoS ONE 2016; 11: e0159421.

3. Noda N, Aida R, Kishimoto S, Ishiguro K, Fukuchi-Mizutani M, Tanaka Y et al. Genetic engineering of novel bluer-colored chrysanthemums produced by accumulation of delphinidin-based anthocyanins. Plant Cell Physiol 2013; 54: 1684–1695.
4. Ono E, Fukuchi-Mizutani M, Nakamura N, Fukui Y, Yonekura-Sakakibara K, Yamaguchi M et al. Yellow flowers generated by expression of the aurone biosynthetic pathway. Proc Natl Acad Sci USA 2006; 103: 11075–11080.
5. Sasaki K, Aida R, Niki T, Yamaguchi H, Narumi T, Nishijima T et al. High-efficiency improvement of transgenic torenia flowers by ion beam irradiation. Plant Biotechnol 2008; 25: 81–89.
6. Ohta M, Matsui K, Hiratsu K, Shinshi H, Ohme-Takagi M. Repression domains of class II ERF transcriptional repressors share an essential motif for active repression. Plant Cell 2001; 13: 1959–1968.
7. Hiratsu K, Ohta M, Matsui K, Ohme-Takagi M. The SUPERMAN protein is an active repressor whose carboxy-terminal repression domain is required for the development of normal flowers. FEBS Lett 2002; 514: 351–354.
8. Hiratsu K, Matsui K, Koyama T, Ohme-Takagi M. Dominant repression of target genes by chimeric repressors that include the EAR motif, a repression domain in Arabidopsis. Plant J 2003; 34: 733–739.
9. Mitsuda N, Umemura Y, Ikeda M, Shikata M, Koyama T, Matsui K et al. FioreDB: a database of phenotypic information induced by the chimeric repressor silencing technology (CRES-T) in Arabidopsis and floricultural plants. Plant Biotechnol 2008; 25: 37–43.
10. Mitsuda N, Takiguchi Y, Shikata M, Sage-Ono K, Ono M, Sasaki K et al. The new FioreDB database provides comprehensive information on plant transcription factors and phenotypes induced by CRES-T in ornamental and model plants. Plant Biotechnol 2011; 28: 123–130.
11. Shikata M, Narumi T, Yamaguchi H, Sasaki K, Aida R, Oshima Y et al. Efficient production of novel floral traits in torenia by collective transformation with chimeric repressors of Arabidopsis transcription factors. Plant Biotechnol 2011; 28: 189–199.
12. Gion K, Suzuri R, Shikata M, Mitsuda N, Oshima Y, Koyama T et al. Morphological changes of Rosa x hybrida by a chimeric repressor of Arabidopsis TCP3. Plant Biotechnol 2011; 28: 149–152.
13. Narumi T, Aida R, Koyama T, Yamaguchi H, Sasaki K, Shikata M et al. Arabidopsis chimeric TCP3 repressor produces novel floral traits in Torenia fournieri and Chrysanthemum morifolium. Plant Biotechnol 2011; 28: 131–140.
14. Shikata M, Ohme-Takagi M. The utility of transcription factors for manipulation of floral traits. Plant Biotechnol 2008; 25: 31–36.
15. Tanaka Y, Yamamura T, Oshima Y, Mitsuda N, Koyama T, Ohme-Takagi M et al. Creating ruffled flower petals in Cyclamen persicum by expression of the chimeric cyclamen TCP repressor. Plant Biotechnol 2011a; 28: 141–147.
16. Sasaki K, Yamaguchi H, Kasajima I, Narumi T, Ohtsubo N. Generation of novel floral traits using a combination of floral organ-specific promoters and a chimeric repressor in Torenia fournieri Lind. Plant Cell Physiol 2016; 57: 1319–1331.
17. Kasajima I, Sasaki K, Tanaka Y, Terakawa T, Ohtsubo N. Large-scale extraction of pure DNA from mature leaves of Cyclamen persicum Mill. and other recalcitrant plants with alkaline polyvinylpolypyrrolidone (PVPP). Sci Hortic 2013; 164: 65–72.
18. Kasajima I. Protocol for DNA extraction from any plant species (alkaline PVPP method). Protocol Exchange 2017;
19. Kasajima I, Ohtsubo N, Sasaki K. Faster, safer, and better DNA purification by ultracentrifugation using GelRed stain and development of mismatch oligo DNA for genome walking. Biosci Biotechnol Biochem 2014; 78: 1902–1905.
20. Oshima Y, Mitsuda N, Nakata M, Nakagawa T, Nagaya S, Kato K et al. Novel vector systems to accelerate functional analysis of transcription factors using chimeric repressor gene-silencing technology (CRES-T). Plant Biotechnol 2011; 28: 201–210.

21. Aida R, Shibata M. Agrobacterium-mediated transformation of torenia (Torenia fournieri). Breed Sci 1995; 45: 71–74.
22. Yamaguchi H, Sasaki K, Shikata M, Aida R, Ohtsubo N. Trehalose drastically extends the in vitro vegetative culture period and facilitates maintenance of Torenia fournieri plants. Plant Biotechnol 2011; 28: 263–266.
23. Kasajima I, Sasaki K. Dichromatism causes color variations in leaves and spices. Color Res Appl 2015; 40: 605–611.
24. Kasajima I. Alexandrite-like effect in purple flowers analyzed with newly devised round RGB diagram. Sci Rep 2016; 6: 29630.
25. Tanaka Y, Yamamura T, Terakawa T. Identification and expression analysis of the Cyclamen persicum MADS-box gene family. Plant Biotechnol 2011b; 28: 167–172.
26. Aida R, Yoshida K, Kondo T, Kishimoto S, Shibata M. Copigmentation gives bluer flowers on transgenic torenia plants with the antisense dihydroflavonol-4-reductase gene. Plant Sci 2000b; 160: 49–56.
27. Bodt SD, Theissen G, de Peer YV. Promoter analysis of MADS-box genes in eudicots through phylogenetic footprinting. Mol Biol Evol 2006; 23: 1293–1303.
28. Cubas P, Lauter N, Doebley J, Coen E. The TCP domain: a motif found in proteins regulating plant growth and development. Plant J 1999; 18: 215–222.
29. Nath U, Crawford BCW, Carpenter R, Coen E. Genetic control of surface curvature. Science 2003; 299: 1404–1407.
30. Cronk QCB. Plant evolution and development in a post-genomic context. Nat Rev Genet 2001; 2: 607–619.
31. Crawford BCW, Nath U, Carpenter R, Coen ES. CINCINNATA controls both cell differentiation and growth in petal lobes and leaves of Antirrhinum. Plant Physiol 2004; 135: 244–253.
32. Bilsborough GD, Runions A, Barkoulas M, Jenkins HW, Hasson A, Galinha C et al. Model for the regulation of Arabidopsis thaliana leaf margin development. Proc Natl Acad Sci USA 2011; 108: 3424–3429.
33. Koyama T, Sato F, Ohme-Takagi M. A role of TCP1 in the longitudinal elongation of leaves in. Arabidopsis Biosci Biotech Biochem 2010; 74: 2145–2147.
34. Hase Y, Tanaka A, Baba T, Watanabe H. FRL1 is required for petal and sepal development in Arabidopsis. Plant J 2000; 24: 21–32.
35. Carland F, Fujioka S, Takatsuto S, Yoshida S, Nelson T. The identification of CVP1 reveals a role for sterols in vascular patterning. Plant Cell 2002; 14: 2045–2058.
36. Hase Y, Fujioka S, Yoshida S, Sun G, Umeda M, Tanaka A. Ectopic endoreduplication caused by sterol alteration results in serrated petals in Arabidopsis. J Exp Bot 2005; 56: 1263–1268.
37. Kasajima I, Yoshizumi T, Ichikawa T, Matsui M, Fujiwara T. Possible involvement of ploidy in tolerance to boron deficiency in Arabidopsis thaliana. Plant Biotechnol 2010; 27: 435–445.
38. Yoshizumi T, Tsumoto Y, Takiguchi T, Nagata N, Yamamoto YY, Kawashima M et al. Increased level of POLYPLOIDY1, a conserved repressor of CYCLINA2 transcription, controls endoreduplication in Arabidopsis. Plant Cell 2006; 18: 2452–2468.
39. Aida R, Kishimoto S, Tanaka Y, Shibata M. Modification of flower color in torenia (Torenia fournieri Lind.) by genetic transformation. Plant Sci 2000a; 153: 33–42.

CHAPTER 17

Differential Physiological and Biochemical Responses Under Variable Culture Conditions in Micro-Propagated *Solenostemon Scutellarioides*: An Important Ornamental Plant

Ranabir SAHU, and Saikat DEWANJEE*

Advanced Pharmacognosy Research Laboratory, Department of Pharmaceutical TechnologyJadavpur UniversityKolkataIndia

ABSTRACT

Solenostemon scutellarioides is a commercially important ornamental plant. In present study, it was attempted to establish a protocol for high frequency *in vitro* regeneration of *S. scutellarioides*. Caulogenesis was found to be significant in solid MS medium supplemented with 0.5 mg dm^{-3} BAP. Combination of GA_3 (0.5 mg dm^{-3}) and BAP (0.5 mg dm^{-3}) induced high frequency regeneration coupled with higher plant height. The plantlets exposed to IBA (1 mg dm^{-3}) exhibited significant root development in terms of first appearance of root and number of roots per shoot. To improve its commercial acceptability in terms of plant architecture and foliage colouration, two weeks old *in vitro* grown plantlets were exposed to different culture conditions namely MS strength, sucrose concentration, pH and light. A variable plant types with diverse ornamental traits were developed under different culture conditions.

Keywords: anthocyanin chlorophyll *in vitro* micropropagation *Solenostemon scutellarioides*

INTRODUCTION

Solenostemon scutellarioides (L.) Codd (syn. Coleus blumei Benth.), commonly known as Coleus or Painted Nettle, is an important ornamental plant valued notably for its vibrant colourful foliage.1 In the last decade, this ornamental plant has gained popularity in the floriculture industry with the introduction of a large number of vegetatively propagated varieties with novel foliage colours. Besides novel foliage colours, opportunities exist to increase their commercial value by manipulating plant architecture. Several hundred commercial cultivars are characterized by their leaf colors, leaf shape, plant architecture and flowering characteristics. The desirable characteristics in S. scutellarioides include improved plant architecture and foliage colour quality. Bright colourful foliages improve its artistic appeal2 and faded colours decrease its product value.3 Considering these traits, commercial Coleus breeding was initiated to develop new cultivars through traditional breeding methods for improved commercial variety. The commercial cultivation of S. scutellarioides is primarily done by cutting method to regenerate similar novel variety. However, to meet with increasing commercial demand high frequency regeneration is desirable and employment of plant tissue culture may serve as an alternative source of S. scutellarioides for commercial purpose.

Plant architecture in S. scutellarioides mainly includes the branching pattern of shoots and leaf morphology. Increasing branching with improved leaf morphology is the main criteria for the selection of better commercial variety. Colouration of different organs such as flower, fruit and leaf is due to the accumulation of chlorophyll, betalains, carotenoids, flavnoids, anthocyanins etc,4 with the latter as the major pigment that impart the red and purple colours observed in S. scutellarioides leaves.5 The attractive colouration of S. scutellarioides leaves is affected by the combination of anthocyanin and chlorophyll content, which is responsible for green colouration.6 Foliage colour may be dependent on factors such as light and pH, analogously to what has been observed in flowers.7,8 Efforts to manipulate plant architecture and foliage colouration to improve commercial acceptability and manipulation of culture condition would serve productive. Therefore the objective of the present work is to establish a protocol for rapid clonal multiplication from limited donor plants and developing different commercial plant types under different culture conditions. The each type with different architecture and foliage colour would serve as the source of explants and mass multiplication achieved by using rapid clonal multiplication protocol.

RESULTS AND DISCUSSION

Nodal segments containing axillary buds have the potential to develop complete plantlets in a nutrient medium containing suitable combinations and concentrations of growth regulators.

To optimize high frequency regeneration, surface sterilized nodal segments of S. scutellarioides were cultured in MS media containing 3% sucrose and pH 5.6. After one week, explants showing axillary buds were transferred into Murashige and Skoog (MS)9 media supplemented with different combinations of phytohormones for organogenesis. Cytokinins are known to stimulate plant cell division and participate in the release of lateral bud dormancy, induction of adventitious bud formation and in growth of lateral buds.10 Different concentrations (0.5 and 1.0 mg dm–3) of cytokinins namely 6-benzylaminopurine (BAP) and kinetin (Kn) were used for the shoot regeneration (caulogenesis). Shoot regeneration response was expressed as number of regenerated shoots per explant and plant height (Figure 1 (a–c), 2a). No detrimental effect was noticed in terms of plant survival with cytokinin treatment. The overall effect of cytokinins on plant height was not promising, which resulted dwarf plantlets with reduced leaf size. The only positive effect was observed in number of shoot generation per explant. BAP (0.5 mg dm–3) was observed to be most beneficial for high frequency regeneration of shoots. No significant difference was observed in terms of plant stature and foliage colour (green with brownish shade) with BAP and Kn treatment. The positive effect of BAP at low concentration was also observed in Coleus forskohlii11 and Plumbago indica.12

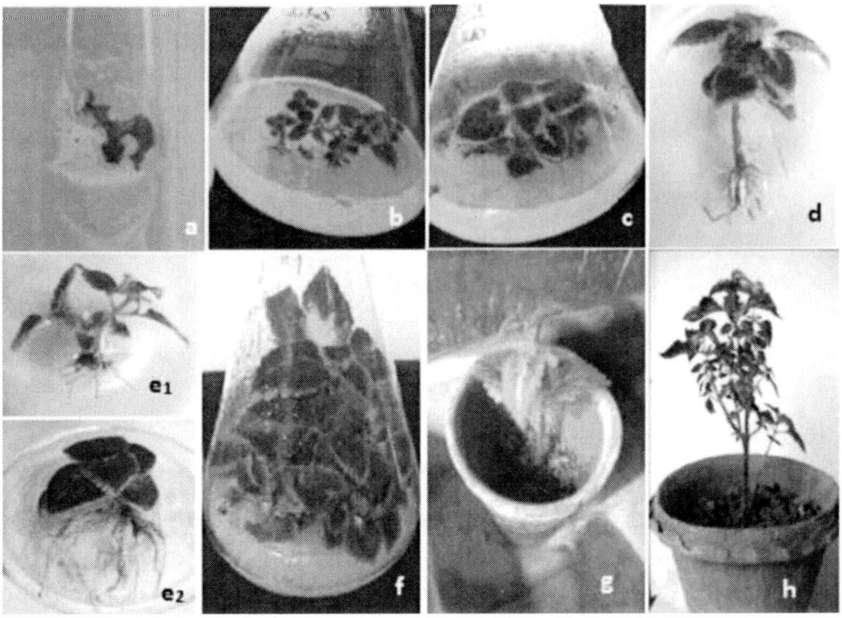

Figure 1. Different stages of micropropagation of S. scutellarioides (a–h): (a) sprouting of shoot buds; (b) initiation of shoot regeneration; (c) emergence of multiple shoot development in BAP (0.5 mg dm−3); (d) enhancement of shoot height with good survival rate in GA3 + BAP (0.5 mg dm−3 each); (e1) initiation of roots at the base of single micro-shoot; (e2) better rooting in media supplemented with IBA (1.0 mg dm−3) at the same period of culture (e1)'; (f) well established in vitro culture; (g) a hardened plant in glass house; (h) micropropagated plant growing under ex vitro environment

Gibberellic acid (GA3) is best known for its role in elongation of axial organs, such as stems, petioles and inflorescences.13 The effect of GA3 on shoot elongation is due to the increase in soluble carbohydrates induced by GA3 and available for metabolic processes.14 In present study, GA3 exhibited significant improvement in height especially to the plants previously treated with BAP (dwarf in nature) (Figure 2b). The negative effect of GA3 was observed on survival percentage after one week, particularly to those plants were exposed with higher concentration of GA3. Severe chlorosis was observed in the plants treated with GA3 (1.0 mg dm−3). But the deleterious effect of GA3 was minimized when it was used in combination with BAP. Combination of GA3 and BAP at low concentration exhibited high frequency regeneration coupled with higher plant height. Combination of GA3 (0.5 mg dm−3) and BAP (0.5 mg dm−3) exhibited best result with almost 1.4 and 1.3 fold increase in shoot numbers and plant height respectively than control (Figure 2b). Plant stature was normal as in control but some changes in plant colour were observed. A pinkish colouration was developed on the upper surface of leaf especially near to mid vein. Leaf area was broader than control and cytokinins treated plant. Leaf margin was serrated which was not common in other treatment. The positive effect of GA3 and BAP in combination was also reported in Ocimum basilicum. 15

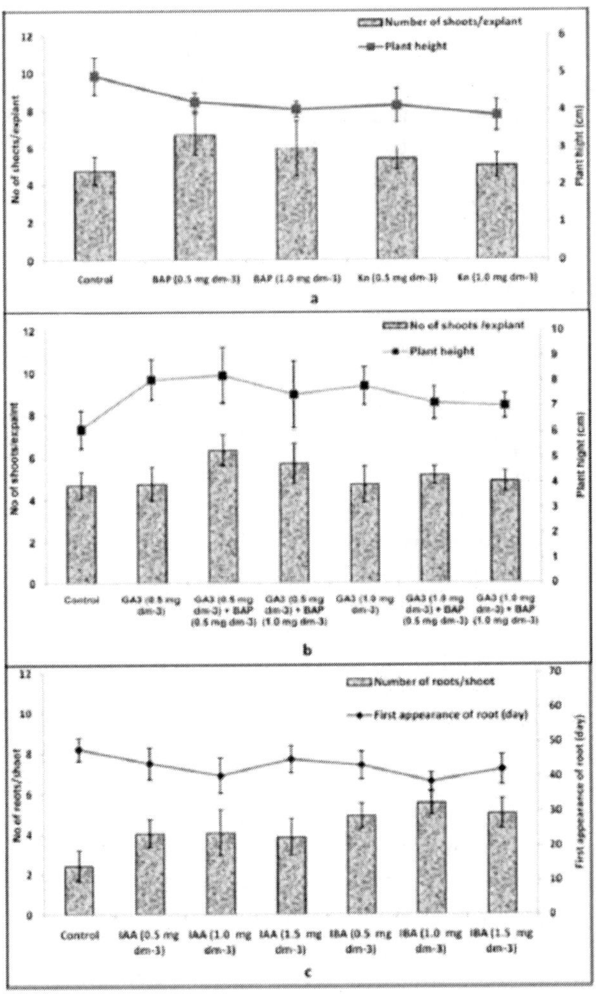

Figure 2. (a) effect of cytokinins on shoot regeneration; (b) effect of GA3 on shoot elongation; (c) effect of auxins on root regeneration from nodal explants of S. scutellarioides

Auxins exert a strong influence in root formation.16 For root generation (rhizogenesis), in vitro regenerated and elongated shoots were transferred into MS medium supplemented with different concentrations (0.5, 1.0 and 1.5 mg dm−3) of indole- 3-acetic acid (IAA) and indole-3-butyric acid (IBA). In course of culture, roots appeared from the base of shoots. The first appearance of root and number of roots per shoot were recorded (Figure 1 (e1, e2), 2c). Both IAA and IBA showed positive effect for early rooting as well increase number of roots per shoot. Plantlets treated with IBA (1 mg dm−1) exhibited best result as roots appeared earliest with almost 2.3 fold increase in number of roots per shoot as compared with control culture. Auxins treatment did not cause any significant morphological changes namely plant architecture and foliage colour. The success of IBA in promoting root induction has been reported in Ceropegia candelabrum17 and Hemidesmus indicus. 18

To study the effect of different external stimuli, two weeks old in vitro grown shoots were exposed with different culture conditions. The regenerated plants maintained in different culture conditions exhibited variable morpho-chemical traits to develop different plant lines with respect to architecture and foliage colouration (Figures 3–6). To study the effect of MS strength, the growth and regeneration potential was found to be improving with MS strength from 1/4 to 1 (Figure 3) but detrimental effect was observed in the cultures treated with 2MS strength. The foliage colours were also significantly varied. The culture with 1/2 and 1 MS strength exhibited beautiful foliage pattern with attractive colouration due to better anthocyanins concentration and selected as developed plant types. The effect of pH of culture media on plant architecture and leaf pattern largely varied (Figure 4). pH did not show significant variation in regeneration potential but it imparted significant difference in leaf morphology and colouration. The plants exhibited diverse attractive foliage colouration between pH 3.6–5.6. The effect of pH on both leaf morphology and foliage colour was found to be optimum at pH 5.6. Higher pH at 7.6 was found to be detrimental for the growth of S. scutellarioides. The plant exposed to pH 4.6 and 5.6 showed highest chlorophyll and anthocyanin content respectively. Amongst the selection, the plants regenerated in pH 3.6, 4.6 and 5.6 were chosen as developed elite types. The effect of sucrose concentration in MS media on plant architecture and foliage pattern was studied (Figure 5). Sucrose concentration did not have significant effect in shoot regeneration per explant but the height improvement was found to be optimum in culture treated with 3% sucrose. Foliage morphology and colouration was found to better in culture treated with 3% sucrose. Chlorophyll concentration in leaves gradually decreased with increasing sucrose concentration. Though sucrose concentrations gave diverse foliage colouration but for the selection of plant type, plants treated with 3% and 4% sucrose were chosen due to better leaf surface and healthy leaf structure. The effect of light on growth and regeneration potential of S. scutellarioides was gradually decreased from white, yellow, green, blue and black exposure (Figure 6). Leaf morphology remained almost similar in white, yellow and green light treatment. Detrimental effect was observed in absence of light. It also significantly reduced ornamental aspect of S. scutellarioides leaves. Chlorophyll and anthocyanin concentrations also largely varied. Green light improved chlorophyll content while white light showed higher anthocyanin content in leaves. Exposure with yellow light exhibited an intermediate concentration of both pigments. Exposure with blue light significantly reduced anthocyanins content.

The high diversity of colour in foliage of S. scutellarioides has been shown to be due variation of anthocyanins and chlorophyll ratio. Anthocyanins are key pigment for attractive pink to purple colouration in S. scutellarioides. Anthocyanins are water soluble pigment having high ecophysiological significance in plant.19 Several hypotheses described that the role of anthocyanins in osmotic regulation to drought and frost stress, antioxidant defense and protection from different light exposure. Based on the hypothesis to develop of new varieties of S. scutellarioides with improved colour quality, the shoots in vitro derived were exposed in vitro with various factors namely MS strength, sucrose concentration, pH and light. In this study, it was observed that selected factors had significant impact on foliage colour development and leaf architecture during developmental stage and the develop plants carried these traits up to ex vitro establishment. Therefore the developed variable plant types with diverse ornamental traits would serve as a secondary source of S. scutellarioides for commercial purpose.

Experimental Section

Collection of Plant Material. S. scutellarioides L. plants were collected from Medicinal Plant Garden of Narendrapur Ramkrishna Mission, South 24 Parganas, India. The plant was authenticated by the Taxonomists of Botanical Survey of India, Shibpur, India. A voucher specimen JU/PT/PC/01/06 was deposited at our laboratory for future reference. The plants were planted in the medicinal plant garden at the Department of Pharmaceutical Technology, Jadavpur University, India.

Preparation and Sterilization of Explants. Nodal segments (0.4 ± 0.1 cm) with internodes were excised from young branches of the plants. The explants containing two buds with oppositely placed axils were used for the experiment. The explants were first washed with running tap water for removing adhering dust particles and then immersed in 1% Teepol

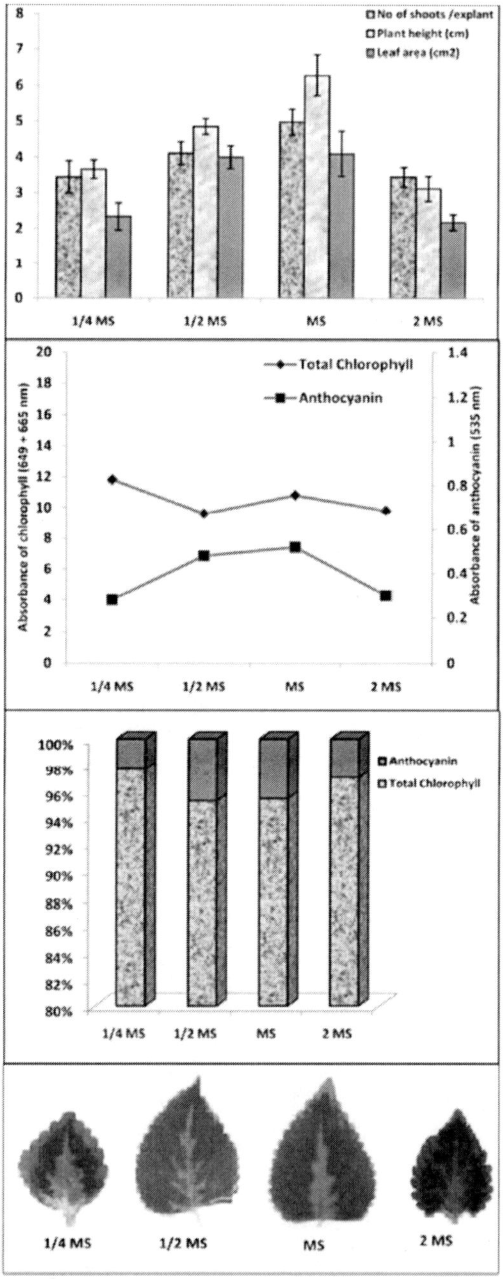

Figure 3. Effect of MS strengths on plant architecture and foliage colour development of S. scutellarioides

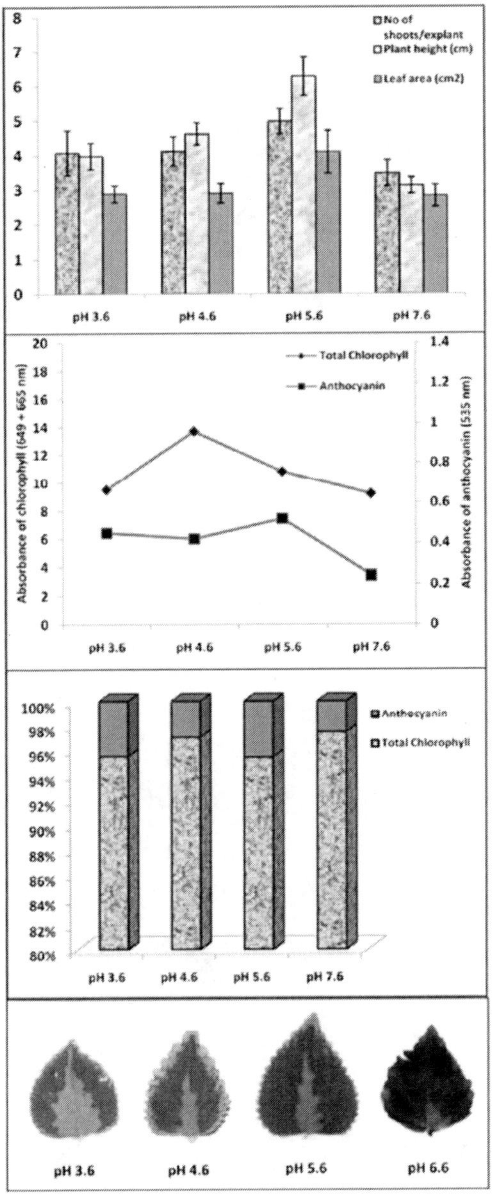

Figure 4. Effect of pH on plant architecture and foliage colour development of S. scutellarioides

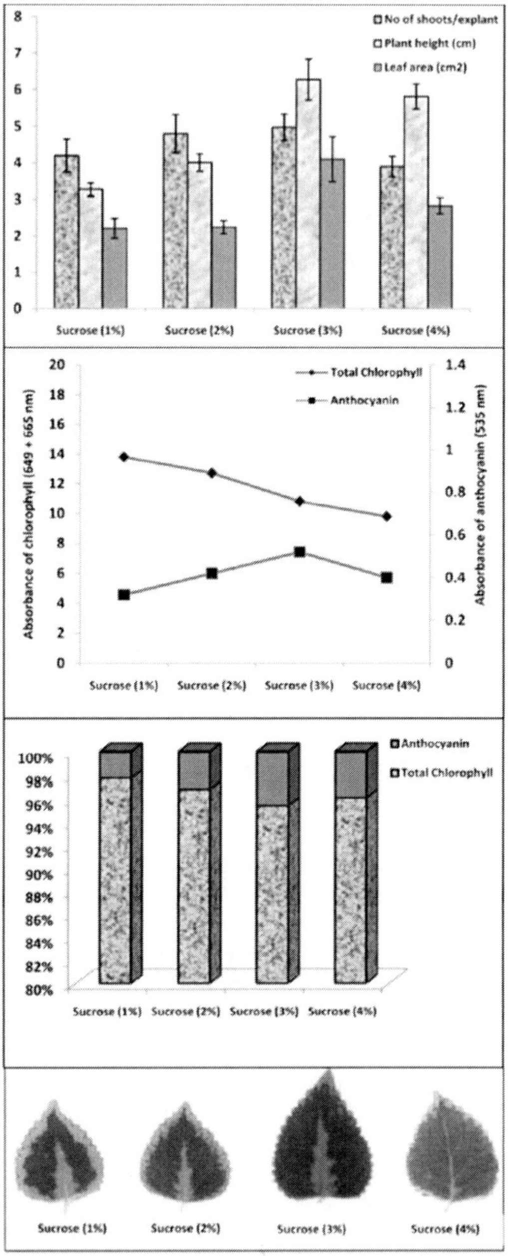

Figure 5. Effect of sucrose concentrations on plant architecture and foliage colour development of S. scutellarioides

detergent (Glaxo India Pvt. Limited, India) for 5 min with continuous shaking in a magnetic stirrer. After gentle washing with sterile distilled water for 4-5 times, the clean segments were surface sterilized with 0.1% mercuric chloride for 1 min, followed by 3-4 times washing with autoclaved distilled water In Vitro Establishment. After surface sterilization, each sterile explant was vertically implanted into single culture tube 150 mm, Borosil, India) contained 25 cm3×(25 sterilized solid MS media. The media was fortified with 3% sucrose (Hi Media, India) gelled with 0.8% Bacto agar (Hi-media, India). The pH was adjusted to 5.6 before autoclaving at 121 ºC for 20 minutes. All cultures were maintained at 22 ± 2 °C and 16 h photoperiod (light intensity 100 μmol m−2 s−1 provided by cool white fluorescent tubes, Philips, India). For adventitious bud induction, 1 week old explants were sub-cultured into 250 cm conical flasks containing 50 cm3 solid MS medium supplemented with different concentrations of two cytokinins (BAP and Kn). After two weeks, the conical flasks exhibited significant caulogenesis were sub-cultured in media containing GA3 singly or in combination with BAP at different concentrations for better shoot development. To study the effect of auxins on rhizogenesis, the healthy in vitro grown plants were transferred in solid MS medium supplemented with IAA and IBA at different concentrations. The best resulting media for rooting was identified on the basis of first appearance of root and number of roots per shoot on day 14.

Ex Vitro Establishment. After two weeks, healthy plantlets with sufficient roots were removed from in vitro condition for hardening. The adhered agar was removed by repeated washing with autoclaved distilled water and the plantlets were transferred to small pots (1 for 1 plantlet), filled with autoclaved soilrite and covered with transparent polythene bags under tissue culture environment. After one week, the plantlets resumed growth were again hardened with autoclaved soilrite-soil mixture (1:1) under same condition for another week. The well hardened plants were transferred to glass house in the large earthen pots for full growth of plants under normal environmental condition for the acclimatization. After two weeks the survived plants were finally transferred in the field.

Manipulation of Media and Culture Conditions to Manipulate Plant Architecture and Foliage Color. To study the effect of different culture conditions two weeks old in vitro grown shoots were exposed with various factors namely MS strength, sucrose concentration, pH and light. One set of culture in MS with 3% sucrose, pH 5.6, under white light was maintained throughout experiment as control. Data were recorded as number of shoots per explant, plant height, leaf surface and foliage chlorophyll and anthocyanin content.

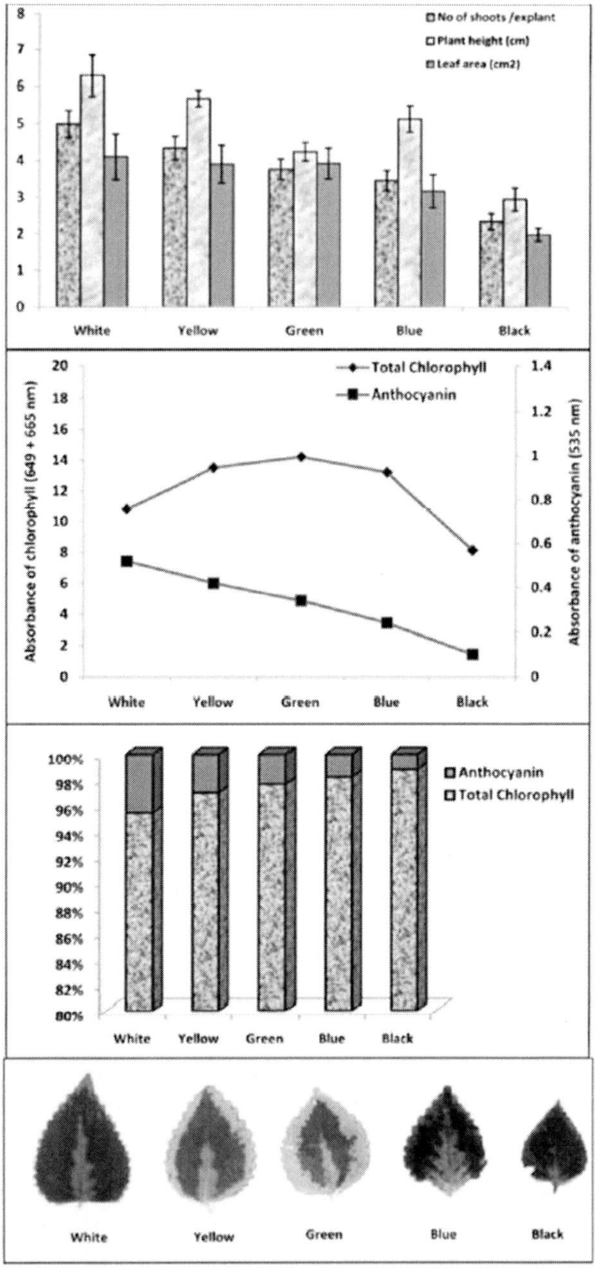

Figure 6. Effect of light on plant architecture and foliage colour development of S. scutellarioides

Total anthocyanins were extracted as previously described20 and absorbance was determined by spectrophotometry at 535 nm. Anthocyanin content was calculated according to the method of Murray and Hackett21 with little modification.4

For chlorophyll determination, collected leaves were frozen in liquid nitrogen. The samples were ground and chlorophyll was extracted. The absorbance was determined at 649 nm and 665 nm. The total chlorophyll a/b concentration was calculated as previously described.22 The ex vitro establishment of different plant lines were done as similar described in previous section.

Statistical Analysis. Each set of experiment consisted of 10 replicates and represented by 10 individual conical flasks. Each experiment was repeated thrice and the data were analyzed statistically using SPSS software (10.0.5, 1999, SPSS Inc).Variability in data was expressed as the mean ± standard deviation.

ACKNOWLEDGMENTS

The authors are thankful to University Grants Commission, New Delhi, India for providing fund to Mr Ranabir Sahu through Rajiv Gandhi National Fellowship, 2011. We are also thankful to Dr. Md. Nasim Ali, Ramkrishna Mission Vivekananda University, Ramkrishna Mission Ashrama Narendrapur, India for providing plant material, S. scutellarioides, for this study.

REFERENCES

1. Garcia, J. F.; O'Neil, R. J. *Biol. Control***2000**, *18*, 225–234.
2. Cantazaro, C.; Bhatti, S.; Muhammad, S.; Abdullah, S. *Hortscience***2005**, *40*, 993–1147.
3. Oren-Shamir, M.; Nissim-Levi, A.; Ovadia, R.; Kagan, S.; Shaked, S. *Acta Horticult.***2003**, *624*, 171–176.
4. Nguyen, P.; Cin, V. D. *Plant Physiol. Bioch.***2009**, *47*, 934–945.
5. Lebowitz, R. *Plant Breeding Review***1985**, *3*, 343–360.
6. Rife, D. C. *J. Hered.***1948**, *39*, 85–91
7. Yoshida, K.; Kawachi, M.; Mori, M.; Maeshima, M.; Kondo, M.; Nishimura, M.; Kondo, T. *Plant Cell Physiol.***2005**, *46*, 407–415.
8. Mol, J.; Grotewold, E.; Koes, R. *Trends Plant Sci.***1998**, *3*, 212–217.
9. Murashige, T.; Skoog, F. *Physiol. Plant.***1962**, *15*, 473–497.
10. Coello, C. Y.; Miceli, C. L.; Orantes, C.; Dendooven, L.; Gutiérrez, F. A. *Gayana Bot.***2010**, *67*, 19–26.

11. Dube, P.; Gangopadhyay, M.; Dewanjee, S.; Ali, M. N. *Indian J. Biotechnol.***2011**, *10*, 228–231.
12. Bhattacharya, R.; Ray, A.; Gangopadhyay, M.; Bhattacharya, S. *Plant Cell Biotechnol. Mol. Biol.***2008**, *8*, 39–46.
13. DeMason, D. A. *Planta***2005**, *222*, 151–166.
14. Bialecka, B.; Kepcznski. J. *Plant Growth Regul.***2007**, *51*, 21–31.
15. Sahoo, Y.; Pattnaik, S. K.; Chand, P. K. *In Vitro Cell. Dev. Biol. Plant***1997**, *33*, 293–296.
16. Gaspar, T. H.; Kevers, C.; Penel, C.; Greppin, H.; Reid D. M.; Thorpe, T. A. *In Vitro Cell. Dev. Biol. Plant***1996**, *32*, 272–289.
17. Beena, M. R.; Martin, K. P.; Kirti, P. B.; Hariharan, M. *Plant Cell Tiss. Org.***2003**, *72*, 285–289.
18. Sreekumar S.; Seeni S.; Pushpangadan P. *Plant Cell Tiss. Org.***2000**, *62*, 211–218
19. Close, D. C.; Beadle, C. I. *The Bot. Rev.***2003**, *69*, 149–161.
20. Singh, A.; Selvi, M.; Sharma, R. *J. Exp. Bot.***1999**, *50*, 1619–1625.
21. Murray, J. R.; Hackett, W. P. *Plant Physiol.***1991**, *97*, 343–351.
22. Goncalves, J.; Antonio, R.; Vieira, M.; Vieira, G. *Rev. Bras. Fisiol. Veg.***2001**, *13*, 149–157.

CHAPTER 18

Sonic Environmental Aesthetics and Landscape Research

Jonathan Prior

Cardiff school of Geography and planning, cardiff University, cardiff, UK

ABSTRACT

The environmental aesthetics literature has primarily focused on the aesthetic qualities and values of landscapes. Within this scholarship, there has been a modest but steady advancement towards explicitly attending to the aesthetic experience of landscape sounds. In this paper, I review the theoretical and applied sonic aesthetics literature pertinent to landscape research, and identify some existing weaknesses. In particular, I demonstrate that there is an ongoing tendency to limit discussions to what is sonically pleasing or displeasing within a given landscape, which, I argue, provides a limited point of entry through which to consider the full scope of landscape sounds. I then turn to offer some ways to address these weaknesses—notably through what I term the development of a 'sensitive ear', and through field recording strategies—in the hope that this will allow scholars to better enfold sonic environmental aesthetics within future theoretical and applied landscape research.

Keywords: Landscape aesthetics, soundscapes, environmental values, field recording, sonic aesthetics

1. INTRODUCTION

Environmental aesthetics 'considers philosophical issues concerning the aesthetic appreciation of the world at large', and so 'the field extends beyond the confines of the artworld and our aesthetic appreciation of works of art' (Carlson & Berleant, 2004, p. 11). While chiefly understood as a subdiscipline of environmental philosophy that emerged in the second half of the twentieth century (Brady, 2003, p. 1), environmental aesthetics research is undertaken under the auspices of various disciplines. So, in addition to theoretical work, there is a strong empirical and applied branch, and there are now numerous attempts to link theoretical and applied work in areas ranging from environmental conservation to landscape planning (see for example Carlson & Lintott, 2008; Porteous, 1996). Across the field, 'aesthetics' is broadly understood as pertaining to what may be aesthetically valued (for instance what is 'beautiful') or disvalued (for instance what is 'ugly' (see Brady, 2011)) in a given environment, as well as other types of aesthetic categories such as the picturesque and the sublime.

The environmental aesthetics literature has given primacy to the aesthetics of landscapes (see among many examples Berleant, 1997; Bourassa, 1991; Gobster, 1999; Parsons, 2008). Within this literature, there have been attempts to demonstrate human psychological reasons as to why certain landscape forms are aesthetically preferred over others (see for example the collection of essays edited by Nasar, 1988). There have also been various examinations of the implications of aesthetic valuations and preferences for the conservation, design and management of landscapes. Here, a normative strand has emerged, which argues that the act of aesthetically valuing landscapes is important for the development of a robust conservation ethic (Berleant, 1997, pp. 36–39; Carlson, 2000), or land ethic (Callicott, 2008; Leopold, 1966). However, it has also been noted that environmental aesthetic values can also conflict with ecological conservation and sustainability goals (Lintott, 2008; Parsons, 1995).

Attendance to the human sensory perception and aesthetic experience of landscape sounds has steadily advanced (however modestly) over the last 20 years or so. In this paper, I reflect on this body of work, and offer some ways in which it could be further developed. I firstly undertake a critical review of this literature, covering both theoretical scholarship chiefly produced by environmental philosophers, and applied work stemming from environmental psychology, urban planning and environmental management researchers. While the review is non-exhaustive, I here examine how sonic aesthetic qualities, experiences and values of landscapes have been broadly conceptualised, and, where relevant, the work that this research has been put to, particularly in regard to the production of landscape policy. As I do so, I identify some existing weaknesses, before turning to offer some ways in which these may be addressed, in the hope that this will allow scholars to better enfold sonic environmental aesthetics within future theoretical and applied landscape research.

2. SONIC AESTHETIC THEORY IN PHILOSOPHICAL LANDSCAPE RESEARCH

The relationship between sonic aesthetics and landscape has been investigated primarily through a concept common to most strands of sonic research—that of the 'soundscape'. 'Soundscape' was first defined by R Murray Schafer as '... any acoustic field of study. We may speak of a musical composition as a soundscape, or a radio programme as a soundscape or an acoustic environment as a soundscape' (Schafer, 1994, p. 7). It is important to note that Schafer deployed the word 'soundscape' in a way that is highly normative: Schafer distinguished the 'hi-fi' from the 'lo-fi' soundscape, where the former 'is one in which discrete sounds can be heard clearly because of the low ambient noise level' (p. 43), and the latter is one where 'there is so much acoustic information that little of it can emerge with clarity' (p. 71). Schafer attributes the hi-fi soundscape to 'the country' and the lo-fi soundscape to post-Industrial Revolution urban settlements, leading Ari Kelman to argue that 'his [Schafer's] notion of "the soundscape" ... is lined with ideological and ecological messages about which sounds "matter" and which do not' (Kelman, 2010, p. 214).

What is important to note here is not that Schafer expresses a sonic preference, but that this preference explicitly rejects the possibility of positive aesthetic experiences in contemporary urbanised landscapes—a judgement that pervades environmental aesthetics (Berleant, 2007). There has since been a shift from a prescriptive to a descriptive formulation of the term; it now usually means the totality of a particular sound environment as perceived by a human listener (Gallagher and Prior, 2014), making the soundscape an apparently useful term for landscape research—so much so that 'soundscape' is often used as a sonic equivalent to the visual 'landscape' (see Porteous & Mastin, 1985; Rodaway, 1994, p. 86).

If reference is made to sounds or soundscapes within the philosophical environmental aesthetics literature, this is normally only in passing, or to support a visual aesthetic argument. For example, Bourassa (1991, pp. 8–9) notes that landscape sounds are relevant to landscape perception and aesthetic experience, but does not develop this idea any further, while Budd (2002, pp. 10–11, 51–52) uses the example of birdsong during a thought experiment about what constitutes the 'natural' (read non-human) in aesthetic experience. Brady (2003) recognises that it is through attentive listening that we can focus our ears on the totality of a soundscape for aesthetic appreciation, and she also notes how sound is at least marginally entangled within historical accounts of the sublime by Alison, Burke, Schopenhauer and Muir (Brady, 2013). Finally, Sepänmaa (2007), in his reflections on the multi-sensory qualities of city environments, briefly states that the sounds of water are an important component of landscape identity, and that sounds characteristic of cities should be afforded protection (Sepänmaa, 2007, p. 94), though neither point is developed.

Aside from these, there are a few instances where the sonic aesthetics of landscapes have been explored more thoroughly. Berleant (2007) states that various sounds within cities—particularly mechanised sounds '... contribute to a three-dimensional auditory texture that is as thick as it is broad' (Berleant, 2007, p. 86), and are thus important to urban landscape experiences. However, Berleant tends toward accounting for urban sounds in a way that is reminiscent of Schafer, conceptualising the totality of 'background drone[s] of traffic', suburban lawnmowers and the hum of air conditioners, as 'inescapable noise', which is a 'kind of environmental oppression' (p. 83). In a more forthright manner he argues that, while there are 'natural' sounds to be heard within urban landscapes, these 'are generally overpowered by mechanical sounds' and that mechanisation within cities 'fill[s] the air with noise and exhaust fumes that are insistent and inescapable, polluting two senses at once' (p. 89).

Jay Douglas Porteous, who has done more than most to extend the scope of environmental aesthetics beyond lines of sight (Porteous, 1982, 1985, 1996), also considers sound to be an important component of urban landscape experience, in tandem with other senses: '... a walk through a landscape garden or a crowded city is a total sensory experience—visual, auditory, olfactory, tactile, kinaesthetic; and in some cities one can taste the air' (Porteous, 1982, p. 84). Porteous also reflects on the 'all-surrounding' qualities of sound in landscapes, stating that sounds have a ubiquitous quality because humans do not possess 'earlids' (Porteous, 1996, p. 33). He then draws a similar conclusion to Berleant, arguing that 'traffic roar increasingly drowns the sounds of nature as well as public music and informational sounds such as foghorns and even sirens' (Porteous, 1996, p. 35).

John Fisher (1998) has provided the most sustained theoretical exploration of the sonic aesthetic qualities of landscapes. Here, he focuses primarily on the extent to which the sounds of (non-human) nature have been routinely ignored within landscape aesthetics research, through examining what he considers to be 'the most significant impediments to including sound in accounts of the aesthetics of nature' (Fisher, 1998, p. 167). These impediments include the apparently habitual suppression or inattention to sounds, which is 'largely learned behaviour' (p. 169), and a Western cultural emphasis on the value of musical over non-musical sounds.

The 'most significant' reason provided by Fisher concerns the apparent subjectivity of experiencing and judging natural sounds. Fisher cites Allen Carlson's 'natural environmental model' approach to environmental aesthetic appreciation, in which Carlson argues that if scientific categories of the natural world are used when we make environmental aesthetic judgments, then objectively correct or incorrect judgments will follow. Fisher then details how the sounds of nature cannot be appreciated objectively, principally by demonstrating that as natural sounds cannot be 'framed' ('which sounds do I pay attention to and for how long?' p. 172), we cannot even agree on what the object of appreciation is—a key requirement of Carlson's model. Nonetheless, as Fisher details these supposed impediments, he dispels each one in turn. Thus, we are still left with the question: why are sonic aesthetics almost wholly neglected within philosophical landscape research, more than 15 years on from when Fisher raised this concern?

3. SONIC AESTHETICS IN APPLIED RESEARCH

Applied landscape researchers have tended to reflect on the sonic aesthetics of landscapes through the lens of environmental psychology in which people's sonic preferences are evaluated. For example, Carles, Bernáldez and de Lucio (1992) undertook an environmental perception survey to understand people's responses to different landscape sounds when combined with different landscape images, concluding that 'natural' sounds 'seem to be preferred' (Carles et al., 1992, p. 55; see also Carles, Barrio, & de Lucio, 1999). Further research suggests that there is a preference for 'natural' over 'human' and 'mechanical' sounds in urban landscapes (Axelsson, Nilsson, Hellström, & Lundén, 2014; Lin & Lam, 2010; Yang & Kang, 2005), and that certain social factors of research participants, such as age, appear to influence the strength of preferences (Yu & Kang, 2010).

The relative 'tranquillity' of different landscape types has been a key area of investigation within the landscape preference literature. Herzog and Bosley (1992) undertook experiments to measure the degree of tranquillity afforded by 66 different 'natural' landscape states, where 'tranquillity' was taken to mean 'how much ... you think this environment would encourage relaxation, peace of mind, escape from the strains of living' (Herzog & Bosley, 1992, p. 117). Despite the clear sonic connotations of this definition, participants were only asked to rate landscape tranquillity based on a series of static images of these landscapes (see also Herzog & Barnes, 1999).

As a way of rectifying the lack of sonic information, Pheasant, Fisher, Watts, Whitaker and Horoshenkov (2010) undertook a two-stage experiment using static photographs and video recordings (including sound) of a range of 'natural' and 'cultural' landscapes. The researchers sought to understand interactions between sound and vision, and how these may modify participants' perception of 'tranquillity'—here taken to mean 'a quiet peaceful place'. The results of this experiment led the authors to argue that sound and vision are both essential when accounting for 'tranquil' landscapes, in part answering Gifford and Fan Ng's (1982) earlier concern that environmental preference research has tended to rely upon single sensory, rather than multi-sensory, environmental representations (see also Hetherington, Daniel, & Brown, 1993).

A series of studies have also been conducted to investigate the impact of 'noise' on people's experiences of non-urban protected landscapes. For example, Benfield, Bell, Troup and Soderstrom (2010) used recordings of different anthropogenic and non-anthropogenic sounds to understand how they affect people's aesthetic ratings of static scenes from five US national parks, concluding that anthropogenic 'noise', including air and ground traffic but also human voices, 'decreased participant ratings of serenity', which chimes with the results of similar experiments (for example Kariel, 1990; Mace, Bell, & Loomis, 1999; Mace, Bell, Loomis, & Haas, 2003; Pilcher, Newman, & Manning, 2009; Tarrant, Haas, & Manfredo, 1995). Such results form the basis of prevailing discussions within applied landscape research over how to best manage unwanted sounds, particularly in national parks and wilderness areas (see for example Dumyahn & Pijanowski, 2011; Lynch, Joyce, & Fristrup, 2011; Mace, Bell, & Loomis, 2004; Miller, 2008; Pepper, Nascarella, & Kendall, 2003). Noise abatement policies are now common across all levels of governance, from supranational institutions to cities, and tools such as noise exposure maps are one prevalent means of gathering data on where anti-noise interventions should take place (Wissmann, 2014).

While a far less cohesive body of work than the environmental psychology literature described, soundscape planning has emerged as a multidisciplinary way of thinking through the interweaving of landscape, sound, and experience:

> Noise management is the current paradigm for management of the outdoor acoustic environment, involving a large body of knowledge, practice, law, policing and control activities ... By contrast, soundscape planning focuses on acoustic environments that are

regarded positively—that people prefer or consider as desirable environments. (Brown & Muhar, 2004, pp. 828–829)

In their work on soundscape planning, Matsinos et al. (2008) confront the framing of soundscape policy as always a question of noise management, arguing that rural landscapes are 'characterised by great spatial and temporal sound variability' (Matsinos et al., 2008, p. 946). The authors go on to outline an approach to the qualitative categorisation, and subsequent quantitative mapping of sound sources in the rural north of Corfu, with the aim of aiding landscape monitoring and management practices that better account for such variability.

Moving from a rural to an urban landscape, 'spatial and temporal sound variability' is understood to be at least a partially positive component of the '24 hour city' by Adams et al. (2006). Through a series of semi-structured interviews, the authors reveal that the assignment of positive or negative aesthetic value by residents to the sounds of Clerkenwell, central London, is highly context-specific. For example, the sounds of all-night parties and the 'hum' of traffic are either tolerated or aesthetically appreciated, when they are understood and interpreted within the social context of their production (see also Raimbault & Dubois, 2005). It is thus concluded that top-down noise abatement policies do not necessarily reflect the actual sonic aesthetic values held by a city's inhabitants, and so: '... the risk is that sound energy levels become the indicator of sustainable sounds and the implied goal becomes a silent city. Where then is the space for the 'buzz' that many people say they come to the city for?' (Adams et al., 2006, p. 2396).

The 'Positive Soundscape Project' also amply demonstrates an appreciation that there is potential discrepancy between the objectives of noise abatement policies—however well intentioned—and the sonic aesthetic values of those who live in urbanised landscapes. As the name suggests, the project sought 'to move away from a focus on negative noise and to identify a means whereby the concept of positive soundscapes can effectively be incorporated into planning' (Davies et al., 2007). One strand of the project involved ascertaining people's 'favourite sounds' from inhabitants of Manchester and London so as to create a compendium of the types of sonic aesthetic qualities that people positively value within urbanised landscapes (Davies et al., 2009). Such a compendium could clearly influence soundscape policy by focusing on what people positively value, as well as what they do not, in the landscapes that they inhabit (see also Porteous & Mastin, 1985).

4. EXTENDING AND BROADENING SONIC AESTHETIC LANDSCAPE RESEARCH

Though there are some notable exceptions aligned with soundscape planning, I have shown that there is a propensity within both the philosophical and applied landscape literature to limit discussions to what is pleasing (which is equated with 'peace' and 'tranquillity'), or what is displeasing (which is equated with 'noise'), mirroring the focus of visual landscape preference studies (Berleant & Carlson, 2007, p. 15).

As such, the literature nearly unanimously ascribes positive aesthetic value to the sounds of landscapes only when there is a relative *absence* of sounds, meaning that amplitude becomes the measure of sonic aesthetic value (though see Zhang & Kang, 2007), rather than the *presence* of other sonic qualities, such as timbre, pitch, texture, resonance, rarity or periodicity. This results in the identification of the majority of anthropogenic sound events in landscapes—especially those landscapes designated as 'natural'—as 'out of place' (Cresswell, 1996) and thus in need of management. This encompasses mechanical sounds and human communicative and bodily sounds, leading to experimental efforts to regulate the amplitude of human speech in parks, as well as the sounds of moving human bodies (Stack, Newman, Manning, & Fristrup, 2011; see also Benfield et al., 2010, p. 109).

While it is not my intention to undermine concerns over noise pollution that can have very real negative effects on human and non-human lives (Blickley & Patricelli, 2010), I think that it is crucial to highlight that making normative claims over the appropriateness of particular sonic geographies (see Matless, 2005), and perhaps most problematically biopolitical claims over sounding human bodies in certain landscapes, raises a series of questions yet to be addressed. For example, is relative silence always appropriate or desirable, or are other ways of sounding within landscapes—for instance with liveliness and high dynamic range (the difference between the quietest and loudest sound as perceived by a human listener in a given sonic environment)—also legitimate? Should we regulate sounds that we may merely find 'annoying', as some research suggests (see for example Pilcher et al., 2009)? Such a proposition appears to reinforce the bifurcation of human and non-human worlds, which rests upon the assumption that humans are negative agents within 'natural' landscapes.

More fundamentally, it seems vital to question the ongoing interpretation of the sounds of landscapes as merely a conduit to human pleasure or displeasure—a stance that is known as the hedonic theory of value (Brady, 2011, p. 96)—which pervades both landscape preference and soundscape planning research. Such a stance demonstrates a relative paucity of thought about how the sounds of landscapes may be experienced and valued. Clearly, many landscape sounds resist being easily categorised as either bringing about pleasure or displeasure, meaning that these sounds are under-represented (if not completely ignored), and so remain under-theorised within research.

Further, using a pleasure-displeasure framework as the basis for soundscape policy deliberations is troubling from an ecological perspective. If we are to take the conservation of soundscapes seriously (Dumyahn & Pijanowski, 2011), then we need to be cautious of managing landscapes with human pleasure as the ultimate goal. Just as landscape theorists have challenged the notion that what brings visual aesthetic pleasure correlates with healthily functioning landscapes (Gobster, 1999), we must also ask whether sonic aesthetic pleasure—as currently conceived within the literature—aligns with landscape conservation and sustainability objectives. If 'tranquillity' and 'peacefulness' are continuously reaffirmed as the ultimate objective of soundscape policy both within 'natural' and 'cultural' landscape settings, what space is there for those natural sounds that do not engender pleasure—the 'unpleasant, raucous, cackling cry of magpies', the 'ugly screech of a jay'; the 'horrible scream of foxes in heat' (Cox, 2015, pp. 88–89)—within such policies? A rejoinder to this may be that all of non-human nature is inherently beautiful—a position within philosophical environmental aesthetics called 'positive aesthetics'—and so what is presumed to be ugly or unpleasurable is in fact beautiful, but this unconvincingly sidesteps the very real existence of negative aesthetic qualities in the natural world (see Brady, 2011).

I now turn to outline some ways in which sonic aesthetics can be more broadly accounted for, and embedded within, theoretical and applied landscape research, beyond the current focus on aesthetic pleasure and displeasure. I want to add the proviso that what follows is tentative and exploratory, and that further reflections could have been offered were it not for a lack of space.

4.1. Listening to the sounds of landscapes

When it comes to making sense of the world, humans are highly adept at doing so through vision (Rodaway, 1994, p. 115). Visual perception dominates over other modes of sensory perception (Posner, Nissen, & Klein, 1976)—certainly in humans with no sensory impairments. The reasons underlying this dominance are debated amongst neuroscientists and psychologists (see for example Spence, Parise, & Chen, 2012), but for the present discussion I am interested in how this dominance has—alongside a general Western cultural prevalence for producing visual representations of landscapes (not only within the arts, including landscape design and architecture, but also landscape research itself (see Lange, 2011))—shaped how landscape perception and aesthetic appreciation are approached by landscape scholars.

In acknowledging such dominance, I am not bemoaning 'ocularcentrism' (Macpherson, 2006); nor am I implicitly arguing for attention to the sonic at the expense of other modes of landscape perception as some sort of corrective. Rather, if we are to take multi-sensory landscape aesthetics seriously, then it appears necessary to critically reflect on current approaches to landscape research that have been constructed with vision and visuality in mind. In the sonic domain, to better account for the multiplicity of landscape sounds and soundscapes, I want to suggest that a 'sensitive ear' to the world is necessary.

A sensitive ear is a listening disposition that is attentive to the diversity of landscape sounds, and the multitudinous ways in which we may aesthetically experience and value them. Rather than beginning (and indeed ending) with trying to discover which sounds within a given landscape arouse human 'pleasure' or 'displeasure', such a disposition is keenly open to the reception of—and our responses to—sounds that are not easily accommodated within such a hedonic theory of aesthetic values.

It is useful at this juncture to note that 'listening' and 'hearing' are distinct, as Handel (1989) explains: 'The physical pressure wave enables perception but does not force it. Listening is active; it allows age, experience, expectation, and expertise to influence perception' (1989, p. 3). Even taking into account the influences of 'age, experience, expectation, and expertise', listening has the benefit of being less discriminatory of the sounds of landscapes than passive hearing, which tends to only draw our attention to a particular sound or group of sounds when there is a change in pitch, volume or so on. Such listening is not necessarily an instinctive practice; rather, it should be regarded as a skill that can be developed through listening exercises (Schafer, 1969).

As a component of a sensitive ear, attentive listening can complicate attempts to locate landscape sounds along a pleasure-displeasure continuum, as it reveals a range of sounds whose aesthetic qualities are more ambiguous than this can hope to contain, including dull, mundane, melancholic and sublime qualities. In its less discriminatory stance toward the diversity of landscape sounds, a sensitive ear is also receptive to those sounds that may be unfamiliar (say, because they are rare within a given landscape), or so overly familiar that they do not register as pleasurable or unpleasurable, but are no less important to the constitution of a sonic landscape character.

My suggestion about the usefulness of a 'sensitive ear' may strike some as banal or condescending, but this is not my intention. While I am cautious of over-generalising any assumed 'lack' of listening by landscape researchers, and receptive to the argument that listening is not inherently positive given the various political and economic power relations between listener and listened (Gallagher, 2013), based on the sounds that have been documented and discussed within the landscape literature, it would appear that attentive listening is not entrenched to the extent that it is within music and sonic arts scholarship (see for example Carlyle & Lane, 2013; Chion, 1994; Oliveros, 2005).

If a sensitive ear is the first step towards better accounting for the diversity of sounds as they actually exist in landscapes (rather than only attending to those that precipitate pleasure or displeasure), and the various ways in which we may aesthetically experience and value them, then the second step is the development of tools for better documenting and communicating what a sensitive ear to the world can reveal.

4.2. Documenting and communicating sonic aesthetics of landscapes

> This day of rain has its own temptation, and the steady downpour gives my brief detour a special quality. The sound on the roof of my car is an auditory pointillism, punctuated by the regular beat of the windshield wipers and supported by the gentle undertone of the engine. They become a thickly textured accompaniment to the sights, movements, and sensations of my drive. (Berleant, 1992, p. 41)

Sensitive to the ways in which aesthetic qualities can be relayed to a reading audience, Berleant (1992) outlines 'descriptive aesthetics' as a convincing mode of doing so: 'descriptive aesthetics combines acute observation with compelling language to encourage the reader toward vivid aesthetic encounters' (Berleant, 1992, p. 26). While Berleant is keen to point out that descriptive aesthetics can go 'beyond a communicative function' (p. 26), this form of writing has become commonplace within the philosophical environmental aesthetics literature, primarily as a means to communicate the particularities of a given landscape, taking note of both formal and expressive aesthetic qualities to do so (see Budd, 2002, pp. 112–118).

Among the examples of rich textual aesthetic accounts of landscapes, Allen Carlson's visual descriptions of modern agricultural landscapes (Carlson, 2000, pp. 185–189), Stan Godlovitch's depiction of the deliberate breaking up of a frozen stream in Alberta (Godlovitch, 1994) and Berleant's (1992) own multi-sensorial account of driving through a Connecticut landscape (excerpted above), are particularly evocative. Berleant's work aside, however, sonic descriptive aesthetics are negligible. Instead, in the literature I have outlined, communication of the sounds of landscapes amounts to cataloguing sound sources and attaching an aesthetic value judgement—normally an adjective such as 'beautiful', 'serene', 'noisy' or 'annoying'. This means that the specific qualities of particular landscape sounds previously mentioned (timbre, pitch and so on) are neglected.

It may be fruitful, then, for landscape scholars to look to other forms of landscape writing to enlarge and enliven their own descriptive work. Nature poems, for instance, 'frequently contain descriptions of sounds' (Fisher, 1999, pp. 39–40), and contemporary nature writing by the likes of Kathleen Jamie and Robert Macfarlane, offers imaginative ways of writing about landscape aesthetics beyond vision. Further, there have been some experimental forms of writing about sound, such as Daniela Cascella's interrogations of the triangulated relationship between listening, reading and writing in—and about— landscapes (Cascella, 2012), as well as Augoyard and Torgue's (2008) approach to formalising different types of sonic qualities that spans music, architecture and landscape. What these texts do is provide landscape researchers with an extended vocabulary that not only captures a range of sonic aesthetic qualities and values beyond the usual list of adjectives, but also (especially in Augoyard & Torgue, 2008) distinguish between, and help to describe, different types of properties of sounds, how they are shaped by different spaces (as sounds do not exist in a spatial vacuum) and how they might be experienced by a listener.

When writing about the sounds of landscapes, it seems necessary to assess the usefulness of prevailing aesthetic classifications. For instance, in their assessment of the role of the category 'natural beauty' in landscape legislation, Selman and Swanwick (2010) posit that a 'modern understanding of natural beauty' includes senses other than the visual (Selman & Swanwick, 2010, p. 22), meaning that it has the potential to be enlarged beyond visual aesthetics. Similarly, the 'picturesque' has been developed to interpret developments in late eighteenth- and early nineteenth-century music that parallel picturesque gardening and landscaping (Richards, 2001). Not all existing aesthetic categories, however, may be so readily enlarged, and so it may well be necessary to create new categories or bring to the fore known but contemporarily neglected ones, as has notably happened in visual aesthetic theory (see for instance Brady, 2013; and Ngai, 2012).

Aside from these rather modest textual-linguistic interventions, I want to go further by proposing field recording—the practice of audio recording outside of the controlled environment of a recording studio—as a legitimate mode of documenting and communicating sonic aesthetic qualities to supplement descriptive textual articulation. Landscape aestheticians are adept at providing photographic and diagrammatic representations in their work (see for instance Porteous, 1996; and Sadler & Carlson, 1982), as a means of documenting visual aesthetic qualities, but there are only a few instances of researchers turning to audio recording practices as a means of landscape documentation, and fewer still that have presented recordings within research outputs (two exceptions being Cusack, 2001, and Prior, 2012).

It is important to highlight that this proposition is not an argument for 'objectivity' compared to any assumed 'subjectivity' of textual documentation; such an assertion must be guarded against. Indeed, the act of field recording, from the initial idea of what to record (and what not to), to the placement of microphones, through to the editing and replay of recordings, is not an objective tool of documentation (see Altman, 1992, p. 40). What it is, however, is an argument in favour of better capturing the experiential richness of the sounds of landscapes—a richness that is difficult to attain through textual description alone.

Audio recordings can act as a representation of the sonic qualities of a particular landscape, while also bringing about different affective, emotional (see Gallagher, 2015) and aesthetic responses for different listeners. Importantly, while a recordist may have their own ideas about the types of responses that a landscape recording may lead to, the latter appears less prescriptive than a written textual description. If I state that a forested landscape is alive with the beautiful, intricate sounds of birds during the spring dawn chorus, or is unnerving during the winter months when passing winds creak leafless branches, I am making declarative statements that afford little space within which a reader can form their own appraisal of the aesthetic qualities of this landscape. An audio recording of the landscape in question, by contrast, would provide such space.

Certain recording techniques allow us to aestheticise sounds that 'are hidden, fleeting, beyond or at the periphery of everyday awareness' (Gallagher and Prior, 2014, p. 271). A listener to a field recording can control the volume during playback; turning up the volume 'one may become aware of a distant rumble of traffic, the flitting of insects, [or] wind whistling around objects' (Gallagher and Prior, 2014, p. 271). Pushing this further, we may deploy a range of different microphones and transducers to reveal and aestheticise those sounds that may not be immediately available to the naked human ear, including hydrophones (underwater microphones), contact microphones (which transduce vibrations moving through solid objects) and geophones (which are sensitive to sub-surface ground movements). As such, audio recording technologies can enable the extension of a sensitive ear to the world.

Audio recordings may also allow us to express what is otherwise challenging to articulate. The work of Cheryl Foster (Foster, 1998) on what she terms the 'narrative' and 'ambient' dimensions of environmental aesthetic valuing, is pertinent to what I am forwarding here. In the former, which Foster argues has come to dominate attempts to articulate environmental aesthetic experiences, certain frameworks— she mentions mythology, social history, but particularly the natural sciences—are used as a means to narrate aesthetic experiences and values, so that they can be communicated, adjudicated, interpreted and so on. Foster (1998, p. 131) states: 'we filter the perceptual properties of nature's surface through a frame of reference that functions as narrative in character, one that contextualises the objects before us'. By contrast, the ambient 'does not rely *in practice* upon any standards, frameworks, or narratives external to the experiencing individual ...' Indeed, 'conceptual frameworks recede() and we encounter nature as an enveloping other' (p. 133; emphasis in original):

> The textures of earth as we move over them, the sounds of the winds and the wildlife and trees, the moistness or dryness of the air, the nascent colors or seasonal mutations—all can melt into a synthesized backdrop for ambient contemplation of both the backdrop itself and the sensuous way we relate to it. (Foster, 1998, p. 134)

Within Foster's formulation of the ambient, there are clear echoes of Berleant's aesthetics of landscape engagement (Berleant, 1997), and Steven Bourassa's 'sensory aesthetics', in which experiences occur 'without regard to any meanings or associations they may arouse' (Bourassa, 1991, p. 22). Because of the particularities of this dimension of aesthetic experience, a communicative dilemma arises as: 'the depth experienced in the ambient dimension of aesthetic value resists straightforward prose' (Foster, 1998, p. 135). Though 'we *require* words' (p. 135; emphasis in original), the difficulty of articulating the ambient dimension through verbal-textual documentation, requires us to 'adjust our methods of accounting for it accordingly' (p. 136). Foster posits art as one strategy, while I would like to propose field recording as a promising mode of landscape aesthetic articulation, as through field recording there is the possibility of at least partially documenting sonic components of the ambient dimensions of aesthetic experience, in addition to narrative dimensions.[1]

Indeed, returning to Berleant's description excerpted above, which I think is an evocative way of attending to the ambient, all sorts of questions are raised about this experience that would be well served by the addition of an audio recording of the event. How exactly, for instance, does the 'gentle undertone of the engine' 'support' the sound of raindrops on the car roof? How does this rain sound compare to the sounds of rain I have experienced? What sounds are deemed extraneous to this recounting, and are thus edited out? Again, while this is not an argument in favour of 'objective' documentation, a field recording may improve a reading/listening audience's ability to critically engage with this landscape experience, and so in turn agree with, modify or refute, an aesthetic judgement of landscape—critically here including those judgements pertaining to its ambient dimensions—in a way that is not easily possible with textual descriptive aesthetics; something that Berleant readily acknowledges (Berleant, 1992, p. 22).

5. CONCLUSION

In this paper, I have critically reviewed the extant theoretical and applied literature that has addressed the aesthetics of landscapes from a sonic perspective. I have shown that the overwhelming majority of this work has been executed with policy-relevance in mind; landscape sounds have nearly unanimously been accounted for so as to make landscape design recommendations, with the intention of either controlling unwanted 'noise' or preserving positively valued soundscapes, which so often equates to the preservation of a relative *absence* of sounds.

I then demonstrated how this approach renders sounds and soundscapes in a markedly instrumental manner, in that it focuses in on human pleasure and displeasure. While there are valid human health reasons for doing so, I argued that this provides a limited point of entry through which to consider the full scope of human and non-human sounds in landscapes, and also the variegated ways in which we aesthetically experience and respond to these sounds. Self-evidently, this approach can say little (if anything) about landscape sounds that fall within aesthetically ambiguous categories, meaning that sounds that do not clearly and directly engender aesthetic pleasure or displeasure are not addressed in the literature. This approach may also, perhaps unwittingly, work against landscape conservation and sustainability objectives. Thus, it would seem judicious to be critically aware of the consequences— however unintentional—of establishing pleasure as the pinnacle aesthetic value of soundscape planning and design.

As a way of extending and broadening sonic aesthetic landscape research beyond a pleasure-displeasure approach, I then considered both attentive listening and different modes of sonic communication. I showed how these are appropriate to the particular qualities of sounds, and are thus well suited to the tasks of accounting for and disseminating the diversity of sounds present within landscapes. While it is important to be sensitively attuned to the demands of sonic aesthetics, I outlined these potential future directions not as a way of demarcating sound as (necessarily) an isolated area for future landscape research. Instead, I see these approaches as a means to enlarge the scope of current sonic environmental aesthetics, in such a way that does not denigrate—but rather complements—other modes of sensing and experiencing landscapes.

NOTE

1. While the act of producing a sonic representation through field recording could be construed as the creation of a type of 'narrative', this differs substantially to a verbal-textual narrative, in that such recordings do not necessarily rely upon an external frame of reference, such as social history or the natural sciences.

DISCLOSURE STATEMENT

No potential conflict of interest was reported by the author.

REFERENCES

1. Adams, M., Cox, T., Moore, G., Croxford, B., Refaee, M., & Sharples, S. (2006). Sustainable soundscapes: Noise policy and the urban experience. *Urban Studies, 43*, 2385–2398.
2. Altman, R. (1992). Introduction: Four and a half film fallacies. In R. Altman (Ed.), *Sound theory/Sound practice* (pp. 35–45). London: Routledge.
3. Augoyard, J.-F., & Torgue, H. (2008). *Sonic experience: A guide to everyday sounds.* Montreal: McGill-Queen's University Press. Axelsson, Ö., Nilsson, M. E., Hellström, B., & Lundén, P. (2014). A field experiment on the impact of sounds from a jet-and-
4. basin fountain on soundscape quality in an urban park. *Landscape and Urban Planning, 123*, 49–60.
5. Benfield, J. A., Bell, P. A., Troup, L. J., & Soderstrom, N. C. (2010). Aesthetic and affective effects of vocal and traffic noise on natural landscape assessment. *Journal of Environmental Psychology, 30*, 103–111.
6. Berleant, A. (1992). *The aesthetics of environment.* Philadelphia, PA: Temple University Press.
7. Berleant, A. (1997). *Living in the landscape: Toward an aesthetics of environment.* Kansas, KS: University Press of Kansas. Berleant, A. (2007). Cultivating an urban

aesthetic. In A. Berleant & A. Carlson (Eds.), *The aesthetics of human environments* (pp. 79–91). Peterborough: Broadview Press.

8. Berleant, A., & Carlson, A. (2007). Introduction: The aesthetics of human environments. In A. Berleant & A. Carlson (Eds.), *The aesthetics of human environments* (pp. 13–46). Peterborough: Broadview Press.
9. Blickley, J. L., & Patricelli, G. L. (2010). Impacts of anthropogenic noise on wildlife: Research priorities for the development of standards and mitigation. *Journal of International Wildlife Law & Policy, 13*, 274–292.
10. Bourassa, S. C. (1991). *The aesthetics of landscape*. London: Belhaven Press.
11. Brady, E. (2003). *Aesthetics of the natural environment*. Edinburgh: Edinburgh University Press.
12. Brady, E. (2011). The ugly truth: Negative aesthetics and environment. *Royal Institute of Philosophy Supplement, 69*, 83–99. Brady, E. (2013). *The sublime in modern philosophy: Aesthetics, ethics, and nature*. Cambridge: Cambridge University Press. Brown, A. L., & Muhar, A. (2004). An approach to the acoustic design of outdoor space. *Journal of Environmental Planningand Management, 47*, 827–842.
13. Budd, M. (2002). *The aesthetic appreciation of nature: Essays on the aesthetics of nature*. Oxford: Clarendon Press.
14. Callicott, J. B. (2008). Leopold's land aesthetic. In A. Carlson & S. Lintott (Eds.), *Nature, aesthetics, and environmentalism: From beauty to duty* (pp. 105–118). New York, NY: Columbia University Press.
15. Carles, J., Bernáldez, F., & de Lucio, J. (1992). Audio-visual interactions and soundscape preferences. *Landscape Research, 17*, 52–56.
16. Carles, J. L., Barrio, I. L., & de Lucio, J. V. (1999). Sound influence on landscape values. *Landscape and Urban Planning, 43*, 191–200.
17. Carlson, A. (2000). *Aesthetics and the environment: The appreciation of nature, art and architecture*. London: Routledge. Carlson, A., & Berleant, A. (2004). Introduction: The aesthetics of nature. In A. Carlson & A. Berleant (Eds.), *The aesthetics of*
18. *natural environments* (pp. 11–42). Toronto: Broadview Press.
19. Carlson, A., & Lintott, S. (Eds.). 2008. *Nature, aesthetics, and environmentalism: From beauty to duty*. New York, NY: Columbia University Press.
20. Carlyle, A., & Lane, C. (Eds.). (2013). *On listening*. Axminster: Uniformbooks.
21. Cascella, D. (2012). *En abîme: Listening, reading, writing: An archival fiction*. Winchester: Zero Books. Chion, M. (1994). *Audio-vision: Sound on screen*. New York, NY: Columbia University Press.
22. Cox, T. (2015). *Sonic wonderland: A scientific odyssey of sound*. London: Vintage Books.
23. Cresswell, T. (1996). *In place/out of place: Geography, ideology, and transgression*. Minneapolis, MN: University of Minnesota Press.
24. Cusack, P. (2001). *Your favourite London sounds*. London: London Musicians Collective.
25. Davies, W. J., Adams, M. D., Bruce, N. S., Cain, R., Carlyle, A., Cusack, P., ... Plack, C. J. (2007). *The positive soundscape project*. Paper presented at the 19th International Congress on Acoustics, Madrid, 2–7 September 2007.
26. Davies, W. J., Adams, M. D., Bruce, N., Marselle, M., Cain, R., Jennings, P., ... Plack, C. J. (2009). The positive soundscape project: A synthesis of results from many disciplines. *Paper presented at the 38th International Congress and Exposition on Noise Control Engineering*, Ottawa, 23–26 August 2009.
27. Dumyahn, S. L., & Pijanowski, B. C. (2011). Beyond noise mitigation: Managing soundscapes as common-pool resources.*Landscape Ecology, 26*, 1311–1326.
28. Fisher, J. A. (1998). What the hills are alive with: In defense of the sounds of nature. *The Journal of Aesthetics and Art Criticism, 56*, 167–179.
29. Fisher, J. A. (1999). the value of natural sounds. *Journal of Aesthetic Education, 33*, 26–42.

30. Foster, C. (1998). The narrative and the ambient in environmental aesthetics. *The Journal of Aesthetics and Art Criticism, 56*, 127–137.
31. Gallagher, M. (2013). Listening, meaning and power. In A. Carlyle & C. Lane (Eds.), *On listening* (pp. 41–44). Axminster: Uniformbooks.
32. Gallagher, M. (2015). Field recording and the sounding of spaces. *Environment and Planning D: Society and Space, 33*, 560–576. Gallagher, M., & Prior, J. (2014). Sonic geographies: Exploring phonographic methods. *Progress in Human Geography, 38*,267–284.
33. Gifford, R. & Fan Ng, C. (1982). The relative contribution of visual and auditory cues to environmental perception. *Journal of Environmental Psychology, 2*, 275–284.
34. Gobster, P. H. (1999). An ecological aesthetic for forest landscape management. *Landscape Journal, 18*, 54–64. Godlovitch, S. (1994). Icebreakers: Environmentalism and natural aesthetics. *Journal of Applied Philosophy, 11*, 15–30. Handel, S. (1989). *Listening: An introduction to the perception of auditory events*. Cambridge: MIT Press.
35. Herzog, T. R., & Barnes, G. J. (1999). Tranquillity and preference revisited. *Journal of Environmental Psychology, 19*, 171–181. Herzog, T. R., & Bosley, P. J. (1992). Tranquility and preference as affective qualities of natural environments. *Journal of Environmental Psychology, 12*, 115–127.
36. Hetherington, J., Daniel, T. C., & Brown, T. C. (1993). Is motion more important than it sounds?: The medium of presentation in environmental perception research. *Journal of Environmental Psychology, 13*, 283–291.
37. Kariel, H. G. (1990). Factors affecting response to noise in outdoor recreational environments. *The Canadian Geographer, 34*, 142–149.
38. Kelman, A. Y. (2010). Rethinking the soundscape: A critical genealogy of a key term in sound studies. *The Senses and Society, 5*, 212–234.
39. Lange, E. (2011). 99 volumes later: We can visualise. Now what? *Landscape and Urban Planning, 100*, 403–406.
40. Leopold, A. (1966). *A sand county almanac: With other essays on conservation from Round River*. Oxford: Oxford University Press.
41. Lin, H., & Lam, K.-C. (2010). Soundscape of urban open spaces in Hong Kong. *Asian Geographer, 27*, 29–42.
42. Lintott, S. (2008). Toward ecofriendly aesthetics. In A. Carlson & S. Lintott (Eds.), *Nature, aesthetics, and environmentalism: From beauty to duty* (pp. 380–396). New York, NY: Columbia University Press.
43. Lynch, E., Joyce, D., & Fristrup, K. (2011). An assessment of noise audibility and sound levels in U.S. National Parks. *Landscape Ecology, 26*, 1297–1309.
44. Mace, B. L., Bell, P. A., & Loomis, R. J. (1999). Aesthetic, affective, and cognitive effects of noise on natural landscape assessment. *Society and Natural Resources, 12*, 225–242.
45. Mace, B. L., Bell, P. A., Loomis, R. J., & Haas, G. E. (2003). Source attribution of helicopter noise in pristine national park landscapes. *Journal of Park and Recreation Administration, 21*, 97–119.
46. Mace, B. L., Bell, P. A., & Loomis, R. J. (2004). Visibility and natural quiet in national parks and wilderness areas: Psychological considerations. *Environment and Behavior, 36*, 5–31.
47. Macpherson, H. (2006). Landscape's ocular-centrism—and beyond? In B. Tress, G. Tress, G. Fry, & P. Opdam (Eds.), *From landscape research to landscape planning: Aspects of integration, education and application* (pp. 95–104). Dordrecht: Springer.
48. Matless, D. (2005). Sonic geography in a nature region. *Social & Cultural Geography, 6*, 745–766.
49. Matsinos, Y. G., Mazaris, A. D., Papadimitriou, K. D., Mniestris, A., Hatzigiannidis, G., Maioglou, D., & Pantis, J. D. (2008). Spatio-temporal variability in human and natural sounds in a rural landscape. *Landscape Ecology, 23*, 945–959.
50. Miller, N. P. (2008). US national parks and management of park soundscapes: A review. *Applied Acoustics*, 77–92.

51. Nasar, J. L. (Ed.). (1988). *Environmental aesthetics: Theory, research, and applications*. Cambridge: Cambridge University Press. Ngai, S. (2012). *Our aesthetic categories: Zany, cute, interesting*. Cambridge: Harvard University Press.
52. Oliveros, P. (2005). *Deep listening: A composer's sound practice*. Lincoln, NE: iUniverse.
53. Parsons, R. (1995). Conflict between ecological sustainability and environmental aesthetics: Conundrum, canard or curiosity.*Landscape and Urban Planning, 32*, 227–244.
54. Parsons, G. (2008). *Aesthetics and nature*. London: Continuum International Publishing Group.
55. Pepper, C. B., Nascarella, M. A., & Kendall, R. J. (2003). A review of the effects of aircraft noise on wildlife and humans, current control mechanisms, and the need for further study. *Environmental Management, 32*, 418–432.
56. Pheasant, R. J., Fisher, M. N., Watts, G. R., Whitaker, D. J., & Horoshenkov, K. V. (2010). The importance of auditory-visual interaction in the construction of 'tranquil space'. *Journal of Environmental Psychology, 30*, 501–509.
57. Pilcher, E. J., Newman, P., & Manning, R. E. (2009). Understanding and managing experiential aspects of soundscapes at Muir woods national monument. *Environmental Management, 43*, 425–435.
58. Porteous, J. D. (1982). Urban environmental aesthetics. In B. Sadler & A. Carlson (Eds.), *Environmental aesthetics: Essays in interpretation* (pp. 67–95). Victoria: University of Victoria.
59. Porteous, J. D. (1985). Smellscape. *Progress in Physical Geography, 9*, 356–378.
60. Porteous, J. D. (1996). *Environmental aesthetics: Ideas, politics and planning*. London: Routledge.
61. Porteous, J. D., & Mastin, J. F. (1985). Soundscape. *Journal of Architectural and Planning Research, 2*, 169–186.
62. Posner, M. I., Nissen, M. J., & Klein, R. M. (1976). Visual dominance: An information-processing account of its origins and significance. *Psychological Review, 83*, 157–171.
63. Prior, J. (2012). *The roles of aesthetic value in ecological restoration: Cases from the United Kingdom*. PhD thesis, The University of Edinburgh, Edinburgh.
64. Raimbault, M., & Dubois, D. (2005). Urban soundscapes: Experiences and knowledge. *Cities, 22*, 339–350. Richards, A. (2001). *The free fantasia and the musical picturesque*. Cambridge: Cambridge University Press. Rodaway, P. (1994). *Sensuous geographies: Body, sense, and place*. London: Routledge.
65. Sadler, B., & Carlson, A. (Eds.). (1982). *Environmental aesthetics: Essays in interpretation*. Victoria: University of Victoria. Schafer, R. M. (1969). *Ear cleaning: Notes for an experimental music course*. Toronto: Berandol Music.
66. Schafer, R. M. (1994). *The soundscape: Our sonic environment and the tuning of the world*. Rochester, MN: Destiny Books. Selman, P., & Swanwick, C. (2010). On the meaning of natural beauty in landscape legislation. *Landscape Research, 35*, 3–26. Sepänmaa, Y. (2007). Multi-sensoriness and the city. In A. Berleant & A. Carlson (Eds.), *The aesthetics of human environments*
67. (pp. 92–99). Peterborough: Broadview Press.
68. Spence, C., Parise, C., & Chen, Y.-C. (2012). The Colavita visual dominance effect. In M. M. Murray & M. T. Wallace (Eds.), *The neural bases of multisensory processes*. (unpaginated). Boca Raton, FL: CRC Press.
69. Stack, D. W., Newman, P., Manning, R. E., & Fristrup, K. M. (2011). Reducing visitor noise levels at Muir woods national monument using experimental management. *The Journal of the Acoustical Society of America, 129*, 1375–1380.
70. Tarrant, M. A., Haas, G. E., & Manfredo, M. J. (1995). Factors affecting visitor evaluations of aircraft overflights of wilderness areas. *Society and Natural Resources, 8*, 351–360.
71. Wissmann, T. (2014). *Geographies of sound*. Farnham: Ashgate.

72. Yang, W., & Kang, J. (2005). Soundscape and sound preferences in urban squares: A case study in Sheffield. *Journal of Urban Design, 10*, 61–80.

73. Yu, L., & Kang, J. (2010). Factors influencing the sound preference in urban open spaces. *Applied Acoustics, 71*, 622–633. Zhang, M., & Kang, J. (2007). Towards the evaluation, description and creation of soundscapes in urban open areas. *Environment and Planning B: Planning and Design, 34*, 68–86.

Index